Advanced Technologies for Managing Insect Pests

Isaac Ishaaya • Subba Reddy Palli
A. Rami Horowitz
Editors

Advanced Technologies for Managing Insect Pests

Springer

Editors
Isaac Ishaaya
Department of Entomology
The Volcani Center
Agricultural Research Organization
Bet Dagan, Israel

Subba Reddy Palli
Department of Entomology
College of Agriculture
University of Kentucky
Lexington, KY, USA

A. Rami Horowitz
Department of Entomology
Gilat Research Center
Agricultural Research Organization
Gilat, Israel

ISBN 978-94-007-4496-7 ISBN 978-94-007-4497-4 (eBook)
DOI 10.1007/978-94-007-4497-4
Springer Dordrecht Heidelberg New York London

Library of Congress Control Number: 2012943161

Printed on acid-free paper

Springer is part of Springer Science+Business Media (www.springer.com)

Preface

Diverse methods to combat insect pests have been contrived throughout modern history: inorganic, botanical and natural pesticides have been mostly exploited during the nineteenth century, and in the twentieth century, significant progress in the synthesis of new chemicals has resulted in a discovery of structures exhibiting insecticidal activity.

Broad-spectrum conventional insecticides, such as organochlorines, organophosphates, carbamates and pyrethroids were developed and used for controlling insect pests over the past five decades, resulting in reduced losses in agricultural yield. However, many of the conventional methods using broad-spectrum insecticides and others have come under scrutiny because of their undesirable effects on man and the environment. In addition, resistance problems reaching crisis proportion have led to stricter regulations and minimize their use in various agricultural systems. The concept of "Integrated Control" based on economic thresholds and injury levels has been implemented within an ecological framework, where chemical and biological controls could thrive together to form the basis for the modern Integrated Pest Management (IPM) concept.

The use of alternatives based on IPM along with novel approaches for developing novel insecticides acting selectively on a specific site in an insect group and compatible with natural enemies and the environment is one of the important challenges of our current book. One of such approaches is based on disrupting the activity of biochemical sites acting on transcription factors such as the basic Helix-Loop-Helix family, anti-juvenile hormone agents that target juvenile hormone (JH) biosynthetic enzymes, G protein coupled receptors (GPCR) and bursicon as targets for insect control. Another approach is the use of biotechnology or the genetic such as gene silencing (RNA-interference) and Bt-crops. Other sections of the book were devoted to the plant's natural products, optical manipulation and the use of nanotechnology for improving insect control methods.

The authors of the various chapters have a wealth of experience and are considered world leaders specializing in novel approaches of insect pest control.

The editors and authors of the various chapters are indebted to the reviewers of the various chapters for valuable suggestions and criticism: Darren Anderson (Canada),

Yves Carriere (USA), H. Czosnek (Israel), Daniel Doucet (Canada), Alberto Fereres (Spain), Murad Ghanim (Israel), Lawrence I. Gilbert (USA), S. Muthukrishnan (USA), Yoshiaki Nakagawa (Japan), Joshua Ogendo (Kenya), Arthur Retnakaran (Canada), Alvaro Romero (USA), David A. Schooley (USA), Gilles-Eric Seralini (France), Les Shipp (Canada), David Stanley (USA), Stephen S. Tobe (Canada), Phyllis Weintraub (Israel), R. R. Youngman (USA), and Zhaolin Zhang (USA).

Contents

Contributors

Yasmin Akhtar Faculty of Land and Food Systems, University of British Columbia, Vancouver, BC, Canada

Yehezkel Antignus Institute of Plant Protection, Agricultural Research Organization, The Volcani Center, Bet Dagan, Israel

Hua Bai EEB, Brown University, Providence, RI, USA

David Ben-Yakir Institute of Plant Protection, Agricultural Research Organization, The Volcani Center, Bet Dagan, Israel

Kavita Bitra Department of Entomology, College of Agriculture, University of Kentucky, Lexington, KY, USA

Michel Cusson Natural Resources Canada, Canadian Forest Service, Laurentian Forestry Centre, Quebec City, QC, Canada

Béla Darvas Department of Ecotoxicology and Environmental Analysis, Plant Protection Institute, Hungarian Academy of Sciences, Budapest, Hungary

Division of Agro-Environmental Safety, Central Food Science Research Institute, Budapest, Hungary

Shengzhang Dong Division of Plant Sciences-Entomology, University of Missouri, Columbia, MO, USA

Murad Ghanim Department of Entomology, Institute of Plant Protection, Agricultural Research Organization, The Volcani Center, Bet Dagan, Israel

Cornelis J.P. Grimmelikhuijzen Center for Functional and Comparative Insect Genomics, Cell and Neurobiology, Department of Biology, University of Copenhagen, Copenhagen, Denmark

Toshiyuki Harada Health and Crop Sciences Research Laboratory, Sumitomo Chemical Company, Tokyo, Japan

Frank Hauser Centre for Functional and Comparative Insect Genomics, Cell and Neurobiology, Department of Biology, University of Copenhagen, Copenhagen, Denmark

Kenneth F. Haynes Department of Entomology, College of Agriculture, University of Kentucky, Lexington, KY, USA

A. Rami Horowitz Department of Entomology, Agricultural Research Organization, Gilat Research Center, MP Negev, Israel

Isaac Ishaaya Department of Entomology, Agricultural Research organization, The Volcani Center, Bet Dagan, Israel

Murray B. Isman Faculty of Land and Food Systems, University of British Columbia, Vancouver, BC, Canada

Adi Kliot Department of Entomology, Institute of Plant Protection, Agricultural Research Organization, The Volcani Center, Bet Dagan, Israel

Moshe Kostyukovsky Department of Food Quality and Safety, Agricultural Research Organization, The Volcani Center, Bet Dagan, Israel

Shlomo Magdassi Casali Institute of Applied Chemistry, The Center for Nanoscience and Nanotechnology, Institute of Chemistry, The Hebrew University of Jerusalem, Jerusalem, Israel

Katrin Margulis-Goshen Casali Institute of Applied Chemistry, Institute of Chemistry, Center for Nanoscience and Nanotechnology, The Hebrew University of Jerusalem, Israel

Yoshiaki Nakagawa Division of Applied Life Sciences, Graduate School of Agriculture, Kyoto University, Kyoto, Japan

Yossi Offir Polysack Plastic Industries, Nir Yithak-Sufa, Israel

Subba Reddy Palli Department of Entomology, College of Agriculture, University of Kentucky, Lexington, KY, USA

Michael F. Potter Department of Entomology, College of Agriculture, University of Kentucky, Lexington, KY, USA

Stephanie E. Sen Department of Chemistry, The College of New Jersey, Ewing, NJ, USA

Eli Shaaya Department of Food Quality and Safety, Agricultural Research Organization, The Volcani Center, Bet Dagan, Israel

Yosepha Shahak Institute of Plant Sciences, Agricultural Research Organization, The Volcani Center, Bet Dagan, Israel

Tetsuro Shinoda Division of Insect Sciences, National Institute of Agrobiological Sciences, Tsukuba, Ibaraki, Japan

Guy Smagghe Department of Crop Protection, Faculty of Bioscience Engineering, Ghent University, Ghent, Belgium

Qisheng Song Division of Plant Sciences-Entomology, University of Missouri, Columbia, MO, USA

Luc Swevers Insect Molecular Genetics and Biotechnology, Institute of Biosciences and Applications, National Centre for Scientific Research "Demokritos", Aghia Paraskevi, Athens, Greece

András Székács Department of Ecotoxicology and Environmental Analysis, Plant Protection Institute, Hungarian Academy of Sciences, Budapest, Hungary

Chapter 1
Advanced Technologies for Managing Insect Pests: An Overview

A. Rami Horowitz and Isaac Ishaaya

1 Introduction

Throughout modern history, various methods to combat insect pests have been contrived; however, inorganic, botanical and natural pesticides mostly exploited during the nineteenth century. In the twentieth century, significant progress in the synthesis of new chemicals has resulted in a discovery of structures and biological activities of various compounds. Broad-spectrum conventional insecticides, such as organochlorines, organophosphates, carbamates and pyrethroids were developed and used to control insect pests over the past five decades, resulting in reduced losses in agricultural yield. However, the severe adverse effects of these pesticides on the environment, problems of resistance reaching crisis proportions and public protests led to stricter regulations and legislation aimed at reducing their use. Since the 1990s, an implementation of Integrated Pest Management (IPM) principles resulted in two advances; one is the development of novel insecticides with selective properties acting on biochemical sites or physiological processes present in specific insect groups but differ from other organisms in their properties (e.g. Casida and Quistad 1998). This process has led to the formation of compounds, which affect the hormonal regulation of molting and developmental processes such as the insect growth regulators (IGR) and the neonicotinoids. The second advance is the exploitation of other non-chemical methods such as biological and cultural controls, use of pheromones and biopesticides (e.g. Horowitz and Ishaaya 2004;

A.R. Horowitz (✉)
Department of Entomology, Agricultural Research Organization,
Gilat Research Center, MP Negev 85280, Israel
e-mail: hrami@volcani.agri.gov.il

I. Ishaaya
Department of Entomology, Agricultural Research Organization,
The Volcani Center, Bet Dagan 50250, Israel
e-mail: vpisha@volcani.agri.gov.il

I. Ishaaya et al. (eds.), *Advanced Technologies for Managing Insect Pests*,
DOI 10.1007/978-94-007-4497-4_1, © Springer Science+Business Media Dordrecht 2012

Rosell et al. 2008), and the substantial advancement in transgenic crops (James 2011).

Attempts to reduce the use of synthetic pesticides, especially broad-spectrum insecticides in plant protection and to use alternatives and novel methods for pest control or ("biorational control") are the challenges of pest control for the twenty-first century (Ishaaya and Horowitz 2009). The term biorational (biological + rational) pest control or approach can be defined as the use of selective means that are compatible with natural enemies and the environment, with minimal effect on non-target organisms. Biorational control is based on a diversity of chemical, biological and physical resources for controlling insect pests, which results in reduced risk to man and the environment, and in accordance with IPM concepts (Horowitz et al. 2009). Biorational agents and approaches will be the key for inspiring IPM strategies to meet our community challenges (Horowitz and Ishaaya 2004; Ishaaya 2003; Ishaaya et al. 2005; Ishaaya and Horowitz 2009). This overview is based on the different chapters of this book, which deals with advanced and novel technologies for managing insect pests. These technologies focus on safer and environmentally friendly (biorational) approaches.

One such approach is based on disrupting the activity of specific biochemical sites serving as targets for insecticide discovery; these sites include transcription factors belonging to the basic Helix-Loop-Helix (bHLH) family (Chap. 2), anti-juvenile hormone (AJH) agents that target JH biosynthetic enzymes (Chap. 3), G protein-coupled receptors (GPCR) (Chap. 4) and bursicon as targets for insect pest control (Chap. 5). Another section is related to screening potential insecticides by cell-based and other advanced screenings (Chaps. 6 and 7). The third segment deals with novel biotechnology control strategies ("the genetic approach"), which exploit the huge development in arthropod genomics (Chap. 8), gene silencing (RNA-interference) (Chap. 9) and Cry toxins (based on the crystal protein produced by *Bacillus thuringiensis – Bt*) usage in insect control (Chap. 10). The last section of the book covers various new aspects of pest control such as the usefulness of plant natural-product mixtures (Chap. 11), optical manipulation for reducing sucking pests (Chap. 12), recent progress in bed bug management (Chap. 13), and insect pests in stored-product food (Chap. 14) and the utilization of nanotechnology for development of potent insecticides (Chap. 15).

2 Identification of New Insect Target-Sites for Discovery and Development of Novel Insecticides

The development of novel insecticides during the past three decades has led to the formation of compounds, which affect the hormonal regulation of molting and developmental processes in insects such as ecdysone agonists, juvenile hormone mimics and chitin synthesis inhibitors (Reviewed in some other books e.g., Ishaaya 2001; Horowitz and Ishaaya 2004; Ishaaya et al. 2007; Ishaaya and Horowitz 2009). In the present volume, the immense advance in molecular biology has resulted in novel studies on insect target sites.

Transcription factors, belonging to the basic Helix-Loop-Helix (bHLH) family play essential roles in a wide range of developmental processes of higher organisms including insects. In a few model insects such as *Drosophila melanogaster*, *Tribolium castaneum*, and *Bombyx mori*, studies on these factors provided information in relation to the importance of transcription factors in such processes as cell proliferation, determination, differentiation, cell cycle maintenance, and more. Chapter 2 by Bitra and Palli describes structure and function of bHLH transcription factors in insects and discussed their potential as target sites for discovering new insecticides. Their study analyzes systematic functional data on bHLH genes in insects and provides an in-depth contribution to the knowledge on the function of insect bHLH transcription factors in insect development and reproduction.

The bHLH transcription factors could be utilized for developing new insecticides for controlling pests of crops and vectors borne-diseases. It is possible that these factors will be used either in small molecule screens, or in the development of RNAi-based pest management methods. One potential target for insecticidal development is the methoprene tolerant (Met) gene found in *D. melanogaster* and other insects that plays an important role in JH signal transduction.

Another target site for developing potential insecticides is JH, which it is produced in the corpora allata (CA) and is considered an important regulator of insect morphogenesis and reproduction. Since the 1960s, intensive research aimed to exploit its properties for developing "third-generation" pesticides has been carried out (Williams 1967). A few JH-analogs have been developed since then (e.g. methoprene, fenoxycarb and pyriproxyfen), which are able to interfere with JH biosynthesis and are still in use against insect pests; in addition, it is unlikely that these agents would affect non-insect organisms. However, serious resistance problems have been detected in various insect pests to these JH-analogs (e.g. Horowitz et al. 1994). Chapter 3 by Cusson et al. presents new insights into identifying and designing AJH agents that target JH biosynthetic enzymes. The advances in the field of molecular biology and structural bioinformatics have resulted in cloning and characterizing JH biosynthetic enzymes. Hence, it makes it possible to design highly specific and effective enzyme inhibitors. This chapter covers JH biosynthetic pathway along with recent development of inhibitors of JH biosynthesis. Allatostatins are insect neuropeptides that have the ability to inhibit JH synthesis (Stay and Tobe 2007). Their discovery has led to the development of peptide analogs that inhibit JH biosynthesis, an alternative way for interfering with JH production (e.g. AJH agents for the control of pests such as cockroaches, see Chap. 3).

Other receptors that could be used, as target sites for insecticide discovery are the G protein-coupled receptors (GPCRs) reviewed in Chap. 4 by Bai and Palli. GPCRs comprise a large protein family that includes seven-transmembrane domains and sense signals outside the cell and transducer them by activating intercellular pathways (e.g. Gilman 1987). GPCRs are found in almost all the eukaryotic organisms (Perez 2005). These proteins contain a large number of structurally diverse receptors intervening diverse functions, such as senses and behavioral regulation (including hormones and pheromones) as well as immune and nervous systems. Since these receptors are involved in many human diseases, they are considered as a key drug target (Filmore 2004).

In their chapter, Bai and Palli review advances that have been made regarding identification, functional characterization and target screening of GPCRs, and their feasibility for developing insecticides. These receptors are involved in different functions of insect including metabolism, development and reproduction. Although many insect GPCRs have been deorphanized (functional characterized), they have not been exploited for insecticide discovery. One of the ways to identify GPCRs pesticide target is by RNAi screening. To identify GPCRs that could be used as target sites for developing new insecticides, the authors performed a large-scale RNAi screening in the red flour beetle, *T. castaneum*, as a model insect for various functional genomics studies. The RNAi screen identified quite a few GPCRS that could be useful as target sites for insecticide development (see also, Bai et al. 2011).

Bursicon is an insect neuropeptide hormone that regulates various processes of cuticle tanning and hardening (sclerotization), wing expansion and the maturation process in different insect orders. Bursicon was first discovered in the 1960s in the blowflies *Calliphora erythrocephala* that regulates cuticle tanning in newly emerged adults (Cottrell 1962; Fraenkel and Hsiao 1962); however, the molecular basis of bursicon was elucidated just in 2005. Bursicon is a heterodimeric cystine-knot protein that activates GPCR (Luo et al. 2005). In Chap. 4, Dong and Song described the progress in bursicon research and developments of various neuropeptides for pest control. They also discuss the possibilities of bursicon as a target for pest control agents. Bursicon and its receptor are well studied in *D. melanogaster* and other insects along with signaling pathway for regulating the maturation process. Furthermore, in RNAi assays with *T. castaneum* and other insects, applying injections of double-stranded (ds) RNAs corresponding to gene products of bursicon, resulted in incomplete wing expansion (e.g. Huang et al. 2007; Arakane et al. 2008).

Dong and Song concluded that bursicon could be used as a potential target site for the design of novel and environmentally friendly insecticides. Like other neuropeptides, bursicon and its receptor have a number of shortcomings that prevent its direct use for controlling insect pests.

However, as a result of using non-peptide small molecules (as mimics for neuropeptides) and the prompt development of RNAi transgenic plants, the authors anticipate its potential use for controlling insect pest in the near future.

3 Progress in Screening New Insecticides

The conventional toxicity bioassays using various insects are still the main method for screening new insecticides. In this section, two chapters review and discuss new methods for rapid screenings of novel controlling agents that could replace the traditional bioassays.

In Chap. 6, Smagghe and Swevers discuss the utility and benefits of insect cell lines, along with high throughput screening procedures for novel pesticides. Nowadays, numerous insect cell lines have been established from diverse insect species and from several different tissue sources (e.g. Smagghe et al. 2009). The authors' recent examples of this approach was demonstrated by various insect growth regulators (IGRs), especially ecdysteroid receptor agonists (e.g. tebufenozide, methoxyfenozide). Other examples are the use of insect cell lines for studying mode of action and for screening inhibitors of chitin synthesis. Furthermore, it is useful for evaluating and screening *Bt* insecticidal proteins, and for detecting resistance mechanisms to insecticides. In addition, insect cell lines could be used successfully for screening of new pesticides. They expect this method will be also exploited to discover new enzyme inhibitors and insect-specific targets and pathways.

Chapter 7 by Nakagawa and Harada summarizes the classical quantitative structure activity relationship (QSAR) of the non-steroidal ecdysone agonists (diacylhydrazines) and ecdysteroids. In addition, they review the application of advanced computer-aided drug design for developing novel insecticides. In drug design, computational approaches such as virtual (*in silico*) screening (including homology modeling and bioinformatics that are required for that screening), and high throughput screening (HTS) are commonly used. The former procedure is classified as ligand-based virtual screening and structure-based virtual screening; both are often used for the design of medical drugs but it is not in use to design agrochemical compounds. The problem with the latter is that often the molecular target is not known. Since both screening types are hardly used in agrochemical design, the authors present instead some pharmaceutical examples (e.g., Walters et al. 1998). The authors assume that bioinformatics and genome technologies will be utilized in the future in drug and pesticide design.

4 Genetic and Biotechnological Approaches as Novel Strategies Against Arthropod Pests

In 1987, the first study demonstrating that transgenic tobacco plants, expressing Bt genes could protect plants from insect damage was published (Vaeck et al. 1987). Since 1996, when Bt-crops were first commercially planted, they have become a great success of applied biotechnology in agriculture, especially to control lepidopteran pests in maize and cotton, and this technology has reduced substantially insecticide applications. In general, biotech crops (they mostly contain an herbicide-tolerance trait that have substantially increased herbicide applications) have been extended to 160 million hectares, and it makes them the fastest adopted crop technology in the history of modern agriculture (James 2011; Brookes and Barfoot 2011).

Chapter 8 by Grimmelikhuijzen and Hauser summarizes more than 50 genome-projects of arthropods that have been conducted during the last 12 years. Insect genomes are particularly valuable in insect research studies to improve beneficial arthropod characteristics as well as to improve our understanding of the biology of agricultural and human pests. In addition, the authors suggest using this knowledge in insect genomes to exploit target sites against insect pests. In this context, they emphasized the GPCRs as promising targets for developing a new generation of environmentally friendly pesticides.

Ghanim and Kliot (Chap. 9) reviewed the progress has been made in understanding the RNAi in model organisms such as *C. elegans* and *D. melanogaster* as well as in non-model organisms. RNAi is a process occurs within living cells that regulates gene expression, resulting in its silencing or inactivation (e.g. Fire et al. 1998; Martinez et al. 2002; Melnyk et al. 2011). The RNAi phenomenon is found in many eukaryotes, but firstly it was studied in plants and in model organisms, and also in human cell lines. The process is mediated by the enzyme Dicer, which cleaves double-stranded RNA (dsRNA) molecules into short nucleotides (siRNAs). Each siRNA is uncoiled into two single-stranded RNA (ssRNAs), and the guide strand is incorporated into the RNA-induced silencing complex (RISC) catalyzed by the protein Argonaute. Gene silencing can occur when the guide strand pairs with a complementary strand of a messenger RNA (known as post-transcriptional gene silencing).

Besides the function of RNAi in model organisms (see above), the authors have summarized the major progress of applying this method in non-model organisms. They emphasize the potential of RNAi for developing new control methods against insect pests, and exemplifying this approach with recent results from their research on the application of the RNAi technology in the whitefly *Bemisia tabaci*, a major pest in various agricultural systems.

Szekacs and Darvas (Chap. 10) reviewed and discussed different approaches and uses of crystalline (Cry) endotoxins from *Bt* as both, bacterial insecticides and expressed proteins in genetically modified plants. They stated some differences between both technologies and concluded that the main advantage of Bt-plants is that they save labor and energy as compared with *Bt* field-application; however, since the former continuously produces Cry toxin during plant growth, without connection to pest's attacks, Bt plants do not comply with the principle of IPM. They summarized that both technologies cannot be considered as equivalent ones; because of the very mild effect of the Bt-based bioinsecticides, it has been accepted as suitable for organic agriculture.

In general, the authors' point of view is in debate; other many reports have demonstrated that there is clear evidence that Bt-crops have significant advantages, including reduction in use of broad-spectrum insecticides that is one of the primary goals of IPM (e.g. Cannon 2000; Romeis et al. 2008; US National Research Council 2010).

5 Other New Advanced Strategies for Management of Insect Pests

This section covers other new aspects of pest control such as the usefulness of plant natural-product mixtures, optical manipulation for controlling sucking pests, recent progress in management of bed bug, insect pests in dry (store) food and the utilization of nanotechnology for development of potent insecticides.

5.1 Usefulness of Plant Natural-Product Mixtures

Natural substances are involved in the plant defensive against herbivores and pathogens. Isman and Akhtar (2007) have reviewed various plant natural products with insecticidal activities. As they are naturally occurring chemicals, the exploitation of such products may be useful for developing ecologically sound pesticides. Akhtar and Isam (Chap. 11) suggested another aspect of plant natural products for pest management: the use of mixtures of two or more natural compounds that are more effective as compared with a single active ingredient. They noted that in plants there is chemical interaction of several compounds that are involved in plant defense against herbivores. In addition, the mixture approach may be useful for prolonging the use of natural products and delay evolution of resistance. On the other hand, other authors (e.g., Denholm et al. 1998) suggested that long use of mixtures of synthetic pesticides had failed to control target pests, apparently because of selection of a new, non-synergizable resistance mechanism or a modification of one already present.

Akhtar and Isman gave numerous examples of mixtures of natural products such as complex plant essential oils and semiochemicals that affect target pests more efficacious and synergically than the pure substances isolated from them. They concluded that the discovery of effective mixtures of natural products might lead to the development of successful insect control agents (toxicants, growth/feeding inhibitors, repellents and attractants) as well as the possible use of reduced amounts of each component in the mixture, to reach acceptable levels of efficacy.

5.2 Optical Manipulation for Reducing Sucking Pests

Many insects are attracted to plants following a series of events, initiating from far-orientation, and ending with landing on plants to feed and lay eggs (Antignus 2010). For host finding and flight orientation, insects use mostly optical or vision cues; thus, manipulation of these cues in insects can interfere with their orientation and impede insect-plant communication (e.g. Antignus 2000).

Chapter 12 by Ben-Yakir et al. reviews the literature regarding vision cues and optical manipulation of insect pests in open field and protected crops. They also include results of their studies, and suggestions for future research with this technology. The authors concentrate on sucking insect pests, such as aphids, whiteflies and thrips that cause serious economic damage to agricultural crops worldwide, both by direct feeding on plant sap and by transmitting viruses to important crops. Most of the aforementioned pests have receptors for UV light (peak at 360 nm) and green-yellow light (peak at 520–540 nm), and preventing or diminishing of these spectra can hamper insect's various activities. This approach is suitable for areas with high intensity of sunlight during the seasons when sucking pests are active. Optical manipulations can be performed by repelling, attracting and camouflaging of vision cues, and they can be achieved by incorporating optical additives to mulches (below plants), to cladding materials (plastic sheets, nets and screens above plants) or to other objects. The authors conclude that this technology would not give sufficient protection in crop production but optical manipulation should be integrated along with other IPM tactics with the hope that their use will reduce the use of insecticides.

5.3 Recent Progress in Bed Bug Management

The bug, *Cimex lectularius* has been known as a human parasite for thousands of years, and its control has been carried out with conventional insecticides, especially pyrethroids. However, in recent years, a world-wide resurgence of bed bug has occurred (Potter 2011). Chapter 13 by Hynes and Potter reviews the recent advances in bed bug management. The authors consider that among other reasons, this resurgence may have resulted from resistance to pyrethroids, which has developed because of the intensive use of this group of insecticides. Hynes and Potter suggest that integrated management with chemical and non-chemical methods is required to better control this pest. Non-chemical methods such as steam, freezing, and vacuuming can be effective but these methods are expensive, restricted to limited space and cannot prevent reintroduction of bed bugs. It is obvious that residual insecticides other than the pyrethroids such as, neonicotinoids and chlorfenapyr are needed to replace the former. Further knowledge of bed bug biology, e.g. mating behavior, is valuable for its detection and for potential new tactics that would manipulate this pest.

5.4 Recent Progress in Management of Insect Pests of Stored Products

Stored-product insects are serious worldwide pests of dried, stored and other durable agricultural commodities (Phillips and Throne 2010). The global losses in stored grain have been estimated at approximately 10% in developed countries, and 20% and more in developing countries. A biorational approach is preferable for controlling stored-product insects because of: the phase-out of very effective insecticides (e.g. methyl bromide), insecticide resistance, and the problem of insecticide residues in food. At present, their control is mainly conducted by fumigation with the organophosphorus toxicant phosphine. Chapter 14, by Kostyukovsky and Shaaya, covers recent progress that was made in management of insect pests in dry food (mainly coleopteran insects) such as *Tribolium castaneum, Sitophilus oryzae* and *Callosobruchus maculates*. In the first section, they describe methods to achieve effective control of these pests by phosphine fumigation. For this aim, they have evaluated a special device containing a heater and a ventilator named “speedbox” (Jakob et al. 2006) that was developed especially for use at low temperatures and shorter treatment time. The use of this device allowed optimizing the Phosphine fumigation by effective control of all developmental stages of major stored product insects at low temperatures and at decreased exposure time.

Besides optimization of phosphine fumigation, Kostyukovsky and Shaaya have evaluated various alternatives to conventional insecticides for controlling stored-product insects such as essential oils as botanical fumigants and diatomaceous earth as a grain protectant. Some essential oils are highly selective to insects, probably because they bind to the insect-selective octopaminergic receptor, a non-mammalian target (Kostyukovsky et al. 2002). A screening of a large number of essential oils from aromatic plants was conducted to isolate effective oils for using as fumigants in grain bins for insect control. The authors reported on two effective essential oils that were examined under laboratory and field conditions with supplementation of CO_2 against *S. oryzae* and *C. maculates*. Their results are the basis for using essential oils as botanical fumigants and alternatives to the toxicant phosphine for controlling stored-product insects.

Diatomaceous earth or diatomite is a chalky, soft siliceous sedimentary rock that is fine-grained into a light-colored powder (Antonides 1997). Among many applications of this compound, it also has been used as a physical pest control agent. Diatomaceous earth was found effective in controlling stored-grain pests, and in addition, the progeny production of some species was highly inhibited.

5.5 The Utilization of Nanotechnology for Development of Potent Insecticides

Nanotechnology is a notion in the field of science and technology that has been greatly developed during the last decade. The studies of nanotechnology exploit the unique behavior of materials and structures (nanomaterials) with dimension of approximately 1–100 nm. The mechanical and physical properties of nanomaterials differ considerably than those of the larger-size bulk materials (or than those of individual atoms and molecules) due to the high surface area of the nanoparticles. This technique is exploited for many applications in the chemical, agriculture, medicine, and cosmetics industries and many more. Although the obvious advantages of nanomaterials, there is concern regarding the safety of these materials for humans and the environment; hence, new safety regulations are needed for this technology (Thomas et al. 2006; Murashov and Geraci 2010).

Chapter 15 by Margulis-Goshen and Magdassi reviews the updated information of nanotechnology for improving insecticide efficacy. Many of the insecticides are organic compounds that are poorly soluble in water, and hence, large amounts of environmentally contaminated organic solvents are usually added to insecticides. Preparation of insecticides as nanoparticles can solve the problem because their utilization allows significantly reduced use of organic solvents. The production the material as nanoparticles results in substantial increase in water solubility, dissolution rate and dispersion uniformity, and their efficacy in the field may be considerably improved. The authors provide an overview of the techniques for making nanoparticles for agricultural use and discuss the risks involved in using those products. They conclude that although the production of nanoparticles is relatively expensive, it is very likely that in the near future this technology will be more widespread.

References

Antignus Y (2000) Manipulation of wavelength-dependent behaviour of insects: an IPM tool to impede insects and restrict epidemics of insect-borne viruses. Virus Res 71:213–220

Antignus Y (2010) Optical manipulations block the spread of *Bemisia tabaci* in greenhouses and the open field. In: Stansly PA, Naranjo SE (eds) *Bemisia*: bionomics and management of a global pest. Springer, Dordrecht, pp 349–356

Antonides LE (1997) Diatomite – 1997. USGS minerals yearbook data 1997, pp 24.1–24.4 http://minerals.usgs.gov/minerals/pubs/commodity/diatomite/250497.pdf

Arakane Y, Li B, Muthukrishnan S, Beeman RW, Kramer KJ, Park Y (2008) Functional analysis of four neuropeptides, EH, ETH, CCAP and bursicon, and their receptors in adult ecdysis behavior of the red flour beetle, *Tribolium castaneum*. Mech Dev 125:984–995

Bai H, Zhu F, Shah K, Palli S (2011) Large-scale RNAi screen of G protein-coupled receptors involved in larval growth, molting and metamorphosis in the red flour beetle. BMC Genomics 12:388

Brookes G, Barfoot P (2011) GM crops: global socio-economic and environmental impacts 1996–2009. PG Economics Ltd., Dorchester, UK

Cannon RJC (2000) *Bt* transgenic crops: risks and benefits. Integr Pest Manag Rev 5:151–173

Casida JE, Quistad GB (1998) Golden age of insecticide research: past, present, or future. Annu Rev Entomol 43:1–16

Cottrell CB (1962) The imaginal ecdysis of blowflies. The control of cuticular hardening and darkening. J Exp Biol 39:395–411

Denholm I, Cahill M, Dennehy TJ, Horowitz AR (1998) Challenges with managing insecticide resistance in agricultural pests, exemplified by the whitefly, *Bemisia tabaci*. Philos Trans R Soc Ser B 353:1757–1767

Filmore D (2004) It's a GPCR world. Mod Drug Discov 11(7):24–28

Fire A, Xu S, Montgomery MK, Kostas SA, Driver SE, Mello CC (1998) Potent and specific genetic interference by double-stranded RNA in *Caenorhabditis elegans*. Nature 391:806–811

Fraenkel G, Hsiao C (1962) Hormonal and nervous control of tanning in the fly. Science 138:27–29

Gilman AG (1987) G proteins: transducers of receptor-generated signals. Annu Rev Biochem 56:615–649

Horowitz AR, Ishaaya I (2004) Biorational insecticides – mechanisms, selectivity and importance in pest management. In: Horowitz R, Ishaaya I (eds) Insect pest management. Springer, Berlin, pp 1–28

Horowitz AR, Ellsworth PC, Ishaaya I (2009) Biorational pest control – an overview. In: Ishaaya I, Horowitz AR (eds) Biorational control of arthropod pests – application and resistance management. Springer, Dordrecht, pp 1–20

Horowitz AR, Forer G, Ishaaya I (1994) Managing resistance in *Bemisia tabaci* in Israel with emphasis on cotton. Pestic Sci 42:113–122

Huang J, Zhang Y, Li M, Wang S, Liu W, Couble P, Zhao G, Huang Y (2007) RNA interference-mediated silencing of the bursicon gene induces defects in wing expansion of silkworm. FEBS Lett 581:697–701

Ishaaya I (2001) Biochemical processes related to insecticide action: an overview. In: Ishaaya I (ed) Biochemical sites of insecticide action and resistance. Springer, Berlin

Ishaaya I (2003) Introduction: biorational insecticides – mechanism and application. Arch Insect Biochem Physiol 54:144

Ishaaya I, Horowitz AR (eds) (2009) Biorational control of arthropod pests – application and resistance management. Springer, Dordrecht

Ishaaya I, Kontsedalov S, Horowitz AR (2005) Biorational insecticides: mechanism and cross-resistance. Arch Insect Biochem Physiol 58:192–199

Ishaaya I, Nauen R, Horowitz AR (eds) (2007) Insecticides design using advanced technologies. Springer, Berlin

Isman MB, Akhtar Y (2007) Plant natural products as a source for developing environmentally acceptable insecticide. In: Ishaaya I, Nauen R, Horowitz AR (eds) Insecticides design using advanced technologies. Springer, Berlin, pp 235–248

Jakob G, Dierks-Lange H, Heck FW, Schmitt S (2006) The speedbox – an innovative application device for the Degesch plates. In: Lorini I, Bacaltchuk B, Beckel H, Deckers D, Sundfeld E, dos Santos JP, Biagi JD, Celaro JC, Faroni LRD'A, Bartolini LdeOF, Sartori MR, Elias MC, Guedes RNC, De-Fonseca RG, Scussel VM (eds) Proceedings of the 9th international working conference on stored product protection, Brazilian Post-harvest Association, Campinas/Sao Paulo, Brazil, 15–18 Oct 2006, pp 564–566

James C (2011) Global status of commercialized biotech/GM crops: 2011, ISAAA brief no 43. ISAAA, Ithaca

Kostyukovsky M, Rafaeli A, Gileadi C, Demchenko N, Shaaya E (2002) Activation of octopaminergic receptors by essential oil constituents isolated from aromatic plants: possible mode of activity against insect pests. Pest Manag Sci 58:1–6

Luo CW, Dewey EM, Sudo S, Ewer J, Hsu SY, Honegger HW, Hsueh AJ (2005) Bursicon, the insect cuticle-hardening hormone, is a heterodimeric cystine knot protein that activates G protein coupled receptor LGR2. Proc Natl Acad Sci USA 102:2820–2825

Martinez J, Patkaniowska A, Urlaub H, Luhrmann R, Tuschl T (2002) Single-stranded antisense siRNAs guide target RNA cleavage in RNAi. Cell 110:563–574

Melnyk CW, Molnar A, Baulcombe DC (2011) Intercellular and systemic movement of RNA silencing signals. EMBO J 30:3553–3563

Murashov V, Geraci C (2010) Regulatory approaches to worker protection in nanotechnology industry in the USA and European Union. National Institute for Occupational Safety and Health, Centers for Disease Control and Prevention, U.S. Department of Health and Human Services. http://ioshic.web.fc2.com/seminar/pdf/20100129_4th_seminar/Murashov_Summary.pdf

Perez DM (2005) From plants to man: the GPCR "tree of life". Mol Pharmacol 67:1383–1384

Phillips TW, Throne JE (2010) Biorational approaches to managing stored-product insects. Annu Rev Entomol 55:375–397

Potter MF (2011) The history of bed bug management with lessons from the past. Am Entomol 57(1):14–25

Romeis J, Shelton A, Kennedy GG (eds) (2008) Integration of insect-resistant genetically modified crops within IPM programs. Springer, Dordrecht

Rosell G, Quero C, Coll J, Guerrero A (2008) Biorational insecticides in pest management. J Pestic Sci 33:103–121

Smagghe G, Goodman CL, Stanley D (2009) Insect cell culture and applications in research and pest control management. In Vitro Cell Dev Biol Anim 45:93–105

Stay B, Tobe SS (2007) The role of allatostatins in juvenile hormone synthesis in insects and crustaceans. Annu Rev Entomol 52:277–299

Thomas T, Thomas K, Sadrieh N, Savage N, Adair P, Bronaugh R (2006) Research strategies for safety evaluation of nanomaterials, part VII: evaluating consumer exposure to nanoscale materials. Toxicol Sci 91(1):14–19

US National Research Council (2010) The impact of genetically engineered crops on farm sustainability in the United States. National Academies Press, Washington, DC

Vaeck M, Reynaerts A, Hofte H, Jansens S, De Beuckeleer M, Dean C, Zabeau M, Van Montagu M, Leemans J (1987) Transgenic plants protected from insect attack. Nature 328:33–37

Walters WP, Stahl MT, Murcko MA (1998) Virtual screening – an overview. Drug Discov Today 3:160–178

Williams CM (1967) Third-generation pesticides. Sci Am 217:13–17

Chapter 2
bHLH Transcription Factors: Potential Target Sites for Insecticide Development

Kavita Bitra and Subba Reddy Palli

1 Introduction

Transcription factors belonging to the basic Helix-Loop-Helix (bHLH) family play a central role in cell proliferation, determination, differentiation, cell cycle maintenance, and homeostasis or stress response pathways (Jan and Jan 1993; Weintraub 1993; Hassan and Bellen 2000). Members of this extensive protein family are characterized by the presence of a basic DNA binding region, which is required for the formation of functional DNA binding complexes (Murre et al. 1989a; Kadesch 1993). The bHLH domain is approximately 60 amino acids in length and comprises a DNA-binding basic region of 15 amino acids followed by two α- helices separated by a variable loop region (Ferre-D'Amare et al. 1993).

Numerous genes coding for bHLH transcription factors have been identified in both vertebrates and invertebrates. These include eight genes in yeast, 39 in *Caenorhabditis elegans,* 39 in *Gallus gallus*, 39 in *Brachydanio rerio*, 46 in *Ciona intestinalis*, 47 in *Xenopus laevis*, 59 in *Drosophila melanogaster*, 53 in *Tribolium castaneum*, 52 in *Bombyx mori,* 87 in *Lagocephalus lagocephalus*, 102 in *Mus musculus*, 118 in *Homo sapiens*, 167 in *Oryza sativa,* 54 in *Acyrthosiphon pisum,* 51 in *Apis mellifera* and 147 in *Arabidopsis* (Ledent et al. 2002; Satou and Satoh 2003; Simionato et al. 2007; Wang et al. 2007b, 2008; Dang et al. 2011). The members of the bHLH superfamily from *D. melanogaster* have been classified into various functional families as defined by additional domains such as bHLH- Per, Arnt, Sim (PAS), Hairy – enhancer of split (HES), Myc/Upstream Transcription Factor (USF), Atonal, Mesp, Hand, p48, Shout and Achaete- scute (AS-C) (Moore et al. 2000). mRNA *in situ* analysis demonstrated that genes encoding some of these bHLH proteins are expressed in several tissue types but are particularly concentrated in

K. Bitra • S.R. Palli (✉)
Department of Entomology, College of Agriculture, University of Kentucky,
Lexington, KY 40546, USA
e-mail: bitrakavita@gmail.com; rpalli@uky.edu

I. Ishaaya et al. (eds.), *Advanced Technologies for Managing Insect Pests*,
DOI 10.1007/978-94-007-4497-4_2, © Springer Science+Business Media Dordrecht 2012

Table 2.1 The insect bHLH members

Family name	Group	bHLH protein	General function
bHLH-PAS	C	Met, GCE, SRC, Tango, Trachealess, Sim1, Sim2, Spineless, Cycle, Clock, Dysfusion, Hypoxia	Play major roles during development
HES	E	H/ESPl, Side1, Side2, Hairy, Deadpan, Hey1, Hey2	Embryogenesis, cell proliferation and tissue differentiation
Myc/USF	B	Myc, Max, Mnt, Mlx, Flocculin like	Transcriptional activators and are also involved in female reproduction
Hand	A	Twist, Hand	Involved in heart formation
Shout	A	Delilah1, Delilah2, Shout	Involved in muscle differentiation, development and attachment during embryonic stage
p48	A	Fer1, Fer2, Fer3	Related to pancreatic exocrine cell transcription factor in mouse. Also involved in embryonic development
NeuroD/Neurogenin	A	Dimmed, Tap, NevroD, Beta3	Mainly involved in nervous system development and neurogenesis
Atonal	A	Cato, Atonal, Amos	Nervous system development
AS-C	A	Scute, Asense, Ash	Involved in neurogenesis
Mesp	A	TF21, Sage	Necessary for mesoderm segmentation initiation
Miscellaneous	A, B, D, F	HLH106, Bigmax, AP-4, HLH3B,Paraxis, HLH4C, Nautilus, Daughterless, Emc, Mitf, Collier	These proteins lack conserved functional domains and are hence classified as miscellaneous group

the developing nervous system and mesoderm (Moore et al. 2000). A phylogenetic analysis based on a sample of 122 bHLH sequences from animals, plants and fungi has led to their classification into four monophyletic groups of proteins named A, B, C, and D (Table 2.1) (Atchley and Fitch 1997).

The bHLH proteins bind to hexanucleotide DNA sequences referred to as "E Boxes" (CANNTG) further grouped into CACCTG or CAGCTG (Group A) and CACGTG or CATGTG (Group B) (Murre et al. 1989b; Van Doren et al. 1991; Dang et al. 1992). Group C contains bHLH-PAS proteins that bind to ACGTG or GCGTG core sequences (Ledent and Vervoort 2001). Group D corresponds to HLH proteins, that lack a basic domain and are hence unable to bind DNA (Ledent and Vervoort 2001). This group includes Extramacrochaete (Emc) protein (Ellis et al. 1990; Garrell and Modolell 1990), which act as antagonists of Group A bHLH proteins (Van Doren et al. 1991, 1992). Additional two groups of bHLH proteins have been described recently. Group E corresponds to the family of bHLH proteins which bind preferentially to sequences typically of N boxes (CACGCG or CACGAG). They also contain an additional orange domain and one WRPW peptide in their carboxyl terminus. Group F contains the family of HLH proteins that have the COE domain which has an additional domain involved in both dimerization and DNA binding (Ledent and Vervoort 2001). Some differences in E-box sequences have been reported for different bHLH dimers in *C. elegans*, (Grove et al. 2009).

The bHLH-PAS family is one of the sub-families of bHLH superfamily of transcription factors. The *D. melanogaster Methoprene-tolerant* (*Met*) gene (also known as *Resistance to juvenile hormone*, *Rst (1) JH*) encodes a transcriptional regulator of the bHLH-PAS domain family (Ashok et al. 1998). Met has been shown to bind Juvenile hormone (Shemshedini et al. 1990; Miura et al. 2005) and, therefore, is a good candidate for the elusive JH receptor (Dubrovsky 2005; Flatt and Kawecki 2004; Godlewski et al. 2006).

In a recent study (Bitra et al. 2009), the *T. castaneum* genome sequence deposited into the National Center for Biotechnology Information (NCBI) and the Human Genome Sequencing Center (HGSC) databases (Richards et al. 2008) was searched and 53 *bHLH* genes were identified. A phylogenetic analysis using Bayesian analysis has classified these 53 *bHLH* genes into ten functional families as defined by the presence of bHLH-PAS, HES, Myc/USF2, Atonal, Mesp, Hand, p48, Shout and Achaete- scute (AS-C) domains. Functional analysis of *bHLH-PAS* and *HES* members showed that knocking-down the expression of seven genes belonging to these families affected the growth or development during larval or pupal stages. Out of the 53 bHLH superfamily transcription factors tested, 31 members play an important role in female reproduction in *T. castaneum* (Bitra and Palli 2010).

Chemical pest control using toxic insecticides is the major management practice used today for the control of most of insect pests. The major concern for using the chemical control is the development of resistance by the pests to the majority of insecticide classes. So, there is an urgent need for development of insecticides working through newer target sites. Since a few juvenile hormone analogues such as hydroprene, methoprene and pyriproxifen targeting putative JH receptor (Met) are already registered for pest management, there is a potential to develop more insecticides targeting various bHLH transcription factors. The powerful *Drosophila* genetics tools have been applied to decipher the biological functions of specific *D. melanogaster* bHLH transcription factors. In the past few years, *T. castaneum* has been utilized as a model insect for various functional genomics studies due to its susceptibility to systemic RNA interference (RNAi) effects after dsRNA injection. bHLH transcription factors have been identified in various insects but very little is known about the function of bHLH transcription factors during the insect development. Previous studies on bHLH transcription factors were mainly focused on identification and understanding the function of bHLH transcription factors in various insects (Wang et al. 2007b, 2008; Bitra et al. 2009; Bitra and Palli 2010;

Dang et al. 2011). Not much has been done to develop them as target sites for insecticide development. In the following pages we will briefly summarize studies on various bHLH transcription factors.

2 bHLH-PAS Family

The PAS domain is named after three proteins, *D. melanogaster* period (Per), the human aryl hydrocarbon receptor nuclear translocator (ARNT) and *D. melanogaster* single-minded (*Sim*) (Zelzer et al. 1997). The members of this family are characterized by the presence of a 260–310 residue-long PAS domain that is involved in dimerization and other functions (Kewley et al. 2004). The PAS proteins are known to heterodimerize with other transcription factors and regulate various developmental processes. Eleven members of the PAS family have been identified in the fruit fly and the red flour beetle (Table 2.1) (Bitra et al. 2009). In the red flour beetle, only one protein related to both germ cell expressed (*gce*) and *Met* of *D. melanogaster* is present. In mosquitoes, a single ortholog of *DmMet* and *Dmgce* is present (Wang et al. 2007a). In *D. melanogaster*, *Met* mutants showed increased resistance to JH or its analogue, methoprene (Wilson and Fabian 1986), also *Met* mutant flies are viable (Wilson and Ashok 1998). RNA interference of *GCE* expression in transgenic flies resulted in lethality prior to the adult stage in the absence of Met (Baumann et al. 2010). *TcMet* is one of the most studied members of this family. Suppression of *Met* expression by the injection of *TcMet* dsRNA in the third or fourth larval instars caused precocious metamorphosis two instars later i.e., (after the fifth or sixth) instead of after the seventh or eighth instar (Konopova and Jindra 2007). Knocking down the expression of *Met* in *T. castaneum* disrupted larval-pupal ecdysis and led to the precocious development of adult structures resulting in the development of adultoid tissues underneath the larval skin (Parthasarathy et al. 2008). Impaired function of *TcMet* renders *T. castaneum* resistant to the ectopic effects of JH or JH analog, hydroprene, irrespective of the time and route of application (Parthasarathy et al. 2008; Parthasarathy and Palli 2009). Knocking down the expression of gene coding for juvenile hormone acetyl methyl transferase (JHAMT), an enzyme involved in JH biosynthesis or *Met* by RNAi in males, decreased the egg and progeny production by untreated females mated with dsRNA injected males by affecting accessory gland protein (Acp) secretions (Parthasarathy et al. 2009). *TcMet* is also required for JH regulation of female reproduction especially vitellogenin synthesis in *T. castaneum* (Parthasarathy et al. 2010). These studies clearly established an important role for *TcMet* in juvenile hormone action. Since JH regulates almost every aspect of an insect's life, it is likely that *TcMet* can be used as a potential target site for insecticide development. Transcription factor Met is present only in insect genomes, hence the insecticides targeting Met should be fairly safe to humans and other animals.

Another well-studied member of bHLH-PAS family is Steroid Receptor Coactivator (*SRC*). Recent studies in JH-responsive Aag-2 cells showed that

Fig. 2.1 Phenotypes observed after injection of bHLH-PAS and HES family genes dsRNAs. Control *malE* or *bHLH* dsRNA was injected into day-one final instar larva. The pictures shown are various phenotypes observed during the larval and pupal stages; (**a, b**) final instar larvae injected with *malE* dsRNA and the pupae formed from the final instar larvae injected with *malE* dsRNA; (**c, d**) *SRC* dsRNA injected larvae died 10–15 days after injection; (**e, f**) *Tango* and *Trachealess* dsRNA injected insects died during the quiescent stage; (**g**) pupae formed with problems in wing development after the injection of *Hypoxia* dsRNA; (**h**) abnormal pupae with defects in wing development after the injection of *E(spl)mgamma1* dsRNA (Reprinted with permission from Bitra et al. 2009)

Aedes aegypti homologues of both *Met* and *SRC* are required for the expression of the JH-response gene, *kr-h1*. In addition, *SRC* is also required for expression of ecdysone-response genes (Zhang et al. 2011). In the same study, all ten bHLH-PAS transcription factors were screened as potential heterodimeric partners of *TcMet* and *TcSRC* was identified as a heterodimeric partner of *TcMet* and played an important role in expression of JH response genes. In the mosquito, *A. aegypti, Met* and *FISC* (mosquito homologue of *SRC*) form a functional complex on the juvenile hormone response element (JHRE) identified in the promoter of early trypsin (*AaET*) gene in the presence of JH and activate transcription of JH-response genes (Li et al. 2011).

In *T. castaneum SRC* dsRNA injected larvae remained small, did not reach critical weight to undergo metamorphosis, became sluggish and eventually died (Fig. 2.1c). Lipid metabolism in *TcSRC* dsRNA injected larvae was disrupted when compared with control larvae (Bitra et al. 2009). The *D. melanogaster* homolog for *SRC* is *Taiman,* a nuclear receptor coactivator, which is shown to be involved in follicular cell migration (Bai et al. 2000). Recent studies of *D. melanogaster* identified a BTB domain transcription factor, Abrupt, that attenuates 20E signaling by interacting directly with the bHLH domain of DmTaiman (Jang et al. 2009). *T. castaneum* (TcSRC) had 14% amino acid sequence identity in the overall protein and 48% in the bHLH region with DmTaiman (Bitra et al. 2009). In contrast, *TcSRC* showed 17% amino acid identity in the overall protein and 71% in the bHLH region

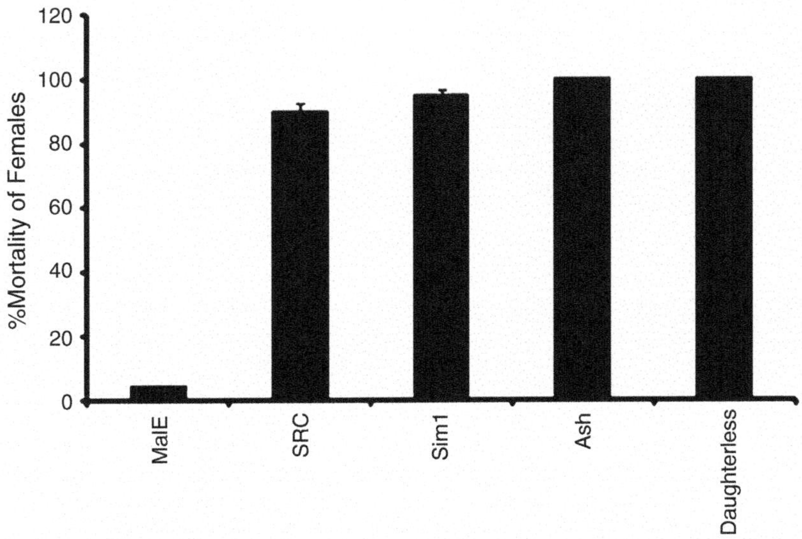

Fig. 2.2 Effect of Knock-down the expression of bHLH superfamily genes on the survival of female beetles. *TcSRC, TcSim1, TcAsh* and *TcDaughterless* dsRNAs were injected into newly emerged female beetles within 24 h after adult emergence and the mortality was observed 7 days after injection of dsRNA. Mean ± SE of three independent replicates are shown (Reprinted with permission from Bitra and Palli 2010)

with human SRC suggesting that this bHLH protein is closer to human SRC than to DmTaiman. The mosquito homologue of the vertebrate SRC is FISC, a p160 coactivator of the ecdysone receptor complex. Protein-protein interactions between the nuclear receptor BetaFTZ- F1 and FISC are required for the stage specific expression of 20E effector genes during mosquito reproduction (Zhu et al. 2006). RNAi studies in *T. castaneum* and previous studies on related proteins in the fruit fly and mosquito suggest that the insect *SRC* plays multiple roles in growth, development and reproduction.

bHLH-PAS family transcription factors, *TcTango*, *TcTrachealess, TcSpineless* and *TcHypoxia* are required for larval to pupal metamorphosis (Bitra et al. 2009). *Tango* and *Trachealess* dsRNA injected final instar *T. castaneum* larvae died during quiescent stage whereas spineless and *hypoxia* dsRNA injected final instar larvae died as pupae and were not able to develop into normal adults (Fig. 2.1). *TcSRC* and *TcSim1* dsRNA injected females died within 1 week after the injection of dsRNA (Fig. 2.2). Six members belonging to this family *TcTango*, *TcTrachealess, TcSpineless, TcDysfusion, TcHypoxia,* and *TcSim2* also affected female reproduction in *T. castaneum* (Bitra and Palli 2010). Interestingly, knocking down the expression of *TcSpineless* in female beetles affected both the egg laying and a few eggs that were laid were not able to develop into first instar larvae (Fig. 2.3). In flies, *Sim* mutants are embryonic lethal (Thomas et al. 1988). *DmSim* is also required for normal development of the female germ line, while it also appears to have the most

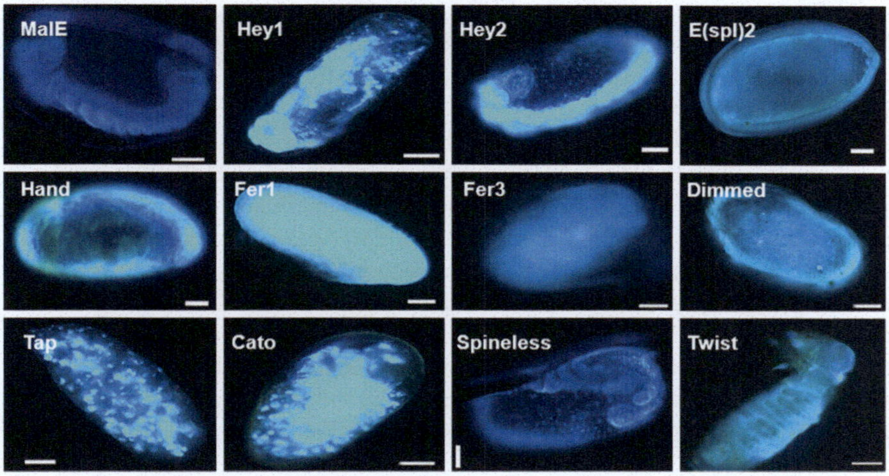

Fig. 2.3 Nuclear staining of embryos of *T. castaneum*. Eggs were collected within 24 h after egg laying. After 5 days of egg laying eggs were fixed overnight in 4% paraformaldehyde. Specimens were washed twice with 1XPBS for 5 min and stained using DAPI for 5 min to observe the stage of arrest of embryogenesis. Scale bar 100 μm (Reprinted with permission from Bitra and Palli 2010)

specific affect on the embryonic patterning (Mayer and Nusslein-Volhard 1988). In *Sim* mutants, the embryonic gut is affected and embryonic anal pad is deformed (Maeda et al. 2007). In *D. melanogaster, Trachealess* mutants display no trachea and this gene product is required for tracheal and salivary gland development (Parrish et al. 2006). Silencing the gene using RNAi caused defects in muscle development and dendrite morphogenesis during the embryonic stage (Parrish et al. 2006). RNAi of *DmTango* resulted in reduced arborization of dendritic neurons, defects in muscle and dendrite morphogenesis during embryonic stage (Parrish et al. 2006). *DmSpineless* mutants display transformation of distal antennae to leg, reduction in size of bristles and deletion of distal leg structures in adults (Duncan et al. 1998) and sterility. Severe loss of function mutation resulted in amplification of sex combs on the first leg in adult flies (Kuzin et al. 1997). *DmSpineless* also regulates dendrite diversity in the dendritic arborization of sensory neurons (Kim et al. 2006).

3 HES Family

HES family members play key roles during embryogenesis, cell proliferation and tissue differentiation and play important roles in Notch signaling pathway (Muskavitch 1994). The members of this family contain additional domains such as orange and the C-terminal WRPW motif. These domains allow these proteins to repress transcription by interacting with proteins such as Groucho (Fisher and

Caudy 1998). Knocking down the expression of *TcHairy, TcDeadpan, and TcSide1* has affected oogenesis while injecting dsRNAs of *TcSide1, TcHey1*, or *TcHey2* into female adults affected embryogenesis (Bitra and Palli 2010) and none of the eggs laid by females were able to develop into normal adults. Hey1-depleted embryos stained with DAPI showed that most of the embryos were blocked at mid-stage of germ band growth and Hey2-depleted embryos were blocked at the completion of germ band growth (Fig. 2.3). *Hey* gene products play important roles during embryogenesis in *D. melanogaster* (Leimeister et al. 1999). During embryogenesis in *D. melanogaster hairy* acts as a pair rule gene in the establishment of segments (Zhang and Levine 1999). In *T. castaneum E(spl)mgamma1* dsRNA injected larvae developed into abnormal pupae and pupal to adult development was impaired (Fig. 2.1h). The pupae that developed from *E(spl)mgamma1* dsRNA injected larvae showed defects in wing development. E(spl) complex of genes show a distinct pattern of expression in wing imaginal discs in flies (Jennings et al. 1994).

4 Myc/USF Family

Myc/USF family of bHLH proteins is widely expressed in many different cell types and it belongs to bHLH-Zip family. Myc, Max and Mxi function in regulation of alternative states of cell activation or quiescence in *D. melanogaster* (Amati and Land 1994). Loss of function alleles of *Mnt* in *D. melanogaster* resulted in flies with increased weight and decreased life span (Loo et al. 2005). *D. melanogaster* females are sterile as a result of defective oogenesis and degeneration of the egg chamber occurs in *DmMyc* mutant females (Gallant et al. 1996). *Dmmyc* mutants failed to attain the normal size and showed arrested development (Pierce et al. 2004). Myc mutant flies are viable, and smaller than the wild type flies (Gallant et al. 1996). In *T. castaneum* females injected with *Myc* and *Max* dsRNA's and mated with uninjected male beetles showed severe defects in egg laying. Ovaries dissected from these beetles were smaller and the oocyte growth was arrested (Bitra and Palli 2010).

5 Hand Family

In *Hand* family, two members are identified in both *D. melanogaster* and *T. castaneum* (Bitra et al. 2009). DmHand is the *D. melanogaster* ortholog of the vertebrate Hand protein and is 69% homologous in the bHLH domain to vertebrate Hand. These proteins are involved in heart formation in flies and vertebrates. *DmH*and expression is detected in bilateral stripes in the ventral mesoderm beginning at stage 10 of embryonic development. This mRNA is also present in the dorsal vessel (heart) and the circular visceral musculature, the two tissues derived from this mesoderm. *DmHand* mRNA is detected in a small subset of cells in the central nervous system at stage 13 (Moore et al. 2000). Most of *hand* mutants in *D. melanogaster* die during late embryonic and early larval stages exhibiting hypoplastic myocardium, and a deficiency of pericardial and lymph gland hematopoietic cells (Han et al. 2006). In *T. castaneum* knocking down the expression of *Hand* gene in adult females, which were subsequently mated with un-injected males, did not affect oogenesis but the eggs were not able to develop into offspring (Fig. 2.3).

In flies, formation of the tracheal tree is severely affected in *twist* mutant embryos and there is only partial development of dorsal and ventral branches (Franch-Marro and Casanova 2000). *DmTwist* also plays an important role in the development of mesoderm during embryonic stage (Arora and Nusslein-Volhard 1992). In *D. melanogaster,* Twist activates Snail, Tinman, Bagpipe, and Mef2 (Moore et al. 2000). Knocking down the expression of *TcTwist* in adult female beetles did not affect oogenesis, but none of the eggs laid by the RNAi beetles was able to develop into larvae (Bitra and Palli 2010). DAPI staining of embryos showed that development was not blocked in the dsRNA-*Twist* injected embryos, but the embryos failed to hatch and emerge as the first instar larvae (Fig. 2.3). In *B. mori*, reverse transcription polymerase chain reaction (RT-PCR) and western blot analyses revealed that *BmTwist* is expressed during all developmental stages in various larval tissues (Guo et al. 2011).

6 Mesp Family

Mesp family proteins are necessary for mesoderm segmentation initiation and have 53% sequence identity with vertebrate proteins in bHLH domain. DmSage (salivary gland-expressed bHLH) is distantly related to the vertebrate Mesp family and is expressed in the salivary glands of *D. melanogaster* (Moore et al. 2000). It is highly expressed during early stages of embryonic development. Its zygotic expression begins in the salivary gland anlage at stage 10 and persists until stage 15 (Moore et al. 2000). Mesp gene family member *HLH54F* is highly expressed in the prothoracic glands of *B. mori* and *D. melanogaster*, an organ that produces the insect steroid hormone, ecdysone (Namiki et al. 2009).

7 Shout Family

Three members (Delilah1 and Delilah2, Shout) of this family have been identified in *T. castaneum* and two members (Delilah, Shout) of this family are identified in *D. melanogaster* (Bitra et al. 2009). DmDelilah protein forms heterodimers with E12 protein that binds to muscle creatine kinase protein and plays an important role in the differentiation of epidermal cells into muscle attachment sites (Armand et al. 1994). In *T. castaneum* knocking down the expression of *Delilah2* gene did not affect egg laying, but has affected egg hatching. The eggs failed to hatch and first instar larvae underneath the chorion membrane were dead (Bitra and Palli 2010).

8 p48 Family

D. melanogaster and *T. castaneum* p48 family members (Fer1, Fer2 and Fer3) (Bitra et al. 2009) are closely related to the bHLH domain of the p48 subunit of PTF1, a pancreatic, exocrine cell-specific transcription factor identified in mouse (Moore et al. 2000). *DmFer1* is expressed in the epidermal cells when they start producing cuticular proteins. *DmFer3* is expressed in the posterior midgut primordial at stage 11 and in the anterior midgut primordial at stage 12 (Moore et al. 2000). In *T. castaneum* knocking down the expression of *TcFer1* and *TcFer3* in female adults blocked embryogenesis and none of the eggs laid by RNAi beetles developed into first instar larvae (Fig. 2.3) and *TcFer2* dsRNA injected females failed to lay eggs (Bitra and Palli 2010).

9 NeuroD/Neurogenin Family

Four genes belong to this family have been identified in *T. castaneum* (*NeuroD, Beta3, Tap and Dimmed*) (Bitra et al. 2009); whereas, in *D. melanogaster* (*Beta3, Tap and Dimmed*) only three members of this family have been identified. *TcNeuroD* is a unique gene present in *T. castaneum* and is absent in all the other insect genomes sequenced. Null mutations in *D. melanogaster Beta3* caused larval lethality due to failed gut function and showed abnormalities such as defective sensory perception (Dettman et al. 2001). Beta3 showed 96% sequence identity between fly and vertebrate proteins in the bHLH domain. It is also required for viability of *D. melanogaster* adults (Kimble et al. 1990). *TcBeta3* dsRNA injected female beetles failed to lay eggs and only few eggs were able to develop into offspring and the ovarian development was blocked (Bitra and Palli 2010). When *Dimmed* gene was misexpressed in embryos of flies most of them died (Hewes et al. 2003). Similar to the effect in *D. melanogaster*, *TcDimmed* dsRNA injected females were able to lay eggs but all of the eggs failed to hatch and develop into offspring. Embryos depleted of Dimmed protein stained with DAPI showed that most of the embryos died at the beginning of the cellular blastoderm stage (Fig. 2.3). *Tap* dsRNA injected females in *T. castaneum* behaved similar to *Dimmed* dsRNA injected females except that the embryos died at early stages of germ band growth (Fig. 2.3).

10 Atonal Family

The members of this family (*Atonal, Amos and Cato*) are mainly involved in the nervous system development. Loss of *Amos* function by RNAi in flies eliminated neuron formation in embryos and mis-expression of *Amos* resulted in ectopic multiple dendritic neurons (Huang et al. 2000). In *T. castaneum Cato* dsRNA injected females laid eggs that failed to develop into larvae and embryos died at the early stages of germ band growth (Bitra and Palli 2010). Functional studies in flies showed that Cato-deficient embryos contain chordotonal neurons that are consistently malformed, appearing longer and often thicker than wild type embryos (Goulding et al. 2000).

11 AS-C Family

Members of the AS-C family are known to be involved in neurogenesis. These proteins include *D. melanogaster* proneural proteins, Achaete, Scute, Lethal of scute (L'sc) and Asense (Ghysen et al. 1993). The AS-C proteins function at several stages of neurogenesis in *D. melanogaster*. Loss of achaete scute homolog (*Ash*) gene function causes neural hypoplasia during embryonic stage (Jimenez and Campos-Ortega 1990) and mutations cause loss of specific cluster of bristles in adult flies (Simpson 1990). In *T. castaneum* only three genes coding for members of this family have been identified (Bitra et al. 2009). RNAi and misexpression studies showed that *TcAsh* is necessary for neural precursor formation and survival in *T. castaneum* (Fig. 2.2) and sufficient for neural precursor formation in *D. melanogaster* (Wheeler et al. 2003; Bitra and Palli 2010). In *B. mori*, four *AS-C* homolog genes referred to as *Bm ASH, Bm ASH2, Bm ASH3,* and *Bmase* have been identified (Tong et al. 2008). RNAi of AS-C genes further indicated that *BmAsh* is necessary for the formation of scales in silkworm wing (Tong et al. 2008).

12 Miscellaneous Genes

Miscellaneous genes are named based on the absence of conserved functional domains (Moore et al. 2000; Bitra et al. 2009). *HLH106, Bigmax, AP-4, HLH3B, HLH4C, Nautilus, Emc, Daughterless, Paraxis* and *Mitf* are classified under this group in *T. castaneum* and *D. melanogaster* (Moore et al. 2000; Bitra et al. 2009). Functional studies showed that *D. melanogaster* larvae lacking HLH106, a homologue of sterol regulatory element binding protein (SREBP), died prior to the third instar stage (Kunte et al. 2006). *HLH106* mutant of *D. melanogaster* larvae showed pronounced growth defects prior to lethality and substantial defects in the genes required for fatty acid synthesis.

In flies, *Daughterless* mutant cells give rise to a narrow posterior scar across the eye and play an important role in the photoreceptor cells (Brown et al. 1996). During embryonic stage, this gene is required for survival of salivary gland cells after invagination and for development of central and peripheral nervous system in both the sexes (Caudy et al. 1988). Hypomorphic daughterless mutant genotypes exhibit dramatic defects during oogenesis including aberrantly defined follicles and loss of interfolicular stalks (Cummings and Cronmiller 1994). It is also required for the complete differentiation of polar and stalk cells (Smith et al. 2002). In *T. castaneum*, knocking down the expression of *daughterless* gene caused 100% mortality of both male and female beetles (Fig. 2.2). In addition, *AP-4* dsRNA injected females showed a decrease in egg number when compared to control adult beetles (Bitra and Palli 2010).

DmEmc functions as a regulator of sensory organ precursor formation and formation of organs such as midgut suggesting a role in regulating morphogenesis (Ellis 1994). It also participates in cell proliferation during vein patterning as well as in vein differentiation (de Celis et al. 1995). The mutant embryos do not hatch and their cuticle displays multiple alterations. *Emc* is involved in the separation of adjacent egg chambers, which contain multiple germ line cysts and lack intervening and polar stalk cells (Adam and Montell 2004).

13 Future Direction

The bHLH transcription factors identified to date could serve as potential insecticide targets. These proteins could be used in small molecule screen, or in the development of RNAi-based pest management methods. One possible target that can be effectively utilized is Met. With the recent availability of information regarding the possible role of Met in JH action, new insecticides could be identified using Met protein in high-throughput screening assays. The insecticides developed targeting Met should be safe to humans and other animals because *Met* gene is present only in insect genomes. Other possible target sites include *Daughterless, SRC*, and *Ash* genes because knocking down the expression of these genes resulted in lethality of *T. castaneum* (Fig. 2.3). One other possible mechanism of utilization of these target sites is through direct ingestion of dsRNA or by oral RNAi using hairpin RNA generating constructs expressed in plants or bacteria (Baum et al. 2007; Zhu et al. 2011).

14 Summary and Conclusions

The functional studies on bHLH transcription factors in few model insects such as *D. melanogaster*, *T. castaneum*, and *B. mori* provide evidence about their essential functions in these insects raising the possibility of using them as new target sites. With the availability of more whole genome sequences in the coming years, more homologues of these transcription factors will be identified from insects especially from those that are economically important. Some of these could be utilized for developing new insecticides for controlling pests of crops and vectors of disease-causing pathogens. Currently, there are only a few commercial insecticides such as JH analogs targeting bHLH transcription factors. Insecticide resistance is a major growing threat, which further increases the need for developing new insecticides targeting new target sites.

In a recent RNAi screen in *T. castaneum*, we identified seven bHLH genes that are required for larval development and 31 genes for female reproduction and survival. Our study provided a systematic functional analysis on bHLH genes in *T. castaneum* and increased our understanding on the function of insect bHLH transcription factors in insect development and reproduction. In the near future, similar studies in pest insects and disease vectors will identify bHLH transcription factors that could be used for developing new pest management methods.

Acknowledgements The research in Palli laboratory was supported by grants from the National Science Foundation (IBN-0421856), the National Institute of Health (GM070559-07), and the National Research Initiative of the USDA-NIFA (2011-67013-30143). This report is contribution number 11-08-075 from the Kentucky Agricultural Experimental Station.

References

Adam JC, Montell DJ (2004) A role for extra macrochaetae downstream of Notch in follicle cell differentiation. Development 131(23):5971–5980. doi:131/23/5971 [pii] 10.1242/dev.01442

Amati B, Land H (1994) Myc-Max-Mad: a transcription factor network controlling cell cycle progression, differentiation and death. Curr Opin Genet Dev 4(1):102–108

Armand P, Knapp AC, Hirsch AJ, Wieschaus EF, Cole MD (1994) A novel basic helix-loop-helix protein is expressed in muscle attachment sites of the Drosophila epidermis. Mol Cell Biol 14(6):4145–4154

Arora K, Nusslein-Volhard C (1992) Altered mitotic domains reveal fate map changes in Drosophila embryos mutant for zygotic dorsoventral patterning genes. Development 114(4):1003–1024

Ashok M, Turner C, Wilson TG (1998) Insect juvenile hormone resistance gene homology with the bHLH-PAS family of transcriptional regulators. Proc Natl Acad Sci USA 95(6):2761–2766

Atchley WR, Fitch WM (1997) A natural classification of the basic helix-loop-helix class of transcription factors. Proc Natl Acad Sci USA 94(10):5172–5176

Bai J, Uehara Y, Montell DJ (2000) Regulation of invasive cell behavior by taiman, a Drosophila protein related to AIB1, a steroid receptor coactivator amplified in breast cancer. Cell 103(7):1047–1058. doi:S0092-8674(00)00208-7 [pii]

Baum JA, Bogaert T, Clinton W, Heck GR, Feldmann P, Ilagan O, Johnson S, Plaetinck G, Munyikwa T, Pleau M, Vaughn T, Roberts J (2007) Control of coleopteran insect pests through RNA interference. Nat Biotechnol 25(11):1322–1326

Baumann A, Barry J, Wang S, Fujiwara Y, Wilson TG (2010) Paralogous genes involved in juvenile hormone action in Drosophila melanogaster. Genetics 185(4):1327–1336. doi:genetics.110.116962 [pii] 10.1534/genetics.110.116962

Bitra K, Palli SR (2010) The members of bHLH transcription factor superfamily are required for female reproduction in the red flour beetle, Tribolium castaneum. J Insect Physiol 56(10):1481–1489. doi:S0022-1910(10)00071-5 [pii] 10.1016/j.jinsphys.2010.03.005

Bitra K, Tan A, Dowling A, Palli SR (2009) Functional characterization of PAS and HES family bHLH transcription factors during the metamorphosis of the red flour beetle, Tribolium castaneum. Gene 448(1):74–87. doi:S0378-1119(09)00436-3 [pii] 10.1016/j.gene.2009.08.003

Brown NL, Paddock SW, Sattler CA, Cronmiller C, Thomas BJ, Carroll SB (1996) Daughterless is required for Drosophila photoreceptor cell determination, eye morphogenesis, and cell cycle progression. Dev Biol 179(1):65–78. doi:10.1006/dbio.1996.0241 S0012-1606(96)90241-9 [pii]

Caudy M, Vassin H, Brand M, Tuma R, Jan LY, Jan YN (1988) Daughterless, a Drosophila gene essential for both neurogenesis and sex determination, has sequence similarities to myc and the achaete-scute complex. Cell 55(6):1061–1067. doi:0092-8674(88)90250-4 [pii]

Cummings CA, Cronmiller C (1994) The daughterless gene functions together with Notch and Delta in the control of ovarian follicle development in Drosophila. Development 120(2):381–394

Dang CV, Dolde C, Gillison ML, Kato GJ (1992) Discrimination between related DNA sites by a single amino acid residue of Myc-related basic-helix-loop-helix proteins. Proc Natl Acad Sci USA 89(2):599–602

Dang CW, Wang Y, Chen KP, Yao Q, Zhang DB, Guo M (2011) The basic helix-loop-helix transcription factor family in the pea aphid, Acyrthosiphon pisum. J Insect Sci 11:84. doi:10.1673/031.011.8401

de Celis JF, Baonza A, Garcia-Bellido A (1995) Behavior of extramacrochaetae mutant cells in the morphogenesis of the Drosophila wing. Mech Dev 53(2):209–221. doi:0925-4773(95)00436-5 [pii]

Dettman RW, Turner FR, Hoyle HD, Raff EC (2001) Embryonic expression of the divergent Drosophila beta3-tubulin isoform is required for larval behavior. Genetics 158(1):253–263

Dubrovsky EB (2005) Hormonal cross talk in insect development. Trends Endocrinol Metab 16(1):6–11. doi:S1043-2760(04)00274-7 [pii] 10.1016/j.tem.2004.11.003

Duncan DM, Burgess EA, Duncan I (1998) Control of distal antennal identity and tarsal development in Drosophila by spineless-aristapedia, a homolog of the mammalian dioxin receptor. Genes Dev 12(9):1290–1303

Ellis HM, Span DR, Posakony JN (1990) Extramacrochaete, a negative regulator of sensory organ development in Drosophila, defines a new class of helix-loop-helix proteins. Cell 61(1):27–38

Ellis HM (1994) Embryonic expression and function of the Drosophila helix-loop-helix gene, extramacrochaetae. Mech Dev 47(1):65–72. doi:0925-4773(94)90096-5 [pii]

Ferre-D'Amare AR, Prendergast GC, Ziff EB, Burley SK (1993) Recognition by Max of its cognate DNA through a dimeric b/HLH/Z domain. Nature 363(6424):38–45. doi:10.1038/363038a0

Fisher A, Caudy M (1998) The function of hairy-related bHLH repressor proteins in cell fate decisions. Bioessays 20(4):298–306

Flatt T, Kawecki TJ (2004) Pleiotropic effects of methoprene-tolerant (Met), a gene involved in juvenile hormone metabolism, on life history traits in Drosophila melanogaster. Genetica 122(2):141–160

Franch-Marro X, Casanova J (2000) The alternative migratory pathways of the Drosophila tracheal cells are associated with distinct subsets of mesodermal cells. Dev Biol 227(1):80–90. doi:10.1006/dbio.2000.9890 S0012-1606(00)99890-7 [pii]

Gallant P, Shiio Y, Cheng PF, Parkhurst SM, Eisenman RN (1996) Myc and Max homologs in Drosophila. Science 274(5292):1523–1527

Garrell J, Modolell J (1990) The Drosophila extramacrochaetae locus, an antagonist of proneural genes that, like these genes, encodes a helix-loop-helix-protein. Cell 61(1):39–48

Ghysen A, Dambly-Chaudiere C, Jan LY, Jan YN (1993) Cell interactions and gene interactions in peripheral neurogenesis. Genes Dev 7(5):723–733

Godlewski J, Wang S, Wilson TG (2006) Interaction of bHLH-PAS proteins involved in juvenile hormone reception in Drosophila. Biochem Biophys Res Commun 342(4):1305–1311. doi:S0006-291X(06)00402-5 [pii] 10.1016/j.bbrc.2006.02.097

Goulding SE, White NM, Jarman AP (2000) cato encodes a basic helix-loop-helix transcription factor implicated in the correct differentiation of Drosophila sense organs. Dev Biol 221(1):120–131. doi:10.1006/dbio.2000.9677 S0012-1606(00)99677-5 [pii]

Grove CA, De Masi F, Barrasa MI, Newburger DE, Alkema MJ, Bulyk ML, Walhout AJ (2009) A multiparameter network reveals extensive divergence between C. elegans bHLH transcription factors. Cell 138(2):314–327. doi:S0092-8674(09)00519-4 [pii] 10.1016/j.cell.2009.04.058

Guo M, Wang Y, Shi J, Kang L, Yao Q, Wang F, Qin L, Chen K (2011) Molecular cloning and characterization of twist gene in Bombyx mori. Mol Cell Biochem 348(1–2):69–76. doi:10.1007/s11010-010-0639-7

Han Z, Yi P, Li X, Olson EN (2006) Hand, an evolutionarily conserved bHLH transcription factor required for Drosophila cardiogenesis and hematopoiesis. Development 133(6):1175–1182. doi:dev.02285 [pii] 10.1242/dev.02285

Hassan BA, Bellen HJ (2000) Doing the MATH: is the mouse a good model for fly development? Genes Dev 14(15):1852–1865

Hewes RS, Park D, Gauthier SA, Schaefer AM, Taghert PH (2003) The bHLH protein Dimmed controls neuroendocrine cell differentiation in Drosophila. Development 130(9):1771–1781

Huang ML, Hsu CH, Chien CT (2000) The proneural gene amos promotes multiple dendritic neuron formation in the Drosophila peripheral nervous system. Neuron 25(1):57–67. doi:S0896-6273(00)80871-5 [pii]

Jan YN, Jan LY (1993) HLH proteins, fly neurogenesis, and vertebrate myogenesis. Cell 75(5): 827–830. doi:0092-8674(93)90525-U [pii]

Jang AC, Chang YC, Bai J, Montell D (2009) Border-cell migration requires integration of spatial and temporal signals by the BTB protein Abrupt. Nat Cell Biol 11(5):569–579. doi:ncb1863 [pii] 10.1038/ncb1863

Jennings B, Preiss A, Delidakis C, Bray S (1994) The Notch signalling pathway is required for Enhancer of split bHLH protein expression during neurogenesis in the Drosophila embryo. Development 120(12):3537–3548

Jimenez F, Campos-Ortega JA (1990) Defective neuroblast commitment in mutants of the achaete-scute complex and adjacent genes of D. melanogaster. Neuron 5(1):81–89. doi:0896-6273(90)90036-F [pii]

Kadesch T (1993) Consequences of heteromeric interactions among helix-loop-helix proteins. Cell Growth Differ 4(1):49–55

Kewley RJ, Whitelaw ML, Chapman-Smith A (2004) The mammalian basic helix-loop-helix/PAS family of transcriptional regulators. Int J Biochem Cell Biol 36(2):189–204. doi:S1357272503002115 [pii]

Kim MD, Jan LY, Jan YN (2006) The bHLH-PAS protein Spineless is necessary for the diversification of dendrite morphology of Drosophila dendritic arborization neurons. Genes Dev 20(20):2806–2819. doi:gad.1459706 [pii] 10.1101/gad.1459706

Kimble M, Dettman RW, Raff EC (1990) The beta 3-tubulin gene of Drosophila melanogaster is essential for viability and fertility. Genetics 126(4):991–1005

Konopova B, Jindra M (2007) Juvenile hormone resistance gene Methoprene-tolerant controls entry into metamorphosis in the beetle Tribolium castaneum. Proc Natl Acad Sci USA 104(25):10488–10493. doi:0703719104 [pii] 10.1073/pnas.0703719104

Kunte AS, Matthews KA, Rawson RB (2006) Fatty acid auxotrophy in Drosophila larvae lacking SREBP. Cell Metab 3(6):439–448. doi:S1550-4131(06)00130-6 [pii] 10.1016/j.cmet.2006.04.011

Kuzin B, Doszhanov K, Mazo A (1997) Interaction between spineless-aristapedia gene and genes from Antennapedia and bithorax complexes of Drosophila melanogaster. Int J Dev Biol 41(6):867–875

Ledent V, Vervoort M (2001) The basic helix-loop-helix protein family: comparative genomics and phylogenetic analysis. Genome Res 11(5):754–770. doi:10.1101/gr.177001

Ledent V, Paquet O, Vervoort M (2002) Phylogenetic analysis of the human basic helix-loop-helix proteins. Genome Biol 3(6):RESEARCH0030

Leimeister C, Externbrink A, Klamt B, Gessler M (1999) Hey genes: a novel subfamily of hairy- and Enhancer of split related genes specifically expressed during mouse embryogenesis. Mech Dev 85(1–2):173–177. doi:S0925-4773(99)00080-5 [pii]

Li M, Mead EA, Zhu J (2011) Heterodimer of two bHLH-PAS proteins mediates juvenile hormone-induced gene expression. Proc Natl Acad Sci USA 108(2):638–643. doi:1013914108 [pii] 10.1073/pnas.1013914108

Loo LW, Secombe J, Little JT, Carlos LS, Yost C, Cheng PF, Flynn EM, Edgar BA, Eisenman RN (2005) The transcriptional repressor dMnt is a regulator of growth in Drosophila melanogaster. Mol Cell Biol 25(16):7078–7091. doi:25/16/7078 [pii] 10.1128/MCB.25.16.7078-7091.2005

Maeda R, Hozumi S, Taniguchi K, Sasamura T, Murakami R, Matsuno K (2007) Roles of single-minded in the left-right asymmetric development of the Drosophila embryonic gut. Mech Dev 124(3):204–217. doi:S0925-4773(06)00216-4 [pii] 10.1016/j.mod.2006.12.001

Mayer U, Nusslein-Volhard C (1988) A group of genes required for pattern formation in the ventral ectoderm of the Drosophila embryo. Genes Dev 2(11):1496–1511

Miura K, Oda M, Makita S, Chinzei Y (2005) Characterization of the Drosophila Methoprene -tolerant gene product. Juvenile hormone binding and ligand-dependent gene regulation. FEBS J 272(5):1169–1178. doi:EJB4552 [pii] 10.1111/j.1742-4658.2005.04552.x

Moore AW, Barbel S, Jan LY, Jan YN (2000) A genomewide survey of basic helix-loop-helix factors in Drosophila. Proc Natl Acad Sci USA 97(19):10436–10441. doi:10.1073/pnas.170301897 170301897 [pii]

Murre C, McCaw PS, Baltimore D (1989a) A new DNA binding and dimerization motif in immunoglobulin enhancer binding, daughterless, MyoD, and myc proteins. Cell 56(5):777–783. doi:0092-8674(89)90682-X [pii]

Murre C, McCaw PS, Vaessin H, Caudy M, Jan LY, Jan YN, Cabrera CV, Buskin JN, Hauschka SD, Lassar AB et al (1989b) Interactions between heterologous helix-loop-helix proteins generate complexes that bind specifically to a common DNA sequence. Cell 58(3):537–544. doi:0092-8674(89)90434-0 [pii]

Muskavitch MA (1994) Delta-notch signaling and Drosophila cell fate choice. Dev Biol 166(2):415–430. doi:S0012-1606(84)71326-1 [pii] 10.1006/dbio.1994.1326

Namiki T, Niwa R, Higuchi A, Yoshiyama T, Mita K, Kataoka H (2009) A basic-HLH transcription factor, HLH54F, is highly expressed in the prothoracic gland in the silkworm Bombyx mori and the fruit fly Drosophila melanogaster. Biosci Biotechnol Biochem 73(3):762–765. doi:JST. JSTAGE/bbb/80737 [pii]

Parrish JZ, Kim MD, Jan LY, Jan YN (2006) Genome-wide analyses identify transcription factors required for proper morphogenesis of Drosophila sensory neuron dendrites. Genes Dev 20(7):820–835. doi:gad.1391006 [pii] 10.1101/gad.1391006

Parthasarathy R, Palli SR (2009) Molecular analysis of juvenile hormone analog action in controlling the metamorphosis of the red flour beetle, Tribolium castaneum. Arch Insect Biochem Physiol 70(1):57–70. doi:10.1002/arch.20288

Parthasarathy R, Tan A, Palli SR (2008) bHLH-PAS family transcription factor methoprene-tolerant plays a key role in JH action in preventing the premature development of adult structures during larval-pupal metamorphosis. Mech Dev 125(7):601–616. doi:S0925-4773(08)00036-1 [pii] 10.1016/j.mod.2008.03.004

Parthasarathy R, Tan A, Sun Z, Chen Z, Rankin M, Palli SR (2009) Juvenile hormone regulation of male accessory gland activity in the red flour beetle, Tribolium castaneum. Mech Dev 126(7):563–579. doi:S0925-4773(09)00038-0 [pii] 10.1016/j.mod.2009.03.005

Parthasarathy R, Sun Z, Bai H, Palli SR (2010) Juvenile hormone regulation of vitellogenin synthesis in the red flour beetle, Tribolium castaneum. Insect Biochem Mol Biol 40(5): 405–414. doi:S0965-1748(10)00082-2 [pii] 10.1016/j.ibmb.2010.03.006

Pierce SB, Yost C, Britton JS, Loo LW, Flynn EM, Edgar BA, Eisenman RN (2004) dMyc is required for larval growth and endoreplication in Drosophila. Development 131(10):2317–2327. doi:10.1242/dev.01108 131/10/2317 [pii]

Richards S, Gibbs RA, Weinstock GM, Brown SJ, Denell R, Beeman RW, Gibbs R, Bucher G, Friedrich M, Grimmelikhuijzen CJ, Klingler M, Lorenzen M, Roth S, Schroder R, Tautz D, Zdobnov EM, Muzny D, Attaway T, Bell S, Buhay CJ, Chandrabose MN, Chavez D, Clerk-Blankenburg KP, Cree A, Dao M, Davis C, Chacko J, Dinh H, Dugan-Rocha S, Fowler G, Garner TT, Garnes J, Gnirke A, Hawes A, Hernandez J, Hines S, Holder M, Hume J, Jhangiani SN, Joshi V, Khan ZM, Jackson L, Kovar C, Kowis A, Lee S, Lewis LR, Margolis J, Morgan M, Nazareth LV, Nguyen N, Okwuonu G, Parker D, Ruiz SJ, Santibanez J, Savard J, Scherer SE, Schneider B, Sodergren E, Vattahil S, Villasana D, White CS, Wright R, Park Y, Lord J, Oppert B, Brown S, Wang L, Weinstock G, Liu Y, Worley K, Elsik CG, Reese JT, Elhaik E, Landan G, Graur D, Arensburger P, Atkinson P, Beidler J, Demuth JP, Drury DW, Du YZ, Fujiwara H, Maselli V, Osanai M, Robertson HM, Tu Z, Wang JJ, Wang S, Song H, Zhang L, Werner D, Stanke M, Morgenstern B, Solovyev V, Kosarev P, Brown G, Chen HC, Ermolaeva O, Hlavina W, Kapustin Y, Kiryutin B, Kitts P, Maglott D, Pruitt K, Sapojnikov V,

Souvorov A, Mackey AJ, Waterhouse RM, Wyder S, Kriventseva EV, Kadowaki T, Bork P, Aranda M, Bao R, Beermann A, Berns N, Bolognesi R, Bonneton F, Bopp D, Butts T, Chaumot A, Denell RE, Ferrier DE, Gordon CM, Jindra M, Lan Q, Lattorff HM, Laudet V, von Levetsow C, Liu Z, Lutz R, Lynch JA, da Fonseca RN, Posnien N, Reuter R, Schinko JB, Schmitt C, Schoppmeier M, Shippy TD, Simonnet F, Marques-Souza H, Tomoyasu Y, Trauner J, Van der Zee M, Vervoort M, Wittkopp N, Wimmer EA, Yang X, Jones AK, Sattelle DB, Ebert PR, Nelson D, Scott JG, Muthukrishnan S, Kramer KJ, Arakane Y, Zhu Q, Hogenkamp D, Dixit R, Jiang H, Zou Z, Marshall J, Elpidina E, Vinokurov K, Oppert C, Evans J, Lu Z, Zhao P, Sumathipala N, Altincicek B, Vilcinskas A, Williams M, Hultmark D, Hetru C, Hauser F, Cazzamali G, Williamson M, Li B, Tanaka Y, Predel R, Neupert S, Schachtner J, Verleyen P, Raible F, Walden KK, Angeli S, Foret S, Schuetz S, Maleszka R, Miller SC, Grossmann D (2008) The genome of the model beetle and pest Tribolium castaneum. Nature 452(7190):949–955. doi:nature06784 [pii] 10.1038/nature06784

Satou Y, Satoh N (2003) Genomewide surveys of developmentally relevant genes in Ciona intestinalis. Dev Genes Evol 213(5–6):211–212. doi:10.1007/s00427-003-0330-z

Shemshedini L, Lanoue M, Wilson TG (1990) Evidence for a juvenile hormone receptor involved in protein synthesis in Drosophila melanogaster. J Biol Chem 265(4):1913–1918

Simionato E, Ledent V, Richards G, Thomas-Chollier M, Kerner P, Coornaert D, Degnan BM, Vervoort M (2007) Origin and diversification of the basic helix-loop-helix gene family in metazoans: insights from comparative genomics. BMC Evol Biol 7:33. doi:1471-2148-7-33 [pii] 10.1186/1471-2148-7-33

Simpson P (1990) Lateral inhibition and the development of the sensory bristles of the adult peripheral nervous system of Drosophila. Development 109(3):509–519

Smith JE 3rd, Cummings CA, Cronmiller C (2002) Daughterless coordinates somatic cell proliferation, differentiation and germline cyst survival during follicle formation in Drosophila. Development 129(13):3255–3267

Thomas JB, Crews ST, Goodman CS (1988) Molecular genetics of the single-minded locus: a gene involved in the development of the Drosophila nervous system. Cell 52(1):133–141. doi:0092-8674(88)90537-5 [pii]

Tong XL, Dai FY, Su MK, Ma Y, Tan D, Zhang Z, He NJ, Xia QY, Lu C, Xiang ZH (2008) Identification and expression of the achaete-scute complex in the silkworm, Bombyx mori. Insect Mol Biol 17(4):395–404. doi:IMB811 [pii] 10.1111/j.1365-2583.2008.00811.x

Van Doren M, Ellis HM, Posakony JW (1991) The Drosophila extramacrochaetae protein antagonizes sequence-specific DNA binding by daughterless/achaete-scute protein complexes. Development 113(1):245–255

Van Doren M, Powell PA, Pasternak D, Singson A, Posakony JW (1992) Spatial regulation of proneural gene activity: auto- and cross-activation of achaete is antagonised by extramacrochaete. Genes Dev 6(12B):2592–2605

Wang S, Baumann A, Wilson TG (2007a) Drosophila melanogaster Methoprene-tolerant (Met) gene homologs from three mosquito species: members of PAS transcriptional factor family. J Insect Physiol 53(3):246–253. doi:S0022-1910(06)00184-3 [pii] 10.1016/j.jinsphys.2006.07.011

Wang Y, Chen K, Yao Q, Wang W, Zhi Z (2007b) The basic helix-loop-helix transcription factor family in Bombyx mori. Dev Genes Evol 217(10):715–723. doi:10.1007/s00427-007-0184-x

Wang Y, Chen K, Yao Q, Wang W, Zhu Z (2008) The basic helix-loop-helix transcription factor family in the honey bee, Apis mellifera. J Insect Sci 8:1–12. doi:10.1673/031.008.4001

Weintraub H (1993) The MyoD family and myogenesis: redundancy, networks, and thresholds. Cell 75(7):1241–1244. doi:0092-8674(93)90610-3 [pii]

Wheeler SR, Carrico ML, Wilson BA, Brown SJ, Skeath JB (2003) The expression and function of the achaete-scute genes in Tribolium castaneum reveals conservation and variation in neural pattern formation and cell fate specification. Development 130(18):4373–4381

Wilson TG, Ashok M (1998) Insecticide resistance resulting from an absence of target-site gene product. Proc Natl Acad Sci USA 95(24):14040–14044

Wilson TG, Fabian J (1986) A Drosophila melanogaster mutant resistant to a chemical analog of juvenile hormone. Dev Biol 118(1):190–201. doi:0012-1606(86)90087-4 [pii]

Zelzer E, Wappner P, Shilo BZ (1997) The PAS domain confers target gene specificity of Drosophila bHLH/PAS proteins. Genes Dev 11(16):2079–2089

Zhang H, Levine M (1999) Groucho and dCtBP mediate separate pathways of transcriptional repression in the Drosophila embryo. Proc Natl Acad Sci USA 96(2):535–540

Zhang Z, Xu J, Sheng Z, Sui Y, Palli SR (2011) Steroid receptor co-activator is required for juvenile hormone signal transduction through a bHLH-PAS transcription factor, methoprene tolerant. J Biol Chem 286(10):8437–8447. doi:M110.191684 [pii] 10.1074/jbc.M110.191684

Zhu J, Chen L, Sun G, Raikhel AS (2006) The competence factor beta Ftz-F1 potentiates ecdysone receptor activity via recruiting a p160/SRC coactivator. Mol Cell Biol 26(24):9402–9412. doi:MCB.01318-06 [pii] 10.1128/MCB.01318-06

Zhu F, Xu J, Palli R, Ferguson J, Palli SR (2011) Ingested RNA interference for managing the populations of the Colorado potato beetle, Leptinotarsa decemlineata. Pest Manag Sci 67(2):175–182. doi:10.1002/ps.2048

Chapter 3
Juvenile Hormone Biosynthetic Enzymes as Targets for Insecticide Discovery

Michel Cusson, Stephanie E. Sen, and Tetsuro Shinoda

1 Introduction

Because of its pivotal roles in insect morphogenesis and reproduction, juvenile hormone (JH) has long been the focus of intensive research aimed at exploiting its properties for the purpose of developing novel pest control products. Under natural conditions, the titers of this hormone are exquisitely controlled and fluctuate in an orchestrated manner in the course of both pre-imaginal and adult life. Whereas its levels must remain low in the eggs for embryogenesis to proceed normally, they increase and remain high during most of larval life, during which JH inhibits the onset of metamorphosis. Soon after the final larval molt, JH hemolymph titers drop to nearly undetectable levels, a condition that is prerequisite for the 20-hydroxyecdysone-induced larval-pupal or nymphal-adult transformation. Although a rise is observed during the prepupal stage (at least in some Lepidoptera), JH titers usually remain low until adult eclosion. From this point on, JH typically acts as a gonadotropin, inducing vitellogenin production in the fat body and its uptake by developing oocytes. In addition, sex pheromone production and sexual receptivity have been shown to be regulated by JH in many species of insects (Cusson 2004). Thus, there are many points in the life history of an insect where interference with JH

M. Cusson (✉)
Natural Resources Canada, Canadian Forest Service, Laurentian Forestry Centre, Quebec City, QC, GIV 4C7, Canada
e-mail: michel.cusson@nrcan-rncan.gc.ca

S.E. Sen
Department of Chemistry, The College of New Jersey, Ewing, NJ 08628-0718, USA
e-mail: sen@tcnj.edu

T. Shinoda
Division of Insect Sciences, National Institute of Agrobiological Sciences, Tsukuba, Ibaraki, Japan
e-mail: shinoda@affrc.go.jp

I. Ishaaya et al. (eds.), *Advanced Technologies for Managing Insect Pests*,
DOI 10.1007/978-94-007-4497-4_3,

endocrinology, either through manipulation of its titers or inhibition of its action at the target site, could severely disrupt development or perturb reproduction.

In a landmark article published in *Scientific American*, Carroll Williams (1967) heralded a new area in insecticide discovery with the development of JH analogs (JHAs), for which he coined the expression "third-generation pesticides". He reasoned that an artificial increase in the levels of JH (simulated by a JHA) at times when such levels are normally low would result in a fatal disruption of insect development. Various JHAs have been developed for this purpose, and some have proven to be effective (e.g., methoprene, pyriproxyfen) for the control of insects whose pest status is restricted to the adult stage (e.g., mosquitoes). However, with a few exceptions (e.g., pyriproxyfen for the control of white flies; Ishaaya and Horowitz 1992), these insecticides do not provide adequate control of crop and forest pests that feed as larvae, as death resulting from failed metamorphosis occurs only at the end of an extended larval stage, during which feeding continues. Thus, a different strategy has to be envisaged for the development of JH-based control products targeting larvae.

A premature, artificially-induced drop in the larval JH titer typically results in precocious metamorphosis (i.e., in a premetamorphic larval stadium), generating miniature adults that are often unviable or cannot reproduce. Different intervention approaches could, at least in theory, lead to this outcome. The principal factor regulating the level of JH in insect hemolymph is the rate of hormone synthesis by the JH producing glands, the corpora allata (CA; Tobe and Stay 1985). Thus, the selective destruction of these glands or the inhibition of JH biosynthesis, either through inhibition of a JH biosynthetic rate-limiting enzyme or interference with one of the factors regulating the expression of these enzymes, will likely induce precocious metamorphosis. In fact, proof of concept has been provided for each of these strategies (Bowers et al. 1976; Minakuchi et al. 2008; Kaneko et al. 2011; these are covered in more detail in the sections below). JH catabolism by JH esterase (JHE) and JH epoxide hydrolase (JHEH) also plays a significant role in the critical lowering of JH titers during the final larval instar of lepidopteran insects (Goodman and Cusson 2012). JHE over-expression in early larval instars has been shown to cause precocious metamorphosis (Tan et al. 2005), although it is unclear how a pest-control product based on this strategy could be developed, in view of the lack of an efficient delivery mechanism. Finally, interference with JH action at the receptor level could have effects similar to those attributable to a drop in JH titer. Unfortunately, the design of such compounds is difficult because little is known about the structure of JH receptors and the design of such compounds necessitates the construction of JH-like structures that may also display JHA activity.

The present chapter focuses on research aimed at identifying anti-JH (AJH) agents that target JH biosynthetic enzymes, but also covers recent work on the design of bio-active mimics of the neuropeptides known to regulate JH biosynthesis in the CA. To give the reader some background on the subject matter, we first provide a brief overview of the JH biosynthetic pathway and of the early work directed at the discovery of AJH agents. This is followed by a review of recent studies aimed at the cloning and characterization of JH biosynthetic enzymes, a prerequisite for the rational design of target-specific inhibitors (for a review of early work on these enzymes, the reader is referred to Schooley and Baker 1985). We conclude with sections on recent progress regarding the development of lead compounds and the future prospects of AJH agents.

2 The JH Biosynthetic Pathway

JH is a lipophilic sesquiterpenoid produced and released into the hemolymph by the CA. To date, eight different forms of JH have been identified from insect tissue and hemolymph. All but one are methyl esters of epoxy-farnesoic acid or of one of its homologs (i.e., with one to several methyl to ethyl substitutions; Fig. 3.1). JH III, the simplest and most ubiquitous one, is the only JH found in the majority of insects (Goodman and Cusson 2012). In the Lepidoptera, five JHs have been reported: JH III and the homologous JH 0, JH I, JH II, and 4-methyl JH I (Röller et al. 1967; Meyer et al. 1968; Judy et al. 1973; Bergot et al. 1980). The CA of cyclorrhaphous Diptera secrete JH III, a bis-epoxide (JHB_3) form of this hormone (Richard et al. 1989; Fig. 3.1), as well as methyl farnesoate (MF), a JH precursor that has been shown to have JH activity in *Drosophila melanogaster* (Harshman et al. 2010; Jones et al. 2010) and in Crustacea (Nagaraju 2007). Finally, another bis-epoxide form of JH III ($JHSB_3$) has recently been identified as the JH of heteropteran insects (Kotaki et al. 2009; Fig. 3.1).

Up to the formation of farnesyl diphosphate (FPP), JH biosynthesis proceeds through the mevalonate pathway (MVP), which insects share with most other organisms; these enzymatic steps may be viewed as the first branch of the JH biosynthetic pathway (Fig. 3.2). Through three sequential steps involving the enzymes acetoacetyl-CoA thiolase (AACT), HMG-CoA synthase (HMGS) and HMG-CoA reductase (HMGR), mevalonate is generated from three units of acetyl-CoA. Mevalonate is then converted to isopentenyl diphosphate (IPP) through three additional steps catalyzed by mevalonate kinase (MVK), phosphomevalonate kinase (PMVK), and mevalonate diphosphate decarboxylase (MDD). IPP is the C_5 isoprene unit used by the short-chain prenyltransferase FPP synthase (FPPS) to generate FPP, the C_{15} "backbone" of JH. FPPS catalyzes two 1′-4 sequential couplings of IPP with dimethylallyl diphosphate (DMAPP; C_5), through the intermediate product geranyl diphosphate (GPP; C_{10}). The chain initiator of this latter reaction, DMAPP, is the allylic isomer of IPP and is formed by an IPP isomerase (IPPI; Fig. 3.2).

The second branch of JH biosynthesis is generally considered to be JH-specific as it comprises enzymatic steps believed to be unique to JH-producing organisms. Conversion of FPP to farnesol is catalyzed by a farnesyl phosphatase (FP), whereafter farnesol undergoes two sequential oxidation reactions that generate farnesal and farnesoic acid (FA). While the first reaction is effected by either a farnesol dehydrogenase (FDH; Mayoral et al. 2009a) or a farnesol oxidase (FO; Sperry and Sen 2001), the enzyme catalyzing the second reaction has yet to be isolated and

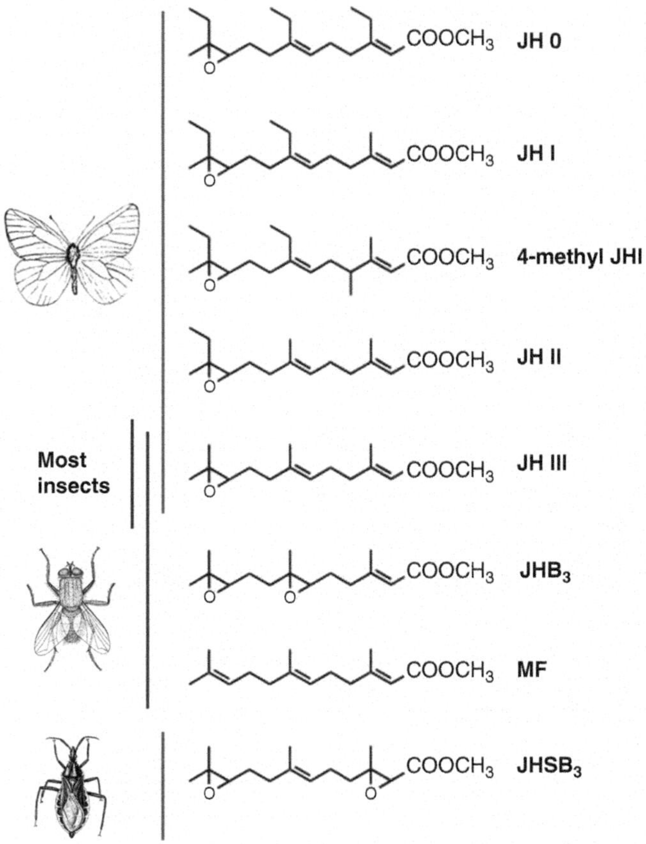

Fig. 3.1 Structures of the eight known juvenile hormones of insects. JH 0, JH I, 4-methyl JH I and JH II have been found only in the Lepidoptera, while JH III is produced by the majority of insects. The higher Diptera produce a bisepoxide form of JH III (JHB_3), in addition to JH III and methyl farnesoate (MF). Finally, a different bisepoxide form of JH III ($JHSB_3$) has been identified as the JH of Heteroptera

characterized. The order of the last two steps, epoxidation and methyl esterification, catalyzed by an epoxidase (FAE or MFE) and a JH acid methyltransferase (JHAMT), may vary between species (Goodman and Cusson 2012; Fig. 3.2).

In the biosynthesis of ethyl-branched JHs by the Lepidoptera, propionyl-CoA substitutes for acetyl-CoA in the initial steps of JH synthesis, leading to the production of the C_6 homolog of IPP (HIPP), which may then be used as substrate, with its allylic isomer (HDMAPP), by FPPS to generate the various FPP homologs that are the precursors of ethyl-branched JHs (Schooley et al. 1973). In the higher Diptera and the Heteroptera, which produce two different bisepoxy forms of JH III, catalytic features of the epoxidases are expected to differ from those found in other insects (Helvig et al. 2004), but characterization of these enzymes must await their isolation.

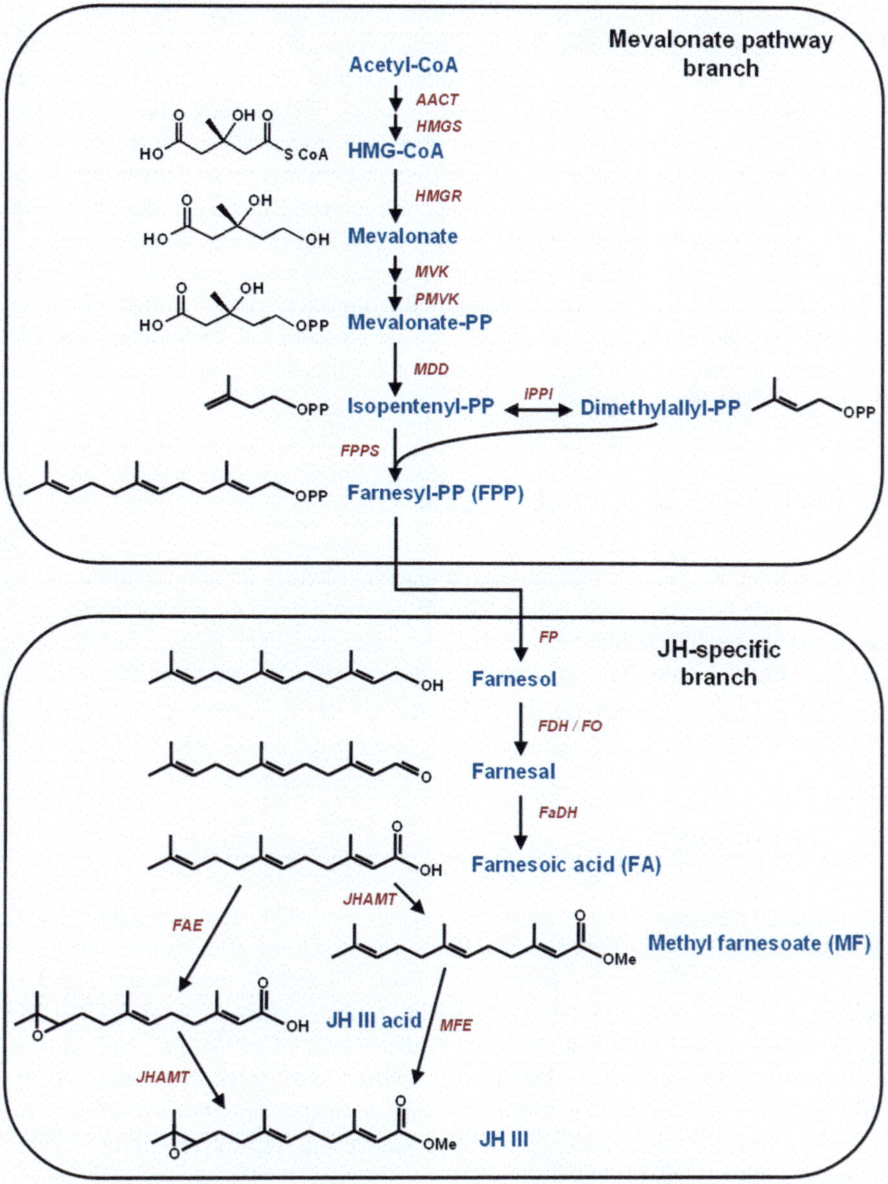

Fig. 3.2 Juvenile hormone biosynthetic pathway. The *first branch* comprises the mevalonate pathway (MVP) enzymes, which insects share with most organisms. The *second branch* features sequential enzymatic steps that are unique to JH biosynthesis. The order of the last two enzymatic steps, methylation and epoxidation, may vary. Abbreviations: *AACT* acetoacetyl-CoA thiolase, *HMGS* HMG-CoA synthase, *HMGR* HMG-CoA reductase, *MVK* mevalonate kinase, *PMVK* phosphomevalonate kinase, *MDD* mevalonate diphosphate decarboxylase, *IPPI* isopentenyl diphosphate isomerase, *FPPS* farnesyl diphosphate synthase, *FP* farnesyl phosphatase, *FDH* farnesol dehydrogenase, *FO* farnesol oxidase, *FaDH* farnesal dehydrogenase, *JHAMT* JH acid methyltransferase, *FAE* farnesoic acid epoxidase, *MFE* methyl farnesoate epoxidase

From the perspective of developing insecticides that target JH biosynthetic enzymes, those of the JH-specific branch have typically been regarded as more suitable, given that inhibitors targeting them are less likely to affect non-insect organisms. As MVP enzymes are shared by most living organisms, it has been surmised that inhibitors directed at them may lack insect-specificity. However, in some groups of insects, MVP enzymes may display unique features that are amenable to the development of target-specific inhibitors. Interestingly, some of the early work on anti-JH agents led to the identification of compounds presumed to target enzymes of the MVP while proving to be specific to the Lepidoptera (see Sect. 3, below). Although this specificity could have resulted from inter-taxonomic differences in inhibitor catabolism, it may also have been the outcome of Lepidoptera-specific enzymatic adaptations evolved to accommodate the homologous precursors of ethyl-branched JHs.

3 Early Work on Anti-JH Agents

For more than 60 years, efforts have been made to identify natural products and to prepare synthetic compounds with AJH activity, using the four general approaches described in the Introduction (for earlier reviews, see Staal 1986; Schooley and Edwards 1996). Figure 3.3 provides the structures of select small-molecule AJHs that have been developed using these strategies. Several of these are discussed in more detail, below.

3.1 Precocenes

These natural products were first discovered in the early 1970s from active insecticidal extracts of the plant *Ageratum houstonianum*. The structures of the active molecules were subsequently identified as precocene I and II (P1 and P2, respectively, Fig. 3.3) and, as their name suggests, are derivatized chromenes (Bowers et al. 1976). The compounds are most interesting because of their mode of action and their varying activity amongst insect species. While often referred to as AJHs, this term is technically inaccurate as they are actually allatotoxins, causing cellular necrosis of the CA.

Precocenes have a fascinating mode of action and an understanding of their pharmacokinetics has provided direction for the design of other topically-active materials. Natural and synthetic precocenes (e.g., acetylenic P2, Feyereisen et al. 1985; Fig. 3.3) are pro-insecticides, first undergoing oxidation to generate the corresponding 3,4-epoxide, which subsequently functions as an alkylating agent. While it was originally suggested that epoxidation within the CA was mediated by MFE, the final step of JH biosynthesis for non-lepidopteran insects, we now know that recombinant CYP15A1, the putative MFE of *Diploptera punctata*, does not metabolize P2 (Helvig et al. 2004). This result suggests that other P-450 enzymes within the CA convert precocene to its reactive form. The mechanism of cellular

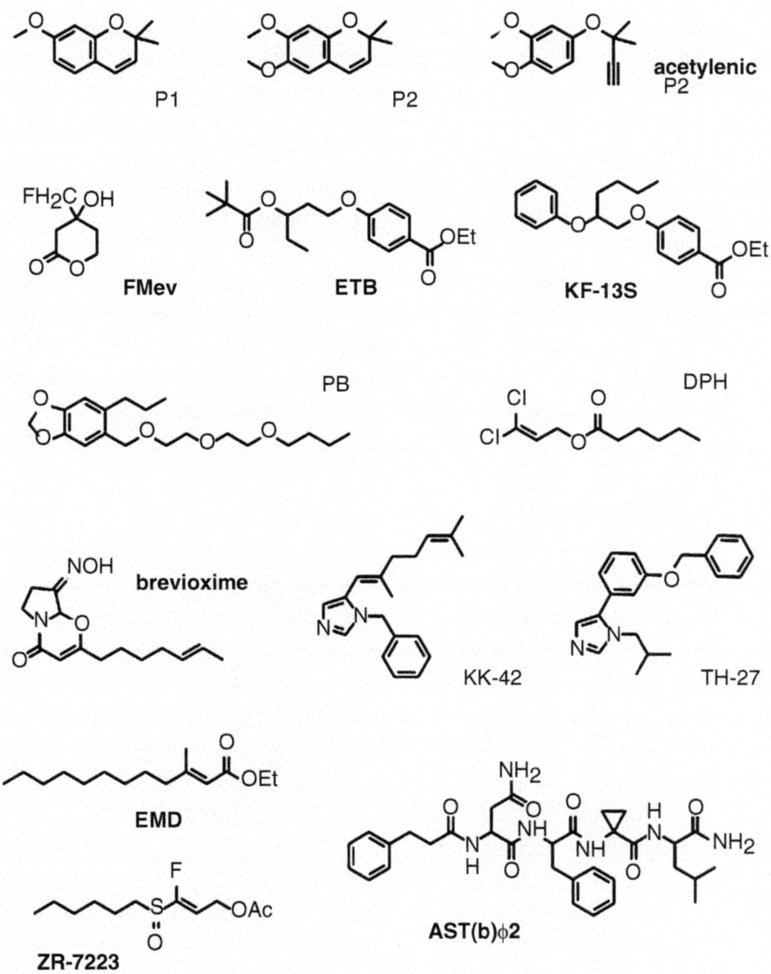

Fig. 3.3 Small molecule agents possessing AJH activity. Abbreviations: *P1* precocene I, *P2* precocene II, *aceylenic P2* 1,2-dimethoxy-4-((2-methylbut-3-yn-2-yl)oxy)benzene, *FMev* fluoromevanolactone, *ETB* ethyl 4-((3-(pivaloyloxy)pentyl)oxy)benzoate, *KF-13S* ethyl (2*S*)-4-((2-phenoxyhexyl)oxy)benzoate, *PB* piperonyl butoxide, *DPH* 3,3-dichloro-2-propenyl hexanoate, *KK-42* (*E*)-1-benzyl-5-(2,6-dimethylhepta-1,5-dien-1-yl)-1*H*-imidazole, *TH-27* 5-(3-(benzyloxy)phenyl)-1-isobutyl-1*H*-imidazole, *EMD* (*E*)-ethyl 3-methyldodec-2-enoate, *ZR-7223* (*E*)-3-fluoro-3-(hexylsulfinyl)allyl acetate, *AST(b)ϕ2* N2-(1-oxo-3-phenylpropyl)-L-asparaginyl-L-phenylalanyl-1-aminocyclopropanecarbonyl- L-leucinamide

necrosis is complex. The 3,4-epoxide may react with incipient nucleophiles present in proteins or other cellular components, including DNA. Research related to the hepatotoxicity of precocene in rats supports metabolic activation by CYPs and macromolecular alkylation, and suggests that toxicity may result from decreased cellular glutathione levels (Hammond and Fry 1996; Ly and Brock 2011).

Precocenes do not exert cytotoxic effects in all insect species, although it appears that most CA are themselves sensitive to precocene exposure. The difference between *in vivo* and *in vitro* activities is due to differences in precocene transport and metabolism within the individual insect species. Studies indicate that several factors are responsible for differences in activity, including levels of epoxide hydrolases, precocene release from the insect cuticle, precocene accumulation and degradation within the fat body, and binding to hemolymph proteins (Brooks and McCaffery 1990).

Many other plant species have been found to contain precocenes, and both natural and unnatural precocenes continue to be studied for their insecticidal properties. The utility of precocenes as naturally-occurring agrochemicals has recently been expanded to the area of food storage. The compounds have anti-fungal activity, making them potentially useful as antimycotoxigenic agents (Jaya and Dubey 2011).

3.2 Fluoromevalonate

The conversion of mevalonate to IPP is catalyzed by three sequential ATP-dependent enzymes: MVK, MVPK and MDD (Fig. 3.2). In relation to JH biosynthesis, these enzymes have been poorly characterized, although they have been cloned from a variety of insects and crystal structures of (non-insect) MVK and MDD are known (Miziorko 2011). Mechanistically, MDD may hold the most promise for AJH development because it is believed to involve a rigid and charged transition state, although selectivity may be difficult, given that the protein is ubiquitous in Nature. As an extension of studies with the MDD dead-end inhibitor, 6-fluoromevalonate 5-diphosphate, fluoromevalonolactone (FMev, Fig. 3.3) was tested for its ability to inhibit JH biosynthesis. Remarkably, FMev is a potent and selective AJH for several lepidopteran species, inhibiting the metabolism of both mevalonate and homomevalonate to JH most effectively in *Manduca sexta* (Quistad et al. 1981; Baker et al. 1986). Although its precise mode of action in insects is unknown, studies with yeast and rat liver MDD would indicate that the compound is hydrolyzed and phosphorylated, and that the corresponding 5-diphosphate 3-phosphate derivative binds to MDD but cannot undergo decarboxylation (Reardon and Abeles 1987). The (*R*)-enantiomer is the active form of FMev. The compound's selectivity for lepidopteran insects is unclear because FMev causes an accumulation of mevalonate-5-diphosphate in *Drosophila* K_c cells (Watson et al. 1985). Despite its early discovery, FMev's limited effectiveness precludes its use as a commercial AJH, and active analogs of FMev have yet to be developed.

3.3 Imidazoles

The final two steps of JH biosynthesis involve the epoxidation of the C-10,11 double bond and *O*-methylation of the C-1 acid. While the order of these steps varies and depends on insect species, the epoxidation is known to be catalyzed by a unique cytochrome P450 enzyme (see Sect. 4.2.4, below). Given that this sesquiterpene epoxidation is specific to insects and is required for JH biosynthesis, inhibition of the epoxidase has long been considered a promising target for AJH design and development.

Unfortunately, because the epoxidase is a cytochrome P450 enzyme, selective inhibition of this protein has been challenging. Although piperonyl butoxide (PB, Fig. 3.3) and its analogs have anti-JH activity, they are broad-spectrum agents, inhibiting MFE, endogenous mixed-function oxidases (MFOs) and, at higher concentrations, even JHAMT (Hammock and Mumby 1978). A more promising class of compounds is the imidazoles, which presumably inhibit the P450 through direct coordination with the iron-heme center. A wide array of imidazole-containing compounds have been prepared and tested and from this several potent anti-JH compounds (e.g., KK-42 and TH-27; Fig. 3.3; Kuwano and Eto 1986; Unnithan et al. 1995) have been obtained. 1,5-Disubstitution of the imidazole moiety appears essential for activity, with alkyl substitution at the N-1 position and *meta*-aryl substitution at C-6 being most effective. In *Bombyx mori* these compounds cause precocious metamorphosis, which is suppressed by the application of methoprene. Because imidazole AJHs decrease total larval duration while increasing cocoon weight and fibroin content, they have economic importance for the silk industry (Yungen and Bharathi 2009).

3.4 ETB

Ethyl 4-[2-(*t*-butylcarbonyloxy)butyloxy]benzoate (ETB; Fig. 3.3) was originally developed in 1975 (Kondo et al. 1977) and discovered to have AJH activity by Staal (1976). This compound is active on *M. sexta* and *B. mori* larvae, causing precocious metamorphosis at lower concentrations, which is partially rescued by the administration of FA (Kiguchi et al. 1984). At high concentrations, ETB functions as a JH agonist and is a weak inducer of JHE activity (Sparks et al. 1979). The activity of ETB is due exclusively to the (−)-enantiomer and is limited to certain lepidopteran species. The dual AJH/JHA activity of ETB and its uncertain mode of action have precluded its development as an effective insecticidal agent. However, recent work on its mode of action and on the development of novel derivatives has created renewed interest in ETB and related compounds (see Sect. 5, below).

3.5 Compounds with Unknown Modes of Action

Many fungal metabolites possess interesting biological activity, including AJH effects (e.g., compactin; Hiruma et al. 1983). Extracts of the fungus *Penicillium brevicompactum*, which displayed anti-JH effects in *Locusta migratoria,* were subfractionated and characterized to yield the novel sesquiterpene-like structure brevioxime (Fig. 3.3, Castillo et al. 1998). This compound caused a dose-dependent inhibition of JH biosynthesis in cultured CA, which could not be restored by the addition of farnesol, farnesoic acid, or mevanolactone. While these results suggest that brevioxime inhibits the final steps of JH biosynthesis, the mode of action of this compound remains unknown. Structures related to brevioxime and possessing AJH activity have been identified (Cantín et al. 1999).

A series of fluorinated vinyl sulfoxides developed in the late 1980s showed promise as potent and selective AJHs against lepidopteran insects (Carney and Brown 1989). The design of these compounds was based on the weakly active EMD (a synthetic competitive epidermal JH receptor ligand, Fig. 3.3) and on analogs of DMAPP (e.g., DPH, Fig. 3.3, Quistad et al. 1985). The most active compounds were acetylated and possessed an unbranched aliphatic chain slightly longer than geraniol (ZR-7223, Fig. 3.3). Topical application to larvae caused premature pupation that was recoverable by co-administration of farnesol. While these results suggest that the compounds target a biosynthetic enzyme related to isoprene construction, the compounds inhibit neither lepidopteran FPPS nor IPPI (Sen, S.E., unpublished results).

4 Recent Work on JH Biosynthetic Enzymes

4.1 Enzymes of the Mevalonate Pathway (MVP)

Among enzymes of the insect MVP, only one, FPPS, has been the focus of directed research in recent years. However, various insect genomics projects have revealed the predicted amino acid sequences of many MVP enzymes. A good example is the work of Noriega et al. (2006), where CA-specific EST libraries from cockroaches and mosquitoes were used to deduce the sequences of many JH biosynthetic enzymes. Another example is the recent sequencing of the genome of the Monarch butterfly, *Danaus plexippus*, in which special attention was given to the entire JH biosynthetic pathway (Zhan et al. 2011). Most significantly, scanning of the genome of the silkworm, *B. mori* (BAG 2004; Mita et al. 2004), led to the identification of all JH-related MVP enzymes, whose cDNAs were then cloned for the purpose of monitoring their transcript levels by q-RT-PCR in several tissues of 4th-instar larvae as well as in CA of larvae, pupae and adults (Kinjoh et al. 2007). With the exception of FPPS, all enzymes were reported to be encoded by single-copy genes. In addition, the transcripts of all MVP enzymes were most abundant in the CA, except for FPPS1, a finding that was corroborated by *in situ* hybridization (Ueda et al. 2009). Overall, age-related variations in CA transcript abundance were found to be relatively well coordinated for most of these enzymes (Kinjoh et al. 2007).

4.1.1 Farnesyl Diphosphate Synthase (FPPS)

This enzyme has now been identified from many insects, where it is typically present as a single-copy gene (for a review, see Vandermoten et al. 2009a). The most striking exception to this rule is the honey bee, *Apis mellifera*, for which sequencing of the

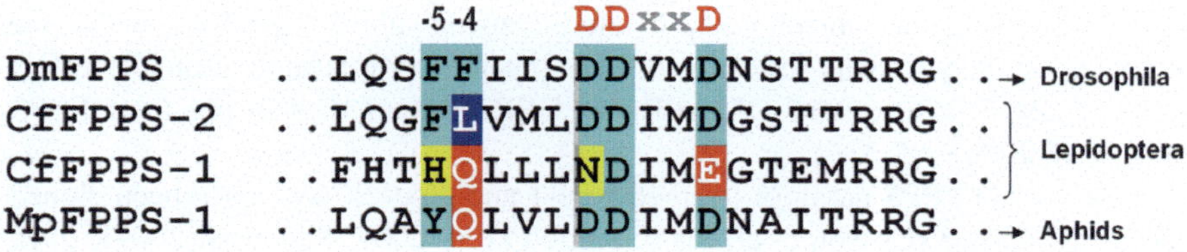

Fig. 3.4 Amino acid alignment of the chain-length determination (CLD) region of selected dipteran, lepidopteran and homopteran FPPSs. The single-copy FPPS of *D. melanogaster* (DmFPPS) features a conventional CLD region, with a conserved first aspartate-rich motif (FARM: DDxxD) and conserved aromatic residues at positions -4 and -5 from the FARM. In comparison, the *C. fumiferana* FPPS-2 (CfFPPS-2) exhibits a non aromatic residue at position -4 from the FARM while CfFPPS-1 displays important substitutions in both the FARM and the residues at positions -4 and -5 from it, one of which (Q) is shared with *Myzus persicae* FPPS

genome revealed the presence of seven putative FPPS paralogs (HGSC 2006) presumably associated with different biochemical pathways. Two other notable exceptions are the FPPSs of Homoptera (aphids) and Lepidoptera. Aphid species examined to date (with one exception: Ma et al. 2010) exhibit two FPPS paralogs that are similar in their primary sequence, and display an important substitution in their "chain-length determination region" (CLD region) relative to most other known eukaryotic FPPSs. The CLD region refers to the amino acid residues located upstream from the first aspartate-rich motif (FARM), with special reference to residues at positions -4 and -5 from the FARM, which are typically aromatic (Phe or Tyr). Although the FARM sequence is conserved in aphid FPPSs (i.e., DDxxD, where x is any amino acid residue), a Gln substitutes for Phe/Tyr at position -4 (Lewis et al. 2008; Vandermoten et al. 2008; Zhang and Li 2008; see Fig. 3.4). Site-directed mutagenesis (Vandermoten et al. 2009b) showed that this substitution was responsible, at least in part, for the bifunctional behavior of the aphid enzyme, as assessed *in vitro* using the recombinant enzyme. Indeed, when the protein was assayed in the presence of DMAPP and [^{14}C]IPP, it produced significant amounts of both GPP and FPP (Vandermoten et al. 2008). In addition, using a linked assay where the aphid FPPS was in the presence of either a monoterpene or sesquiterpene synthase, both the monoterpene and sesquiterpene could be generated (Lewis et al. 2008). As no additional putative GPPS or FPPS could be found in the pea aphid genome, it appears that aphids have evolved an enzyme that can generate both GPP and FPP, which they use as precursors for sex pheromone (GPP) and alarm pheromone/JH (FPP) biosynthesis, respectively (Vandermoten et al. 2008). It should be noted here that only one of the two recombinant aphid FPPSs displayed activity *in vitro*; however, both putative proteins exhibit the same CLD region sequence and are therefore expected to show the same bifunctional behavior (Vandermoten et al. 2008).

In the Lepidoptera, FPPS has long been suspected of exhibiting structural features allowing it to accommodate the bulkier homologous substrates and products used as precursors of ethyl-branched JHs (Cusson et al. 1996). Not surprisingly, prenyltransferase assays conducted using *M. sexta* CA homogenates pointed to a selectivity of the native enzyme towards homologous substrates (Sen et al. 1996).

This finding was later corroborated by studies using substrate analogs, where the lepidopteran enzyme was shown to display greater steric latitude around the C-3 and C-7 alkyl positions of DMAPP and GPP, compared with pig liver FPPS (Sen et al. 2006).

Two distinct FPPS paralogs (termed type-1 and type-2) have since been cloned and characterized (Cusson et al. 2006; Kinjoh et al. 2007; Sen et al. 2007). In *B. mori*, a third paralog has been identified (BmFPPS3), which is very similar to BmFPPS2 (Kinjoh et al. 2007), and both may thus be considered type-2 FPPSs. Interestingly, we have not been able to find a third FPPS paralog in the genome of *Choristoneura fumiferana*, for which we now have a draft assembly (Cusson, M. et al., unpublished results). Similarly, only two FPPSs were found in the recently published genome of the Monarch butterfly (Zhan et al. 2011). Since BmFPPS3 does not appear to be the product of alternative splicing, it may have resulted from a gene duplication event that did not occur in all lepidopteran species.

Type-1 FPPSs were observed to display several active site substitutions when compared with "conventional" eukaryotic FPPSs, particularly in the FARM region, where NDxxE substitutes for DDxxD (Fig. 3.4). The FARM is known to be involved in allylic substrate binding and catalysis and, based on site-directed mutagenesis work done on other FPPSs (Joly and Edwards 1993; Song and Poulter 1994), these substitutions are expected to have an impact on catalysis. In addition, His and Gln substitute for the Phe/Tyr residues at positions -5 and -4 relative to the FARM (Fig. 3.4). In comparison, the type-2 protein displays a more conventional active site, although it exhibits a variable non-aromatic residue at position -4 from the FARM. In addition, its N-terminus exhibits clearly distinct features (Cusson et al. 2006). Although molecular modeling and docking studies pointed to the type-1 (FPPS1) as being the better suited paralog for binding homologous substrates and products, transcriptional analysis of both FPPS1 and FPPS2 in various *B. mori* tissues revealed a ubiquitous distribution for the former and a confinement to the CA for the latter (Cusson et al. 2006; Kinjoh et al. 2007). This observation led to the conclusion that FPPS2 was likely the principal prenyltransferase generating FPP substrates for JH biosynthesis. In subsequent work, the two *C. fumiferana* FPPSs (CfFPPS1 and CfFPPS2) were produced as recombinant proteins using a bacterial expression system and used to conduct enzyme assays to assess their ability to couple various allylic substrates with [^{14}C]IPP. CfFPPS1 was inactive with all substrates tested while CfFPPS2 displayed significant activity with all of them. Interestingly, however, when the two proteins were combined in equal amounts, a synergistic effect was observed, particularly with homo- and bishomo-geranyl diphosphate as allylic substrates, precursors of JH II and JH I, respectively (Sen et al. 2007). Whether this phenomenon is of any functional significance *in vivo* remains to be determined, but transcripts of both FPPS types are present in *B. mori* CA (1 : 20 FPPS1 : FPPS2 ratio in larval CA; Cusson et al. 2006; see also Kinjoh et al. 2007), suggesting that they could form heteromers in this tissue. However, FPPS2 transcript levels were near the q-RT-PCR detection limit in tissues where FPPS1 transcripts were abundant, such as the Malpighian tubules (MTs; Cusson et al. 2006; Kinjoh et al. 2007), and where the formation of FPPS1-FPPS2 heteromers is therefore unlikely. The strikingly high abundance of FPPS1 transcripts in MTs has recently been confirmed for two other lepidopteran species (Barbar, A., Cusson, M., Béliveau, C., Sen, S.E., unpublished results), but the role of this protein in this tissue has yet to be elucidated.

Clearly, both homopteran and lepidopteran FPPSs display taxon-specific active-site features that show promise for the development of target-specific inhibitors. In addition, in these two insect groups, FPPS paralogs appear to be involved in other important pathways besides JH biosynthesis, suggesting that their inhibition could result in multiple physiological disturbances.

4.1.2 Isopentenyl Diphosphate Isomerase (IPPI)

Lepidopteran IPPI has been suspected of displaying substrate selectivity towards the C_6 substrate HIPP. Results supporting this hypothesis were provided by assays using *M. sexta* CA homogenates (Baker et al. 1981) and a partially purified IPPI from *B. mori* (Koyama et al. 1985). In both cases, HIPP was converted to the correct isomer (HDMAPP), whereas pig liver IPPI generated products that could not be used for synthesis of homologous JHs (Koyama et al. 1973). IPPI has now been cloned from many insects and has been observed to be present as a single-copy gene in the genomes of *B. mori* (Kinjoh et al. 2007), *D. plexipus* (Zhan et al. 2011) and *C. fumiferana* (Cusson, M. et al., unpublished). Interestingly, the primary sequences of lepidopteran IPPIs are very similar to those of other insects and other eukaryotes, and sequence comparisons do not reveal obvious lepidopteran-specific substitutions that could account for the ability of this enzyme to isomerize HIPP to HDMAPP better than other eukaryotic IPPIs. Thus, the relevant differences may be subtle and difficult to identify. In this perspective, it is unclear whether IPPI will prove to be a suitable target for the development of lepidopteran-specific inhibitors.

4.2 Enzymes of the JH-Specific Branch

cDNAs encoding enzymes from the JH-specific portion of the pathway have been cloned and characterized only recently. As of the end of 2011, clones had been obtained from at least one insect species for all of these enzymes except for farnesal dehydrogenase.

4.2.1 Farnesyl Phosphatase (FP)

To isolate an insect farnesyl phosphatase, Cao et al. (2009) screened the *Drosophila melanogaster* genome for phosphatases and identified eight genes, three of which were observed to be expressed in the ring gland, using an RT-PCR assay. The cDNAs of two of these were used to produce the recombinant proteins, which were submitted to phosphatase assays. Both proteins displayed phosphatase activity in the presence of para-nitrophenyldiphosphate (p-NPP), but only one displayed high activity against FPP; the same protein was inactive in the presence of the diterpene (C_{20}) precursor geranylgeranyl diphosphate (GGPP). Transcript levels in whole animals correlated well with previously published JH titers for the developmental stages that were examined (2nd and 3rd instar larvae, pupae, and male and female adults). It remains to be determined whether these transcripts were produced predominantly by the ring gland or by other tissues.

4.2.2 Farnesol Dehydrogenase (FDH)/Farnesol Oxidase (FO)

The first demonstration of farnesal production in the CA was provided by Baker et al. (1983) who used *M. sexta* CA homogenates to monitor the formation of both farnesal and farnesoic acid (FA) from [^{3}H]farnesol. These authors observed accumulation of farnesal in the absence of NAD^+ while they noted a shift towards FA accumulation in its presence. This work suggested that the sequential oxidation of farnesol to farnesal and FA was catalyzed by two distinct enzymes, the second one requiring NAD^+ as cofactor. Subsequent work by Sen and Garvin (1995) showed that farnesol metabolism in *M. sexta* CA could be attributed to highly specific enzymes (as opposed to non-specific oxidative enzymes), for which optimal activity was observed with substrates composed of at least three isoprene units, suggesting that these enzymes would be a suitable target for the development of anti-JH agents.

Although alcohol oxidation in various organisms is typically catalyzed by nicotinamide-dependent dehydrogenases, work on *M. sexta* CA homogenates indicated that the enzyme responsible for the oxidation of farnesol to farnesal in this species was not a dehydrogenase but a specific metal-dependent alcohol oxidase, given that the conversion was oxygen-dependent (Sperry and Sen 2001). However, it has not yet been possible to isolate this alcohol oxidase from any insect species. Mayoral et al. (2009a) reported the cloning and characterization of a farnesol dehydrogenase from *Aedes aegypti* CA, for which an EST candidate had been identified in an earlier study (Noriega et al. 2006). The recombinant protein was active as a homodimer and oxidized farnesol to farnesal in the presence of $NADP^+$. However, the protein displayed no stereospecificity and was more reactive towards secondary alcohols. This protein has a typical short-chain dehydrogenase (SDR) fold, with orthologs found in other species, including *B. mori*. Transcript abundance was highest in the midgut and the brain, as opposed to the CA, where transcript levels were relatively low. On the other hand, transcript levels in adult female CA were well correlated with *in vitro* JH biosynthesis (Mayoral et al. 2009a). It remains to be determined whether mosquitoes and moths use different enzymes (i.e., an alcohol dehydrogenase *versus* an alcohol oxidase) to convert farnesol to farnesal.

4.2.3 JH Acid Methyltransferase (JHAMT)

In the Lepidoptera, methylation of the precursor acid is effected by an *O*-methyltransferase that uses JH acid as substrate; in Coleoptera, Orthoptera, Diptera, and Dictyoptera, FA is the putative substrate of this enzyme. While enzymatic studies, including structure-activity relationship studies with CA homogenates, had been performed (Hamnett et al. 1981), it is only recently that structural information on the methyltransferase has become available. The lepidopteran enzyme was first cloned from *B. mori* using a fluorescent differential display approach (Shinoda and Itoyama 2003). One of three PCR amplicons that displayed developmental changes matching those expected for this enzyme was submitted to sequence analysis. The cDNA was found to encode a 278 amino-acid protein with no clear homolog but containing a conserved motif found on several *S*-adenosyl methionine (SAM)-dependent methyltransferases (LLDIGCGSG). Northern blot analysis showed that this transcript could be detected almost exclusively in the CA, and a q-RT-PCR analysis indicated that its transcription starts declining at the spinning stage of last-instar larvae, concomitant with the documented loss of JHAMT activity at this stage. Its levels remain very low until the pharate adult stage, when they start rising again to reach high levels in adult female CA (Kinjoh et al. 2007). The recombinant His-tagged enzyme could methylate both JH acids and FA to the expected products, although it showed greater conversion rates with JH I acid and JH II acid as substrates as compared with JH III acid and FA (at 100 μM). No conversion was observed with several saturated and unsaturated fatty acids. This enzyme is clearly rate-limiting in *B. mori.* Orthologs of JHAMT have now been cloned and characterized in *Tribolium castaneum* (Minakuchi et al. 2008), *D. melanogaster* (Niwa et al. 2008) and *Aedes aegypti* (Mayoral et al. 2009b). In all three species, the enzyme is expressed predominantly in the CA and the recombinant protein can methylate JH III acid and FA at similar rates, with high stereospecificity for (10*R*)-JH III. In *T. castaneum*, RNAi-mediated knockdown of JHAMT in 3rd instars induced a precocious metamorphosis, pointing to the key regulatory role of this enzyme in the red flour beetle (Minakuchi et al. 2008). A similar approach used in *D. melanogaster* did not disrupt metamorphosis but whole body JHAMT overexpression caused a pharate adult lethal phenotype (Niwa et al. 2008). In *A. aegypti*, JHAMT transcript levels increased prior to *in vitro* JH biosynthesis, and its levels were similar for blood-fed and nutrient-deficient mosquitoes, suggesting that it may not be a rate-limiting enzyme in this species (Mayoral et al. 2009b). From the above work, it appears that the enzymes used by Lepidoptera to methylate JH acid and by other groups of insects to methylate FA belong to the same family and display similar catalytic properties. As such, this enzyme is unrelated to the putative FA methyltransferase (FAMeT) isolated from crustacean mandibular organs (e.g., Holford et al. 2004). Interestingly, while most recombinant FAMeT orthologs isolated from either Crustacea or insects have failed to display methyltransferase activity and *D. melanogaster* strains with a deficiency in FAMeT showed no significant decrease in MF, JH III and JHB_3 biosynthesis compared with wild-types (Burtenshaw et al. 2008), a JHAMT ortholog has recently been identified in the water flea *Daphnia pulex* (Hui et al. 2010), suggesting that the crustacean enzyme responsible for the conversion of FA into MF may in fact be a JHAMT ortholog.

Using *O*-methyltransferases from the cyanobacterium *Anabaena variabilis* and the wildflower *Clarkia breweri* as templates, homology models for several insect JHAMTs have been developed (Defelipe et al. 2011). The structures obtained were found to display a typical S-adenosylmethionine transferase fold composed of alternating six β-sheets and nine α-helices. Docking studies with SAM, FA, JH III, palmitic acid, and lauric acid were subsequently performed using the homology model developed for *A. aegypti* JHAMT (AeJHAMT). Consistent with substrate activity measurements, FA and JH III formed low-energy ligand-receptor complexes, with sesquiterpenoid chains interacting with a defined hydrophobic pocket and the carboxyl moiety of each substrate forming hydrogen bonds with the indole nitrogen of Trp-120 (AeJHAMT numbering) and amide nitrogen of Gln-14. The stereospecificity of the enzyme is believed to be due to hydrogen-bonding interactions that are possible between the (10*R*)-epoxide oxygen and Ser-176 and Tyr-178. The fact that these residues are not conserved in all insect JHAMTs may explain why the substrate specificity for (10*R*)-JH III is not the same in all species.

4.2.4 Methyl Farnesoate Epoxidase (MFE)/Farnesoic Acid Epoxidase (FAE)

With the possible exception of what is observed in the Lepidoptera (Bhaskaran et al. 1986) and in *Drosophila* (Moshitzky and Applebaum 1995), the last step of JH biosynthesis is the epoxidation of MF. This epoxidase was known from early studies to be a microsomal cytochrome P450 enzyme in cockroach and locust (Feyereisen et al. 1981; Hammock 1975) but its cDNA was cloned only recently from a cockroach, following the construction of a *Diploptera punctata* CA cDNA library and the 5′ end sequencing of 1,056 clones from it (Helvig et al. 2004). The recombinant protein, CYP15A1, showed high affinity for MF and converted it to JH III. The enzyme could not metabolize other MF-related compounds such as farnesol, farnesoic acid and farnesyl methyl ether as substrates, and it showed selectivity for the natural geometric isomer as well as for the 10*R* enantiomer. CYP15A1 was observed to be expressed only in the CA and only during peak JH production. Surprisingly, no clear orthologs of CYP15A1 could be found in *D. melanogaster*, perhaps because this fly produces a bisepoxide of JH III, the production of which may require a different epoxidase. However, an ortholog of this enzyme was identified in an *A. aegypti* CA EST collection (Noriega et al. 2006) and in the desert locust *Schistocerca gregaria*, in which *in vivo* RNAi targeting CYP15A1 caused MF accumulation in the CA (Marchal et al. 2011). Interestingly, a mutation in the CA-specific CYP15A1 gene of *B. mori* has recently been shown to be responsible for the *dimolting* mutant phenotype, where larvae always undergo a precocious metamorphosis (Daimon et al. 2012). The *B. mori* enzyme uses FA as substrate (as opposed to MF) and must therefore be considered an FA epoxidase (FAE). This study points to the fact that disruption of epoxidase function in the CA can also result in precocious metamorphosis, making this enzyme a suitable target for AJH agent development.

5 Recent Efforts to Develop Inhibitors of JH Biosynthesis

Since Staal's review in 1986, progress in the development of AJH agents has been modest. With respect to inhibitors of JH biosynthetic enzymes, some studies have focused on the assessment and mode of action of new derivatives of previously reported AJH compounds, while on-going work aimed at the development of new chemistries targeting specific enzymes has not yet reached the publication stage. However, there is growing interest in the design of lipophilic, breakdown-resistant analogs of the allatostatins (ASTs), the neuropeptides that inhibit JH production in the CA (reviewed in Stay and Tobe 2007; Audsley et al. 2008; Weaver and Audsley 2009). Presumably, these peptidomimetics interact with cell-surface AST receptors, which then modulate the expression of enzymes involved in JH biosynthesis and/or precursor supply (i.e., upstream of the MVP; Sutherland and Feyereisen 1996).

The AJH activity of 1,5-disubstituted imidazoles was first reported by Kuwano et al. (1983). More recently, 17 of them were assayed for their ability to inhibit JH biosynthesis by cockroach (*D. punctata*) CA *in vitro* (Unnithan et al. 1995), and some exhibited very high potency, with IC_{50} values <100 nM. Four of these compounds also displayed *in vivo* activity following topical application to 2-day-old mated females at 100 μg/insect, inhibiting JH biosynthesis (as determined *in vitro* on isolated CA) and causing an accumulation of MF in the CA. In a subsequent study, 1,5-disubstituted imidazoles displayed parallel effectiveness in inhibiting the activity of a recombinant CYP15A1 and *in vitro* JH biosynthesis by isolated cockroach CA (Helvig et al. 2004), clearly demonstrating that these compounds are effective inhibitors of the MFE. In spite of their high *in vitro* potency, interest in these compounds now appears to have waned, perhaps because of their limited *in vivo* activity and potential reactivity towards other cytochrome P-450 enzymes.

ETB-related compounds have also been the focus of several recent investigations. These molecules have been known for some time to display both AJH and JHA activity depending on the dose used, and to be selective against the Lepidoptera (Staal 1986). Significant efforts have been made to modify the ETB structure to eliminate the compound's undesirable agonistic effects (Ishiguro et al. 2003; Fujita et al. 2005; Furuta et al. 2006). The result of these efforts was the development of ethyl (2*S*)-4-(2-benzylhexyloxy)benzoate (also known as KF-13S; Fig. 3.3), which has no JH agonistic activity and is more potent than ETB. The mode of action of KF-13S and ETB was recently elucidated (Kaneko et al. 2011). Both compounds cause a decrease in JH biosynthesis *in vitro*, as indicated by loss of [^{14}C]labeled JH production by excised CA glands, and KF-13S caused reversible inhibition of JH biosynthesis *in vivo* for insects topically treated during the 3rd larval instar. Transcript levels of several early JH biosynthetic enzymes were found to be lowered by KF-13S treatment, including AACT, HMGS, HMGR, MVK, PMVK, MDD, IPPI, FPPS3, and JHAMT, with HMGS and HMGR transcript levels being the most affected (Kaneko et al. 2011). Thus, ETB-related compounds do not work as enzyme inhibitors but as transcriptional regulators of JH biosynthetic enzymes. In this respect, their overt effect may be seen as being similar to that of ASTs.

Because the CA do not store JH, control of JH biosynthesis must in part involve the regulation of total glandular synthetic activity (Tobe and Pratt 1974). Several neuropeptides have been found to reversibly inhibit JH production and have been studied as potential AJHs. ASTs were first identified in cockroach brains (Woodhead et al. 1989) and were subsequently found in many other insect orders (Tobe and Bendena 2006). As commonly seen for neuropeptides, ASTs occur as multiple sequence-related copies derived from a single gene. ASTs are 6–18 amino acids long and comprise two structural types, the FGLa ASTs and the PISCF ASTs (Coast and Schooley 2011). FGLa ASTs, also known as cockroach ASTs, have a characteristic C-terminal Y/FXFGL-amide sequence, which is essential for activity. A PISCF AST was first identified in the moth *M. sexta* and is distinct from the first AST type in having an N-terminal pyroglutamate, a free C-terminus, and two cysteines that form an intramolecular disulfide bond.

As multiple allatostatins of varying types can be produced within a single insect species, the mode of action of these peptides appears to be complex. In fact, only cockroaches and crickets have endogenous ASTs that inhibit JH biosynthesis, although cross-species reactivity (e.g., *D. melanogaster* AST inhibition of moth JH biosynthesis) has been seen. The other effects of these peptides are varied and relate to myotropic activity and neurotransmission. Several AST receptors, including a G-protein coupled receptor in *D. melanogaster*, have been identified (Nässel 2002; Kreienkamp et al. 2002). The current view is that the original function of the ASTs was not the inhibition of JH production but rather the activation of ion channels leading to reduced membrane potential and input resistance.

Because of their ability to inhibit JH biosynthesis, cockroach-type ASTs have been considered potential targets for AJH development. These compounds, being oligopeptides, have significant pharmacokinetic challenges, including poor absorption, transport, and degradation by endogenous peptidases, plus high production costs. Structure-activity relationship studies indicate that the C-terminal pentapeptide Y/FXFGL-NH_2, which adopts a turn structure in solution, is essential for activity, with the FGL sequence being more essential than Y/FX (Nachman et al. 1998). To increase proteolytic stability, amide substitution by ketomethylene and methyleneamine peptidomimetics, and by sidechain substitution with indane and cyclopropyl groups has been investigated (Piulachs et al. 1997). Modifications of Dippu-AST 1 and Dippu-ASTs 4–6 have yielded peptidomimetics with enhanced peptidase resistance and good *in vitro* activity, as exemplified by allatostatin analog AST(b)ϕ2 (Fig. 3.3), which was highly resistant to hemolymph, brain, and midgut proteases while displaying an IC_{50} for JH biosynthesis of *D. punctata* CA of 1.55 nM (Nachman et al. 1999). Recent work by Kai et al. (2010, 2011) suggests that a bioactive AST analog must contain an aromatic group, an appropriate linker in the Y/FX region and the FGLamide portion. Clearly, significant progress has been made in the rational design of AST analogs and some of them show potential as AJH agents for the control of pests such as cockroaches.

6 Future Prospects

From the foregoing, it is clear that recent efforts in the field of AJH research have focused on the cloning and characterization of JH biosynthetic enzymes, as opposed to the search for and development of novel AJH agents, with the exception of the work on the development of ETB derivatives and the design of AST analogs. This should not be interpreted as a drop in interest in an AJH control strategy. Rather, the change in focus is a reflection in recent technological developments that have enabled the cloning of the genes encoding JH biosynthetic enzymes and their detailed characterization as recombinant proteins. It is hoped that such characterization, combined with structure-activity relationship studies, will pave the way to the rational design of target-specific enzyme inhibitors. Indeed, because the crystal structures of many homologous proteins are known, a computational approach can now be taken to provide new directions for AJH design and evaluation. Already, this research points to promising avenues for future investigations. For example, JHAMT has been shown to be a rate-limiting enzyme in the biosynthesis of JH in *B. mori* (Shinoda and Itoyama 2003; Kinjoh et al. 2007) and *T. castaneum* (Minakuchi et al. 2008), and RNAi targeting this enzyme has provided a remarkable proof of concept for the potential usefulness of agents capable of suppressing this enzyme. Efforts in developing JHAMT-specific inhibitors should therefore be rewarding. Similar comments apply to FDH in mosquitoes, where this enzyme is the rate-limiting step (as opposed to JHAMT) in the JH-specific branch of the pathway (Mayoral et al. 2009a). With respect to MVP enzymes, aphid and moth FPPSs display structural features that should lend themselves to the design of taxon-specific inhibitors. In addition, FPPSs are involved in biochemical pathways other than JH biosynthesis, increasing the potential for physiological disruption. Not surprisingly, FPPS has been shown to be an essential enzyme in several organisms, and inhibitors targeting this enzyme have been developed both as drugs (Oldfield 2010) and herbicides (Cromartie et al. 1991). In this respect, recent advances in the design of lipophilic bisphosphonate inhibitors of FPPS for the treatment of cancer, osteoporosis, and malaria (Zhang et al. 2009; Oldfield 2010) should provide guidance in the development of related molecules as inhibitors of insect FPPSs. Finally, the feasibility of using an RNAi-based pest-control strategy targeting JH biosynthetic enzymes needs to be carefully examined. Indeed, recent work suggests that dsRNA administered *per os* could offer effective and highly species-specific control of certain pests (e.g., Zhu et al. 2011).

Acknowledgments We thank two anonymous reviewers for their constructive comments on an earlier version of the manuscript. Some of the research reviewed in this chapter was supported by grants from the Natural Sciences and Engineering Research Council of Canada and Natural Resources Canada to M.C. and from the National Science Foundation to S.E.S.

References

Audsley N, Matthews HJ, Price NR, Weaver RJ (2008) Allatoregulatory peptides in Lepidoptera, structures, distribution and functions. J Insect Physiol 54:969–980

[BAG] Biology Analysis Group (2004) A draft sequence for the genome of the domesticated silkworm (*Bombyx mori*). Science 306:1937–1940

Baker FC, Lee E, Bergot BJ, Schooley DA (1981) Isomerization of isopentenyl pyrophosphate and homoisopentenyl pyrophosphate by *Manduca sexta* corpora cardiaca-corpora allata homogenates. In: Pratt GE, Brooks GT (eds) Juvenile hormone biochemistry. Elsevier, Amsterdam

Baker FC, Mauchamp B, Tsai LW, Schooley DA (1983) Farnesol and farnesal dehydrogenase(s) in corpora allata of the tobacco hornworm moth, *Manduca sexta*. J Lipid Res 24: 1586–1594

Baker FC, Miller CA, Tsai LW, Jamieson GC, Cerf DC, Schooley DA (1986) The effect of juvenoids, anti juvenile hormone agents, and several intermediates of juvenile hormone biosynthesis on the in vivo juvenile hormone levels in *Manduca sexta* larvae. Insect Biochem 16:741–747

Bergot BJ, Jamieson GC, Ratcliff MA, Schooley DA (1980) JH zero: new naturally occurring insect juvenile hormone from developing embryos of the tobacco hornworm. Science 210:336–338

Bhaskaran G, Sparagana SP, Barrera P, Dahm KH (1986) Change in corpus allatum function during metamorphosis of the tobacco hornworm *Manduca sexta*: regulation at the terminal step in juvenile hormone biosynthesis. Arch Insect Biochem Physiol 3:321–338

Bowers WS, Ohta T, Cleere JS, Marsella PA (1976) Discovery of insect anti-juvenile hormones in plants. Science 193:542–547

Brooks GT, McCaffery AR (1990) The precocene antijuvenile hormones (allatotoxins): a case history in insect toxicology. In: McCaffery AR, Wilson ID (eds) Chromatography and isolation of insect hormones and pheromones. Plenum Press, New York

Burtenshaw SM, Su PP, Zhang JR, Tobe SS, Dayton L, Bendena WG (2008) A putative farnesoic acid *O*-methyltransferase (FAMeT) orthologue in *Drosophila melanogaster* (CG10527): relationship to juvenile hormone biosynthesis? Peptides 29:242–251

Cantín Á, Moya P, Castillo MA, Primo J, Miranda MA, Primo-Yúfera E (1999) Isolation and synthesis of *N*-(2-methyl-3-oxodec-8-enoyl)-2-pyrroline and 2-(hept-5-enyl)-3-methyl-4-oxo-6,7,8,8a-tetrahydro-4*H*-pyrrolo[2,1-*b*]1,3-oxazine – two new fungal metabolites with in vivo anti-juvenile-hormone and insecticidal activity. Eur J Org Chem 1:221–226

Cao L, Zhang P, Grant DF (2009) An insect farnesyl phosphatase homologous to the N-terminal domain of soluble epoxide hydrolase. Biochem Biophys Res Commun 380:188–192

Carney RL, Brown TL (1989) US Patent 4,877,899

Castillo M, Moya P, Couillaud F, Garcerá MD, Martínez-Pardo R (1998) A heterocyclic oxime from a fungus with anti-juvenile hormone activity. Arch Insect Biochem Physiol 37:287–294

Coast GM, Schooley DA (2011) Toward a consensus nomenclature for insect neuropeptides and peptide hormones. Peptides 32:620–631

Cromartie TH, Fisher KJ, Grossman JN (1991) Discovery of a novel site of action for herbicidal bisphosphonates. Pestic Biochem Physiol 63:114–126

Cusson M (2004) Juvenile hormone. In: Capinera JL (ed) Encyclopedia of entomology. Kluwer, Norwell, MA, USA

Cusson M, Le Page A, McNeil JN, Tobe SS (1996) Rate of isoleucine metabolism in lepidopteran corpora allata: regulation of the proportion of juvenile hormone homologs released. Insect Biochem Mol Biol 26:195–201

Cusson M, Béliveau C, Sen SE, Vandermoten S, Rutledge RJ, Stewart D, Francis F, Haubruge É, Rehse P, Huggins DJ, Dowling APG, Grant GH (2006) Characterization and tissue-specific expression of two lepidopteran farnesyl diphosphate synthase homologs: implications for the biosynthesis of ethyl-substituted juvenile hormones. Proteins 65:742–758

Daimon T, Kozaki T, Niwa R, Kobayashi I, Furuta K, Namiki T, Uchino K, Banno Y, Katsuma S, Tamura T, Mita K, Sezutsu H, Nakayama M, Itoyama K, Shimada T, Shinoda T (2012) Prococious metamorphosis in the juvenile hormone-deficient mutant of the silkworm, *Bombyx mori*. PLoS Genet 8:e1002486

Defelipe LA, Dolghih E, Roitberg AE, Nouzova M, Mayoral JG, Noriega FG, Turjanski AG (2011) Juvenile hormone synthesis: "esterify then epoxidize" or "epoxidize then esterify"? Insights from the structural characterization of juvenile hormone acid methyltransferase. Insect Biochem Mol Biol 41:228–235

Feyereisen R, Pratt GE, Hamnett AF (1981) Enzymic synthesis of juvenile hormone in locust corpora allata: evidence for a microsomal cytochrome P-450 linked methyl farnesoate epoxidase. Eur J Biochem 118:231–238

Feyereisen R, Farnsworth DE, Prickett KS, de Montellano PR Ortiz (1985) Suicidal destruction of cytochrome P-450 in the design of inhibitors of insect juvenile hormone biosynthesis. In: Hedin PA, Cutler HG, Hammock BD, Menn JJ, Moreland DE, Plimmer JR (eds) Bioregulators for pest control, vol 276, ACS symposium series. American Chemical Society, Washington, DC

Fujita N, Furuta K, Shirahashi H, Hong S, Shiotsuki T, Kuwano E (2005) Synthesis and anti-juvenile hormone activity of ethyl 4-[2-(6-methyl-3-pyridyloxy)alkyloxy]benzoates. J Pestic Sci 30:192–198

Furuta K, Shirahashi H, Ashibe K, Yamashita H, Nishikawa M, Fujita N, Yamada N, Kuwano E (2006) Synthesis and anti-juvenile hormone activity of alkyl 4-(2-phenoxyalkyloxy)benzoates and related compounds. J Fac Agric Kyushu Univ 51:303–308

Goodman WG, Cusson M (2012) The juvenile hormones. In: Gilbert LI (ed) Insect endocrinology. Elsevier, London

Hammock BD (1975) NADPH dependent epoxidation of methyl farnesoate to juvenile hormone in the cockroach *Blaberus giganteus* L. Life Sci 17:323–328

Hammock BD, Mumby SM (1978) Inhibition of epoxidation of methyl farnesoate to juvenile hormone III by cockroach corpus allatum homogenates. Pestic Biochem Physiol 9:39–47

Hammond AH, Fry JR (1996) Effects of culture duration, cytochrome P-450 inhibition and glutathione depletion on toxicity of diverse xenobiotics. Toxicol In Vitro 10:315–321

Hamnett AF, Pratt GE, Stott KM, Jennings RC (1981) The use of radio HRLC in the identification of the natural substrate of the *O*-methyl transferase and substrate utilization by the enzyme. Dev Endocrinol 15:93–105

Harshman LG, Song KD, Casas J, Shuurmans A, Kuwano E, Kachman SD, Riddiford LM, Hammock BD (2010) Bioassays of compounds with potential juvenoid activity on *Drosophila melanogaster:* juvenile hormone III, bisepoxide JH III and methyl farnesoates. J Insect Physiol 56:1465–1470

Helvig C, Koener JF, Unnithan GC, Feyereisen R (2004) CYP15A1, the cytochrome P450 that catalyzes epoxidation of methyl farnesoate to juvenile hormone III in cockroach corpora allata. Proc Natl Acad Sci USA 101:4024–4029

[HGSC] Honeybee Genome Sequencing Consortium (2006) Insights into social insects from the genome of the honeybee *Apis mellifera*. Nature 443:931–949

Hiruma K, Yagi S, Endo A (1983) ML-236B (Compactin) as an inhibitor of juvenile hormone biosynthesis. J Appl Entomol Zool 18:111–115

Holford KC, Edwards KA, Bendena WG, Tobe SS, Wang Z, Borst DW (2004) Purification and characterization of a mandibular organ protein from the American lobster, *Homarus americanus*: a putative farnesoic acid *O*-methyltransferase. Insect Biochem Mol Biol 34:785–798

Hui JHL, Hayward A, Bendena WG, Takahashi T, Tobe SS (2010) Evolution and functional divergence of enzymes involved in sesquiterpenoid hormone biosynthesis in crustaceans and insects. Peptides 31:451–455

Ishaaya I, Horowitz AR (1992) Novel phenoxy juvenile hormone analog (pyriproxyfen) suppresses embryogenesis and adult emergence of sweetpotato whitefly (Homoptera: Aleyrodidae). J Econ Entomol 85:2113–2117

Ishiguro H, Fujita N, Kim IH, Shiotsuki T, Kuwano E (2003) Ethyl 4-[2-(6-methyl-3-pyridyloxy) butyloxy]benzoate, a novel anti-juvenile hormone agent. Biosci Biotechnol Biochem 67: 2045–2047

Jaya PB, Dubey NK (2011) Evaluation of chemically characterised essential oils of *Coleus aromaticus*, *Hyptis suaveolens* and *Ageratum conyzoides* against storage fungi and aflatoxin contamination of food commodities. Int J Food Sci Technol 46:754–760

Joly A, Edwards PA (1993) Effect of site-directed mutagenesis of conserved aspartate and arginine residues upon farnesyl diphosphate synthase activity. J Biol Chem 268:26983–26989

Jones G, Jones D, Li X, Tang L, Ye L, Teal P, Riddiford LM, Sandifer C, Borovsky D, Martin JR (2010) Activities of natural methyl farnesoids on pupariation and metamorphosis of *Drosophila melanogaster*. J Insect Physiol 56:1456–1464

Judy KJ, Schooley DA, Dunham LL, Hall MS, Bergot J, Siddall JB (1973) Isolation, structure, and absolute configuration of a new natural insect juvenile hormone from *Manduca sexta*. Proc Natl Acad Sci USA 70:1509–1513

Kai Zp, Huang J, Xie Y, Tobe SS, Ling Y, Zhang L, Yc Z, Xl Y (2010) Synthesis, biological activity, and hologram quantitative structure–activity relationships of novel allatostatin analogues. J Agric Food Chem 58:2652–2658

Kai Zp, Xie Y, Huang J, Tobe SS, Jr Z, Ling Y, Zhang L, Yc Z, Xl Y (2011) Peptidomimetics in the discovery of new insect growth regulators: studies on the structure–activity relationships of the core pentapeptide region of allatostatins. J Agric Food Chem 59:2478–2485

Kaneko Y, Furuta K, Kuwano E, Hiruma K (2011) An anti-juvenile hormone agent, ethyl 4-(2-benzylhexyloxy)benzoate, inhibits juvenile hormone synthesis through the suppression of the transcription of juvenile hormone biosynthetic enzymes in the corpora allata in *Bombyx mori*. Insect Biochem Mol Biol 41:788–794

Kiguchi K, Mori T, Akai H (1984) Effects of anti-juvenile hormone "ETB" on the development and metamorphosis of the silkworm, *Bombyx mori*. J Insect Physiol 30:499–506

Kinjoh T, Kaneko Y, Itoyama K, Mita K, Hiruma K, Shinoda T (2007) Control of juvenile hormone biosynthesis in *Bombyx mori*: cloning of the enzymes in the mevalonate pathway and assessment of their developmental expression in the corpora allata. Insect Biochem Mol Biol 37:807–818

Kondo K, Terada O, Oshima K, Mori Y, Mochida K (1977) β-Phenoxy- or substituted phenoxyethanol compounds. US Patent 4016186

Kotaki T, Shinada T, Kaihara K, Ohfune Y, Numata H (2009) Structure determination of a new juvenile hormone from a heteropteran insect. Org Lett 11:5234–5237

Koyama T, Ogura K, Seto S (1973) Studies on isopentenyl pyrophosphate isomerase with artificial substrates. J Biol Chem 248:8043–8051

Koyama T, Matsubara M, Ogura K (1985) Isoprenoid enzyme systems of silkworm. II. Formation of the juvenile hormone skeletons by farnesyl pyrophosphate synthase II. J Biochem 98:457–463

Kreienkamp HJ, Larusson HJ, Witte I, Roeder T, Birgül N, Hönck HH, Harder S, Ellinghausen G, Buck F, Richter D (2002) Functional annotation of two orphan G-protein-coupled receptors, Drostar1 and −2, from *Drosophila melanogaster* and their ligands by reverse pharmacology. J Biol Chem 277:39937–39943

Kuwano E, Eto M (1986) Anti-juvenile hormone effects of an imidazole compound (KK-42) in different larval instars of *Bombyx mori*. Agric Biol Chem 50:2919–2920

Kuwano E, Takeya R, Eto M (1983) Terpenoid imidazoles: new anti-juvenile hormones. Agric Biol Chem 47:921–923

Lewis MJ, Prosser IM, Mohib A, Field LM (2008) Cloning and characterisation of a prenyltransferase from the aphid *Myzus persicae* with potential involvement in alarm pheromone biosynthesis. Insect Mol Biol 17:437–443

Ly VT, Brock B (2011) Effects of CYP inhibitors on precocene I metabolism and toxicity in rat liver slices. Chem Biol Interact 193:109–118

Ma JY, Sun XF, Zhang YL, Li ZX (2010) Molecular cloning and characterization of a prenyltransferase from the cotton aphid, *Aphis gossypii*. Insect Biochem Mol Biol 40:552–561

Marchal E, Zhang JR, Badisco L, Verlinden H, Hult EF, Wielendaele PV, Yagi KJ, Tobe SS, Broeck JV (2011) Final steps in juvenile hormone biosynthesis in the desert locust, *Schistocerca gregaria*. Insect Biochem Mol Biol 41:219–227

Mayoral JG, Nouzova M, Navare A, Noriega FG (2009a) $NADP^+$-dependent farnesol dehydrogenase, a *corpora allata* enzyme involved in juvenile hormone synthesis. Proc Natl Acad Sci USA 106:21091–21096

Mayoral JM, Nouzova M, Yoshiyama M, Shinoda T, Hernandez-Martinez S, Dolghih E, Turjanski AG, Roitberg AE, Priestap H, Perez M, Mackenzie L, Li Y, Noriega FG (2009b) Molecular and functional characterization of a juvenile hormone acid methyltransferase expressed in the corpora allata of mosquitoes. Insect Biochem Mol Biol 39:31–37

Meyer AS, Schneiderman HA, Hanzmann E, Ko J (1968) The two juvenile hormones from the cecropia silk moth. Proc Natl Acad Sci USA 60:853–860

Minakuchi C, Namiki T, Yoshiyama M, Shinoda T (2008) RNAi-mediated knockdown of juvenile hormone acid O-methyltransferase gene causes precocious metamorphosis in the red flour beetle *Tribolium castaneum*. FEBS J 275:2919–2931

Mita K, Kosahara M, Sasaki S et al (2004) The genome sequence of silkworm, *Bombyx mori*. DNA Res 11:27–35

Miziorko HM (2011) Enzymes of the mevalonate pathway of isoprenoid biosynthesis. Arch Biochem Biophys 505:131–143

Moshitzky P, Applebaum SW (1995) Pathway and regulation of JH III-bisepoxide biosynthesis in adult *Drosophila melanogaster* corpus allatum. Arch Insect Biochem Physiol 30:225–237

Nachman RJ, Moyna G, Williams HJ, Tobe SS, Scott AI (1998) Synthesis, biological activity, and conformational studies of insect allatostatin neuropeptide analogs incorporating turn-promoting moieties. Bioorg Med Chem 6:1379–1388

Nachman RJ, Garside CS, Tobe SS (1999) Hemolymph and tissue-bound peptidase-resistant analogs of the insect allatostatins. Peptides 20:23–29

Nagaraju GPC (2007) Is methyl farnesoate a crustacean hormone? Aquaculture 272:39–54

Nässel DR (2002) Neuropeptides in the nervous system of *Drosophila* and other insects: multiple roles as neuromodulators and neurohormones. Prog Neurobiol 68:1–84

Niwa R, Niimi T, Honda N, Yoshiyama M, Itoyama K, Kataoka H, Shinoda T (2008) Juvenile hormone acid *O*-methyltransferase in *Drosophila melanogaster*. Insect Biochem Mol Biol 38:714–720

Noriega FG, Ribeiro JMC, Koener JF, Valenzuela JG, Hernandez-Martinez S, Pham VM, Feyereisen R (2006) Genomic endocrinology of insect juvenile hormone biosynthesis. Insect Biochem Mol Biol 36:366–374

Oldfield E (2010) Targeting isoprenoid biosynthesis for drug discovery: bench to bedside. Acc Chem Res 43:1216–1226

Piulachs MD, Vilaplana L, Bartolome JM, Carreno C, Martin D, Gonzalez-Muniz R, Herranz R, Garcia-Lopez MT, Andreu D, Belles X (1997) Ketomethylene and methyleneamino pseudopeptide analogs of insect allatostatins inhibit juvenile hormone and vitellogenin production in the cockroach *Blattella germanica*. Insect Biochem Mol Biol 27:851–858

Quistad GB, Cerf DC, Schooley DA, Staal GB (1981) Fluoromevalonate acts as an inhibitor of insect juvenile hormone biosynthesis. Nature 289:176–177

Quistad GB, Cerf DC, Kramer SJ, Bergot BJ, Schooley DA (1985) Design of novel insect anti-juvenile hormones: allylic alcohol derivatives. J Agric Food Chem 33:47–50

Reardon JE, Abeles RH (1987) Inhibition of cholesterol biosynthesis by fluorinated mevalonate analogs. Biochemistry 26:4717–4722

Richard DS, Applebaum SW, Sliter TJ, Baker FC, Schooley DA, Reuter CC, Henrich VC, Gilbert LI (1989) Juvenile hormone bisepoxide biosynthesis *in vitro* by the ring gland of *Drosophila*

melanogaster – a putative juvenile hormone in the higher Diptera. Proc Natl Acad Sci USA 86:1421–1425
Röller H, Dahm DH, Sweeley CC, Trost BM (1967) The structure of the juvenile hormone. Angew Chem Int Ed 6:179–180
Schooley DA, Baker FC (1985) Juvenile hormone biosynthesis. In: Kerkut GA, Gilbert LI (eds) Comprehensive insect physiology, biochemistry and pharmacology, vol 7. Pergamon Press, Oxford
Schooley DA, Edwards JP (1996) Anti juvenile hormones: from precocenes to peptide. Brighton Crop Prot Conf Pests Dis 3:1029–1038
Schooley DA, Judy KJ, Bergot BJ, Hall MS, Siddall JB (1973) Biosynthesis of juvenile hormones of *Manduca sexta*: labeling pattern from mevalonate, propionate and acetate. Proc Natl Acad Sci USA 70:2921–2925
Sen SE, Garvin GM (1995) Substrate requirements for lepidopteran farnesol dehydrogenase. J Agric Food Chem 43:820–825
Sen SE, Ewing GJ, Thurston N (1996) Characterization of lepidopteran prenyltransferase in *Manduca sexta* corpora allata. Arch Insect Biochem Physiol 32:315–332
Sen SE, Hitchcock JR, Jordan JL, Richard T (2006) Juvenile hormone biosynthesis in *M. sexta*: substrate specificity of insect prenyltransferase utilizing homologous diphosphate analogs. Insect Biochem Mol Biol 36:827–834
Sen SE, Cusson M, Trobaugh C, Béliveau C, Richard T, Graham W, Mimms A, Roberts G (2007) Purification, properties and heteromeric association of type-1 and type-2 lepidopteran farnesyl diphosphate synthases. Insect Biochem Mol Biol 37:819–828
Shinoda T, Itoyama K (2003) Juvenile hormone acid methyltransferase: a key regulatory enzyme for insect metamorphosis. Proc Natl Acad Sci USA 100:11986–11991
Song L, Poulter CD (1994) Yeast farnesyl-diphosphate synthase: site-directed mutagenesis of residues in highly conserved prenyltransferase domains I and II. Proc Natl Acad Sci USA 91:3044–3048
Sparks TC, Wing KD, Hammock BD (1979) Effects of the anti-hormone-hormone mimic ETB on the induction of insect juvenile hormone esterase in *Trichoplusia ni*. Life Sci 25:445–450
Sperry AE, Sen SE (2001) Farnesol oxidation in insects: evidence that the biosynthesis of insect juvenile hormone is mediated by a specific alcohol oxidase. Insect Biochem Mol Biol 31:171–178
Staal GB (1976) Insect control with insect growth regulators based on insect hormones. In: Marini-Bettólo GB (ed) Natural products and the protection of plants. Pontifica Academia Scientiarum, Rome
Staal GB (1986) Anti juvenile hormone agents. Annu Rev Entomol 31:391–429
Stay B, Tobe SS (2007) The role of allatostatins in juvenile hormone synthesis in insects and crustaceans. Ann Rev Entomol 52:277–299
Sutherland TD, Feyereisen R (1996) Target of cockroach allatostatin in the pathway of juvenile hormone biosynthesis. Mol Cell Endocrinol 120:115–123
Tan A, Tanaka H, Tamura T, Shiotsuki T (2005) Precocious metamorphosis in transgenic silkworms overexpressing juvenile hormone esterase. Proc Natl Acad Sci USA 102:11751–11756
Tobe SS, Bendena WG (2006) Allatostatins in the insects. In: Kastin A (ed) Handbook of biologically active peptides. Academic, New York
Tobe SS, Pratt GE (1974) Dependence of juvenile hormone release from corpus allatum on intraglandular content. Nature 252:474–476
Tobe SS, Stay B (1985) Structure and function of the corpus allatum. Adv Insect Physiol 18:303–438
Ueda H, Shinoda T, Hiruma K (2009) Spatial expression of the mevalonate enzymes involved in juvenile hormone biosynthesis in the corpora allata in *Bombyx mori*. J Insect Physiol 55:798–804
Unnithan GC, Andersen JF, Hisano T, Kuwano E, Feyereisen R (1995) Inhibition of juvenile hormone biosynthesis and methyl farnesoate epoxidase activity by 1,5-disubstituted imidazoles in the cockroach, *Diploptera punctata*. Pestic Sci 43:13–19

Vandermoten S, Charloteaux B, Santini S, Senm SE, Béliveau C, Vandenbol M, Francis F, Brasseur R, Cusson M, Haubruge É (2008) Characterization of a novel aphid prenyltransferase displaying dual geranyl/farnesyl diphosphate synthase activity. FEBS Lett 582:1928–1934

Vandermoten S, Haubruge É, Cusson M (2009a) New insights into short-chain prenyltransferases: structural features, evolutionary history and potential for selective inhibition. Cell Mol Life Sci 66:3685–3695

Vandermoten S, Santini S, Haubruge É, Francis F, Brasseur R, Cusson M, Charloteaux B (2009b) Structural features conferring dual GPP/FPP synthase activity to an aphid prenyltransferase. Insect Biochem Mol Biol 39:707–716

Watson JA, Have CM, Lobos DV, Baker FC, Morrow CJ (1985) Isoprenoid synthesis in isolated embryonic Drosophila cells. Sterol-independent regulatory signal molecule is distal to isopentenyl 1-pyrophosphates. J Biol Chem 260:14083–14091

Weaver RJ, Audsley N (2009) Neuropeptide regulators of juvenile hormone synthesis. Trends Comp Endocrinol Neurobiol 1163:316–329

Williams CM (1967) Third-generation pesticides. Sci Am 217:13–17

Woodhead AP, Stay B, Seidel SL, Khan MA, Tobe SS (1989) Primary structure of four allatostatins: neuropeptide inhibitors of juvenile hormone biosynthesis. Proc Natl Acad Sci USA 86:5997–6001

Yungen M, Bharathi D (2009) Effect of anti-juvenile hormone agent, KK-42 on the incorporation rate of radioactive glycine and biosynthesis of silk proteins of silkworm, *Bombyx mori* L. Toxicol Environ Chem 91:485–491

Zhan S, Merlin C, Boore JL, Reppert SM (2011) The Monarch butterfly genome yields insights into long-distance migration. Cell 147:1171–1185

Zhang YL, Li ZX (2008) Two different farnesyl diphosphate synthase genes exist in the genome of the green peach aphid, *Myzus persicae*. Genome 51:501–510

Zhang Y, Cao R, Yin F et al (2009) Lipophilic bisphosphonates as dual farnesyl/geranylgeranyl diphosphate synthase inhibitors: an x-ray and NMR investigation. J Am Chem Soc 131:5153–5162

Zhu F, Xu J, Palli R, Ferguson J, Palli SR (2011) Ingested RNA interference for managing the populations of the Colorado potato beetle, *Leptinotarsa decemlineata*. Pest Manag Sci 67:175–182

Chapter 4
G Protein-Coupled Receptors as Target Sites for Insecticide Discovery

Hua Bai and Subba Reddy Palli

1 Introduction

G protein-coupled receptors (GPCRs) are proteins that contain seven-transmembrane domains and sense external signals. Binding of external signals to these proteins results in the activation of the heterotrimeric G protein complex. G protein-coupled receptors are members of the largest membrane receptor family, which are present in almost all the eukaryotes (Perez 2005). There are about 800 GPCRs annotated from the human genome and more than 1,000 from the mouse and worm genomes (Fredriksson and Schioth 2005). Despite of its evolutionary success, GPCR family contains a large number of structurally diverse receptors. GPCRs are typically divided into five groups based on their protein structures and ligand-binding properties (Kolakowski 1994; Horn et al. 2003; Fredriksson et al. 2003). Rhodopsin-like GPCR family is the largest family among five GPCR families. Insect GPCRs are generally classified as four families: Rhodopsin-like GPCR, Secretin receptor-like GPCR, Metabotropic glutamate receptor-like GPCR and Atypical GPCR (Frizzled/Smoothened). Within these families, odorant receptors and gustatory receptors (so called chemosensory GPCRs) are a large number of GPCRs that play crucial roles in sensing environmental cues (see ref. Hallem et al. 2006 for a review). Chemosensory GPCRs are excellent targets for developing chemosensory-based pest control tools such as insect repellents and traps. Most of research on this class of GPCRs has been recently reviewed (Touhara and Vosshall 2009). Therefore, we will mainly focus on insect GPCRs other than chemosensory GPCRs in this chapter.

H. Bai (✉)
EEB, Brown University, Providence, RI 02912-G, USA
e-mail: hua_bai@brown.edu

S.R. Palli (✉)
Department of Entomology, College of Agriculture, University of Kentucky, Lexington, KY 40546, USA
e-mail: rpalli@uky.edu

I. Ishaaya et al. (eds.), *Advanced Technologies for Managing Insect Pests*,
DOI 10.1007/978-94-007-4497-4_4, © Springer Science+Business Media Dordrecht 2012

It is remarkable that GPCRs mediate diverse functions, such as the regulation of vision, smell, taste, behavioral and mood regulation, immune system and nervous system. This is also why GPCRs have drawn the most attention in the pharmaceutical industry (Pierce et al. 2002; Gether 2000). About 40% of therapeutic drugs target human GPCRs (Filmore 2004). Insect development, reproduction, metabolism and various behaviors such as feeding and ecdysis, are also under the control of GPCRs (Park and Adams 2005; Broeck 2001). Several GPCRs including ecdysis triggering hormone receptor (ETHR), crustacean cardioactive peptide receptor (CcapR) and bursicon receptor regulate ecdysis behaviour (Truman 2005; Arakane et al. 2008; Loveall and Deitcher 2010). Disrupting or over-activating GPCRs that mediate vital biological functions could lead to dramatic mortality, development arrest and reduced reproduction of insect pests. Therefore, insect GPCRs become a group of potential targets for developing insecticides (Grimmelikhuijzen et al. 2007). Very few efforts have been made to use knowledge on insect GPCRs for the development of insect control agents (Nachman et al. 2001; Predel and Nachman 2001). Based on the mode of action classification form the Insecticide Resistance Action Committee (IRAC), no synthetic pesticides currently used target insect GPCR (http://www.irac-online.org). To discover GPCRs as pesticide targets, a combinatorial approach employing bioinformatics, functional analysis and high-throughput screening for agonists or antagonists is needed.

In the past decade, genomes from more than 20 insect species have been sequenced. More and more insect genomes are being sequenced currently (Robinson et al. 2011). Genome annotation efforts have identified about 200 GPCRs in *Drosophila melanogaster* (Brody and Cravchik 2000), 276 GPCRs in *Anopheles gambiae* (Hill et al. 2002), 56 neurohormone GPCRs in *Apis mellifera* (The HoneyBee Genome Sequencing Consortium 2006), and so on. Functional GPCR assays have been developed to deorphanize the ligand for each orphan GPCR. These assays include bioluminescence assay for monitoring Ca^{2+} mobilization (Stables et al. 1997; Knight et al. 2003; Lu et al. 2011; Staubli et al. 2002) or a luciferase reporter assay for measuring cellular cAMP levels (Chen et al. 1995; Durocher et al. 2000). It is also important to dissect the biological function of GPCR signaling pathways, which will provide further understanding of the mode of action of each GPCR and help to select potential candidate ligands for future pesticide screen. In this chapter, we will cover recent progress made on identification, functional characterization and target screening of GPCRs for developing insecticides.

2 Identification of Insect GPCRs: With a Focus on Neurohormone GPCRs

GPCR is one of the largest protein families in all animals studied. GPCR is recognized by the presence of seven transmembrane (7TM) helices topology. Another feature of GPCR is involvement of G-protein subtypes in the signal-transduction. However, there are G-protein-independent signaling pathways involved in some cases. Except for conserved 7TM structures, there are often large variation in amino acid sequences and length in extracellular and intracellular domains among all GPCRs. The diversity of these domains, including changes in the conformation of 7TM core domain, provides specific properties to each GPCR for recognition and activation by a variety of ligands (Bockaert and Pin 1999). The ligand for GPCR can be light, Ca^{2+}, pheromones, nucleotides, amino acids, peptides, proteins, etc. GPCR classification is therefore based on the type of ligand, receptor sequence homology, as well as functional similarity. GPCRs are typically divided into five main families (Kolakowski 1994; Horn et al. 2003; Fredriksson et al. 2003): Rhodopsin, Secretin, Adhesion, Metabotropic glutamate and Frizzled/Taste2. Rhodopsin-like GPCR family is the largest family among five families. Insect GPCRs are generally classified as four families: Rhodopsin-like GPCR, Secretin receptor-like GPCR, Metabotropic glutamate receptor-like GPCR and Atypical GPCR (Frizzled/Smoothened).

Since the first insect genome sequencing project (*Drosophila melanogaster*) was completed a decade ago (Adams et al. 2000), more than 20 genomes of economically and ecologically important insect species have also been sequenced. With the great advances in genome sequencing technology and dramatically reduced cost of whole genome sequencing, it is not surprising that more and more insect genomes are being sequenced and will be sequenced in the near future. A recent launch of genome initiative, "i5k" (also called "the Manhattan Project of Entomology"), aims to sequence and analyze the genomes of 5,000 species of insects and related arthropods during the next 5 years (Robinson et al. 2011). Tremendous amount of genome sequence information will provide great opportunities for the discovery of potential pesticide targets, including those that belong to GPCR family.

Neurohormone (biogenic amine, neuropeptide and protein hormone) GPCRs, which belong to Rhodopsin and Secretin families, are especially important and play crucial roles in regulating many physiological processes (Hauser et al. 2006). Therefore, neurohormone GPCRs are attractive pesticide targets among the GPCRs. Interestingly, out of 46 human GPCRs targeted by pharmaceutical drugs, about 50% of the targeted GPCRs are neuropeptide and protein hormone receptors, while 26% interact naturally with biogenic amines (Lagerstrom and Schioth 2008). As shown in Table 4.1, similar number of neurohormone GPCRs have been annotated from several sequenced insect genomes. Notably, there are many species-specific neurohormone GPCRs. For example, among all sequenced insect species, inotocin (oxytocin/vasopressin-like) receptors are only found in *Tribolium castaneum, Nasonia vitripennis, Camponotus floridanus* and *Harpegnathos saltator* (Bonasio et al. 2010; Stafflinger et al. 2008). Between two ant species, GPCRs for tachykinin, myosuppressin and FMRFamide were only found in *C. floridanus*, but not in *H. saltator* (Bonasio et al. 2010). GPCRs for neuropeptide-F and corazonin are found in *H. saltator*, but not in *C. floridanus*. Recently, a novel neuropeptide that is structurally similar to adipokinetic hormone (AKH) and corazonin, named ACP (AKH/corazonin-related peptide), was identified as a ligand for orphan GPCRs

Table 4.1 Number of biogenic amine, neuropeptide and protein hormone GPCRs identified through insect genome sequencing projects

Common name	Scientific name	Biogenic amine GPCR	Neuropeptide and protein hormone GPCR	References
Fruit fly	*Drosophila melanogaster*	21	49	Brody and Cravchik (2000), Hauser et al. (2006)
Africa malaria mosquito	*Anopheles gambiae*	18	37	Hill et al. (2002)
Honey bee	*Apis mellifera*	19	37	The Honey Bee Genome Sequencing Consortium (2006), Hauser et al. (2006)
Yellow fever mosquito	*Aedes aegypti*	26	45	Nene et al. (2007)
Domesticated silkworm	*Bombyx mori*	20	46	The International Silkworm Genome Consortium (2008)
Red flour beetle	*Tribolium castaneum*	20	52	Richards et al. (2008), Hauser et al. (2008a)
Pea aphid	*Acyrothosyphon pisum*	18	42	The International Aphid Genomics Consortium (2010)
Parasitoid wasps	*Nasonia vitri-pennis, N. giraulti, N. longicornis Pediculus*	15	39	Werren et al. (2010)
Human body louse	*Humanus humanus*	21	32	Kirkness et al. (2010)
Florida carpenter ant	*Camponotus floridanus*	16	32	Bonasio et al. (2010)
Jerdon's jumping ant	*Harpegnathos saltator*	18	25	Bonasio et al. (2010)

that are closely related to AKH and corazonin GPCRs (Hansen et al. 2010). Interestingly, ACP receptors are found in the mosquitoes (*Anopheles gambiae, Aedes aegypti*, and *Culex pipiens*), the silkworm *Bombyx mori*, the red flour beetle *T. castaneum*, the parasitoid wasp *N. vitripennis*, and the bug *Rhodnius prolixus*. However, the ACP GPCR is not present in 12 *Drosophila* species, the honeybee *A. mellifera*, the pea aphid *Acyrthosiphon pisum* and the body *louse Pediculus humanus* (Hansen et al. 2010). Finally, proctolin receptors are missing from *B. mori* and *A. mellifera* (Roller et al. 2008; Hummon et al. 2006).

In the past decade, many neurohormone GPCRs have also been experimentally deorphanized i.e. a ligand has been identified for the GPCR activation, especially those from *D. melanogaster*. In the following sections, we will summarize deorphanized insect neurohormone GPCRs.

2.1 Biogenic Amine GPCRs

Biogenic amines, such as dopamine, serotonin, octopamine, tyramine and acetylcholine, are neurotransmitters and neurohormones that regulate a variety of behavioral and physiological processes, including locomotion, emotions, circadian rhythms, cardiovascular control, learning and memory (Blenau 2005; Blenau and Baumann 2001). All biogenic amine GPCRs belong to Rhodopsin-like GPCR family.

Dopamine receptors are grouped into D1-like and D2-like receptors based on their ligand binding and signaling transduction properties. Activation of D1-like receptors stimulates cAMP synthesis via Gs protein alpha subunit (Gαs), while activation of D2-like receptors inhibits cAMP formation via Gi protein alpha subunit (Gαi) (Clark and Baro 2007). Several dopamine receptors have been identified and characterized in *D. melanogaster* (Gotzes et al. 1994; Sugamori et al. 1995; Feng et al. 1996; Han et al. 1996; Hearn et al. 2002; Draper et al. 2007), *A. mellifera* (Blenau et al. 1998; Beggs et al. 2005; Ebert et al. 1998), *B. mori* (Ohta et al. 2009; Mitsumasu et al. 2008) and *Papilio xuthus* (Ono and Yoshikawa 2004). A recent study identified a GPCR (DmDopEcR) from *D. melanogaster*, which can be activated by both dopamine and ecdysteroids. This study suggests that beside the regulation of transcription via its nuclear receptor, ecdysteroids can also trigger rapid and non-genomic actions through a membrane receptor (Srivastava et al. 2005).

Serotonin (5-Hydroxytryptamine; 5-HT) receptors are another major group of biogenic amine GPCRs and belong to the diverse family of neurotransmitter receptors. Several serotonin receptors have been identified in *D. melanogaster* (Corey et al. 1994; Witz et al. 1990; Saudou et al. 1992; Saudou and Hen 1994), *A. mellifera* (Thamm et al. 2010; Schlenstedt et al. 2006; Ebert et al. 1998) and *A. aegypti* (Pietrantonio et al. 2001).

Octopamine and tyramine, the decarboxylation products of tyrosine, are the invertebrate counterparts of the vertebrate adrenergic transmitters (Roeder 2005). Octopamine plays key roles in modulating muscle activity in locust, learning and memory in honeybee and fruit fly. Octopamine receptors have been found in

D. melanogaster (Han et al. 1998; Maqueira et al. 2005; Balfanz et al. 2005; Arakawa et al. 1990), *A. mellifera* (Grohmann et al. 2003; Ebert et al. 1998), *B. mori* and *Heliothis virescens* (von Nickisch-Rosenegk et al. 1996), *Schistocerca gregaria* (Verlinden et al. 2010), while tyramine receptors have been identified in *D. melanogaster* (Saudou et al. 1990; Cazzamali et al. 2005a), *A. mellifera* (Blenau et al. 2000), *B. mori* (Huang et al. 2009) and *Periplaneta americana* (Rotte et al. 2009).

In addition, two muscarinic acetylcholine receptors have been cloned and characterized in *D. melanogaster* (Onai et al. 1989; Shapiro et al. 1989). Muscarinic acetylcholine receptors, so called mAChRs, are stimulated by acetylcholine and play important roles in autonomic nervous system.

2.2 Neuropeptide GPCRs

Neuropeptides are small peptides secreted from brain and endocrine glands that are involved in many crucial physiological processes, e.g. feeding behavior, learning and memory, fluid secretion, muscle activity/locomotion, ecdysis behaviors. Most of neuropeptide GPCRs belong to Rhodopsin-like GPCR family, while diuretic hormone receptors and pigment dispersing factor (PDF) belong to Secretin receptor-like GPCR family.

Insect FMRFamides, myosuppressins, sulfakinins and neuropeptide Y-like (NPF and sNPF) are all FMRFamide-related peptides. FMRFamides possibly act at skeletal muscle, and muscle of the heart and intestine. A FMRFamide receptor (CG2114) has been cloned and functionally characterized in *D. melanogaster* (Cazzamali and Grimmelikhuijzen 2002; Meeusen et al. 2002). However, the physiological functions regulated by FMRFamides and their receptors are largely unknown. Recently, a study suggests that FMRFamide-related peptides (FaRPs) may regulate circadian locomotor activity rhythms in the cockroach *Leucophaea maderae* (Soehler et al. 2008). Another study showed that one of abundant FMRFamides, DPKQDFMRFamide enhanced synaptic transmission through activation of FMRFamide receptor (FR) and myosupressin receptor-2 (DmsR-2) in *D. melanogaster*, which leads to enhanced fictive locomotion (Klose et al. 2010).

Myosuppressins were originally identified as a neuropeptide with a FLRFamide car-boxy terminus that can inhibit hindgut contractions of in cockroaches (Holman et al. 1986).

Two receptors activated by *Drosophila* myosuppressin that block both muscle and nervous system activities have been found in *D. melanogaster* (Egerod et al. 2003a; Johnson et al. 2003a).

Insect sulfakinins have been isolated originally as neuropeptides with myotropic activity. Then sulfakinins were found to inhibit feeding behavior in several insect species (Wei et al. 2000; Maestro et al. 2001; Meyering-Vos and Muller 2007; Downer et al. 2007; Wicher et al. 2007). In *D. melanogaster*, sulfakinins are also involved larval odor preference and locomotion activity (Nichols et al. 2008). Two sulfakinin receptors have been identified and functional characterized in *D. melanogaster* (Kubiak et al. 2002).

Short neuropeptides F (sNPFs) is a neuropeptide with the C-terminal consensus sequence, PX1RLRX2amide, and it regulates food intake and body size. Short neuropeptides F receptor has been identified in *D. melanogaster* (Lee et al. 2004). NPF (long NPF) is another neuropeptide Y (NPY)-like peptides with a C-terminus RVRFa. Functions of NPF in *D. melanogaster* have been found in controlling foraging, feeding, alcohol sensitivity and aggression (Chen et al. 2008; Lingo et al. 2007; Wu et al. 2003, 2005).

Pyrokinins are neuropeptides with a C-terminal FXPRLamide that regulate insect sex pheromone production and contractions of visceral muscles (Nachman et al. 2001; Predel and Nachman 2001). Pyrokinin receptors have been identified characterized in *D. melanogaster* and *A. gambiae* (Park et al. 2002; Olsen et al. 2007; Cazzamali et al. 2005b; Rosenkilde et al. 2003).

Proctolin was the first insect neuropeptide to be sequenced (Brown 1975; Starratt and Brown 1975). Proctolin acts as a neuromodulator and a neurohormone to stimulate the contract of both somatic and visceral muscles. Proctolin is also a cotransmitter in glutamatergic motoneurons to enhance neuromuscular transmission (Orchard et al. 1989). Proctolin receptor has been identified and characterized from *D. melanogaster* in two parallel studies (Egerod et al. 2003b; Johnson et al. 2003b).

Capa peptides can stimulate their fluid secretion of malpighian tubules in several dipteran insects, e.g. *D. melanogaster, A. aegypti, A. stephensi* and *G. morsitans* (Pollock et al. 2004). Capa receptor was also found highly expressed in malpighian tubules. Several capa receptors have been identified and functionally examined in *D. melanogaster* and *A. gambiae* respectively (Iversen et al. 2002a; Olsen et al. 2007; Park et al. 2002).

SIFamides with conserved sequence AYRKPPFNGSIFamide are thought to regulate adult courtship behavior (Terhzaz et al. 2007). One SIFamide receptor has been identified in *D. melanogaster* (Terhzaz et al. 2007; Jorgensen et al. 2006).

Insect tachykinins share sequence homology to mammalian tachykinins and are intestinal myotropic peptides found in locust (Schoofs et al. 1990). Other than myostimulatory effects, tachykinins have found to regulate AKH release, secretion in malpighian tubules, chemosensory and locomotion (Winther et al. 2006; Nassel 2002; Ignell et al. 2009). Two tachykinin receptors have been identified and characterized recently in *D. melanogaster* (Birse et al. 2006; Poels et al. 2009).

Insect allatostatins were originally isolated as inhibitory neuropeptides that can block the production of juvenile hormone (JH) in the corpora allata (CA) (Woodhead et al. 1989). However, many other functions of allatostatins were identified in later studies, including myoinhibitory and cardioinhibitory activities (Blackburn et al. 2001; Price et al. 2002). Insect allatostatins can be divided into three different groups based on their distinct consensus amino acid sequences: allatostatin-A, allatostatin-B (or MIP, myoinhibitory peptides) and allatostatin-C. Several allatostatin receptors have been identified in *D. melanogaster* (Kreienkamp et al. 2002; Johnson et al. 2003a; Larsen et al. 2001; Lenz et al. 2000a, b, 2001; Birgul et al. 1999).

Leucokinins, as well as many other myokinins can stimulate spontaneous muscle contractions (Schoofs et al. 1992). A leucokinin receptor has been identified in *D. melanogaster* that is related to neurokinin receptors of mammals (Hauser et al. 2008b).

Insect corazonin was first isolated as a cardioaccelerating peptide from the corpora cardiac of the cockroach *Periplaneta americana* (Veenstra 1989). But the biological functions of corazonin vary among different insects. Corazonin has been associated with polyphenism in migratory locusts, silk production in silkworms and ecdysis behavior in *M. sexta* (Veenstra 2009). Corazonin receptor has been identified and functionally characterized in *D. melanogaster*, *M. sexta* and *A. gambiae* (Park et al. 2002; Cazzamali et al. 2002; Belmont et al. 2006).

AKH is an insect neuropeptide produced by corpora cardiac and regulates carbohydrate and lipid metabolism in the fat body (adipose tissues) of *D. melanogaster*, *A. gambiae* and *T. castaneum* (Park et al. 2002; Hansen et al. 2010). A novel neuropeptide named ACP was recently identified as a ligand for orphan GPCRs that are closely related to AKH and corazonin GPCRs (Hansen et al. 2010). However, the biological function of ACP signaling remains unknown.

Vasopressin and oxytocin are the earliest characterized neuropeptides. They were discovered from the mammalian posterior pituitary, with the activities of antidiuresis and increasing in blood pressure (vasopressin) or promoting contractions of the estrogen-treated uterus preparations and the ejection of milk from the mammary glands (oxytocin). Insect oxytocin/vasopressin-like peptide, named inotocin, and its receptor were recently identified and functional characterized in *T. castaneum* (Stafflinger et al. 2008; Aikins et al. 2008). However, RNAi (RNA interference) for inotocin and its receptor has resulted in no mortality or abnormal phenotype in *T. castaneum* (Aikins et al. 2008).

Several neuropeptides such as CCAP (Cazzamali et al. 2003; Arakane et al. 2008), ETH (Arakane et al. 2008; Iversen et al. 2002b; Park et al. 2003) and their receptors are involved in the regulation of insect ecdysis behavior. Since ecdysis behavior is essential during insect development and growth, ecdysis-regulating GPCRs could be promising targets for developing novel pest control agents.

Diuretic hormones, along with capa peptides and leucokinin, are neuropeptides that stimulate fluid secretion in malpighian tubules in *D. melanogaster*. Two diuretic hormones (DH31 and DH44) share sequence similarities to calcitonin (DH31) and corticotrophin-releasing peptide. Diuretic hormones receptors have been cloned and characterized in *D. melanogaster* (Johnson et al. 2004, 2005; Hector et al. 2009).

The first insect pigment dispersing factor (PDF) was isolated from the grasshopper *Romalea microptera* (Rao et al. 1987). In *D. melanogaster*, PDF expressed in subsets of clock neurons that mainly regulates circadian rhythms and geotactic behaviors (Hyun et al. 2005; Mertens et al. 2005; Lear et al. 2009).

2.3 Protein Hormone GPCRs

Insect glycoprotein hormone GPCRs belong to the family of leucine-rich repeat-containing G protein-coupled receptors (LGRs). The major LGRs in insects are bursicon and the GPA2/GPB5 receptors. The involvement of bursicon receptor in cuticle tanning and wing expansion was discovered in *D. melanogaster* a decade ago (Baker and Truman 2002). Then, the ligand bursicon was identified as a heterodimer of two cystine knot peptides encoded by CG13419 (named burs) and CG15284 (named pburs) (Luo et al. 2005; Mendive et al. 2005). Bursicon receptor has also been characterized in several other insect species, such as *B. mori* (Huang et al. 2007), *T. castaneum* (Arakane et al. 2008; Bai and Palli 2010) and *Manduca sexta* (Dai et al. 2008). In addition to bursicon receptor, there is another LGR in *D. melanogaster* that can be activated by thyrostimulin, a heterodimer of two glycoprotein hormone subunits (GPA2 and GPB5) (Sudo et al. 2005; Hauser et al. 1997). A recent study suggests that GPA2/GPB5 may play important roles in development and anti-diuresis in *D. melanogaster* (Sellami et al. 2011).

2.4 GPCRs in Other Families

Amino acids are known to be the ligands of GPCRs (Bockaert and PhilippePin 1999). These GPCRs are activated by the neurotransmitter glutamate, and play important roles in the central nervous system function of *D. melanogaster* (Pan and Broadie 2007). Mitri et al. (2004) identified homologs of mammalian GluRs in *A. gambiae*, *A. mellifera*, and *D. melanogaster* genomes and named them as AmXR, HBmXR, and DmXR. These studies also showed that the DmXR is activated by a ligand that contains an amino group, extracted from heads of fruit flies, mosquitoes and locusts. Interestingly, orthlogues of these receptors are so far identified only from insects. Recently, DmXr has been deorphanized to show that L-canavanine is a ligand of this receptor (Mitri et al. 2009). Since the presence of this receptor is limited to insects, it could be an excellent candidate for development of insecticides.

GPCRs that belong to other families, e.g. Atypical GPCR (Frizzled/Smoothened), also play crucial roles in regulating insect development and growth. We will discuss the possibility of developing pesticides to target Atypical GPCRs in Sects. 3.2 and 4.

3 Deciphering GPCR Ligand and Functional Analysis

3.1 GPCR Deorphanization

To identify the ligand(s) for each orphan GPCR, two assays are mainly used: (1). Ca^{2+} mobilization assay (Stables et al. 1997; Knight et al. 2003; Lu et al. 2011; Staubli et al. 2002) and (2). cAMP luciferase reporter assay (Chen et al. 1995; Durocher et al. 2000). In Ca^{2+} mobilization assay (as shown in Fig. 4.1), an orphan GPCR, a promiscuous Gα16 protein and apoaequorin are co-expressed in either frog oocytes or mammalian cells, such as Chinese hamster ovary (CHO) cells. Three hours before the assay, the co-factor of apoaequorin, coelenterazine is added to the culture medium. Activation of the receptor would initiate an IP3/Ca^{2+} cascade,

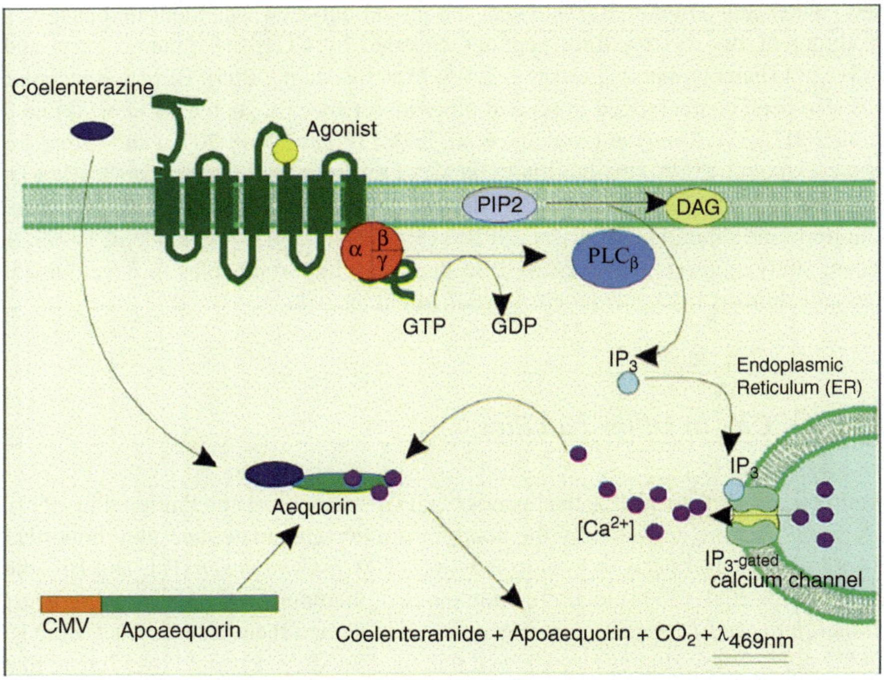

Fig. 4.1 Schematic diagram showing one of the GPCR function assays, Ca^{2+} mobilization assay in CHO cells. An orphan GPCR and Gα16 protein are expressed in a CHO cell line. Upon receptor activation, the increased Ca^{2+} concentration stimulates aequorin to emit light

leading to a strong bioluminescence response (Stables et al. 1997; Cazzamali and Grimmelikhuijzen 2002). Emitting luminescence levels indicate the calcium concentration. As the GPCR signals through the release of intracellular calcium, the intensity of the signal is related to the potency of the ligand. In a cAMP luciferase-reporter assay, human embryonic kidney (HEK) 293 cells are regularly used. An orphan GPCR, together with the appropriate luciferase reporter gene (e.g. 6xcAMP-luciferase) and a proper G protein are transfected into 293 cells. Twenty-four hours post-transfection, cells are exposed to ligand for 3 h in serum-free medium. Then cells are lysed, and luciferase activities are quantified using a plate reader. Ligand potencies are determined by stimulated luciferase activities (cellular cAMP levels) with increasing concentrations of ligand.

3.2 GPCR Functional Analysis

Most functional studies on insect neurohormones were done by injecting peptides synthesized in vitro. However, very little is known about in vivo functions of neurohormones and their receptors. The physiological and behavioral roles of *Drosophila* neuropeptides have been recently reviewed (Nassel and Winther 2010). Although more and more insect GPCRs have been deorphanized, relatively few efforts have been made to apply knowledge on insect GPCRs to the development of novel pest control-methods such as the disruption of neuropeptide signaling systems using neuropeptide mimics (Nachman et al. 2001; Predel and Nachman 2001). Furthermore, it is the critical step to identify the GPCRs mediating normal insect physiology in order to conduct a high-throughput small molecular screen. Based on the classification of mode of actions, issued by IRAC, there are no commercial synthetic-pesticides available using insect GPCRs as targets (http://www.irac-online.org).

The powerful *Drosophila* genetics tools have been applied to decipher the biological functions of *Drosophila* GPCRs. Broad ranges of physiological and behavioral functions regulated by diverse GPCRs have been identified, e.g. feeding behavior, locomotion activity, metabolism and circadian rhythm. Other *Drosophila* GPCRs are found to play important roles in development, molting and reproduction. Many atypical GPCRs such as Frizzle/smoothened family are found to be involved in imaginal disc development during the larval-pupal metamorphosis. Both Frizzled (Fz) and Smoothened (smo) are tissue polarity genes. In *D. melanogaster*, frizzled mutants show strong wing-hair disorientation and negligible segment-polarity of homozygous embryos (Park et al. 1994; Schulte and Bryja 2007). A recently study discovered bursicon receptor is required not only for cuticle tanning, but also plays an important role in the development of the epidermis and imaginal discs during prepupal stage (Loveall and Deitcher 2010). Using a genome-wide RNAi screen approach, sex peptide (SP), one of the male accessory gland proteins (Acps), was found to bind to a GPCR (sex peptide receptor) and mediated female post-mating behaviors in *D. melanogaster* (Yapici et al. 2008). Interestingly, myoinhibitory peptides (MIPs) was recently identified as a second family of SPR ligands, indicating SPR may be involved in functions other than female reproduction (Kim et al. 2010).

In the past few years, *T. castaneum* has been utilized as a model insect for various functional genomics-studies, due to its systemic RNA interference (RNAi) effects through dsRNA injection. Eclosion hormone (EH), ecdysis triggering hormone (ETH), crustacean cardioactive peptide (CCAP) and bursicon are four critical neuropeptides involved in ecdysis behavior. Function analysis of these neuropeptides and their receptors were studied using RNAi in *T. castaneum* (Arakane et al. 2008). As summarized in Fig. 4.2, RNAi of EH and ETH disrupted preecdysis behavior and prevented subsequent ecdysis behavior, while RNAi for CCAP interrupted ecdysis behavior. RNAi of genes encoding receptors for those peptides produced phenocopies comparable to those of their respective cognate neuropeptides.

4 RNAi Screening to Identify GPCR Pesticide Targets: A Case Study

Recently, 111 non-sensory GPCRs [including 72 that have been previously reported as biogenic amine and neuropeptide GPCRs (Hauser et al. 2008a)] were annotated in the *T. castaneum* genome by searching databases (Bai et al. 2011). These 111 GPCRs were classified into four families using the conserved domain prediction program.

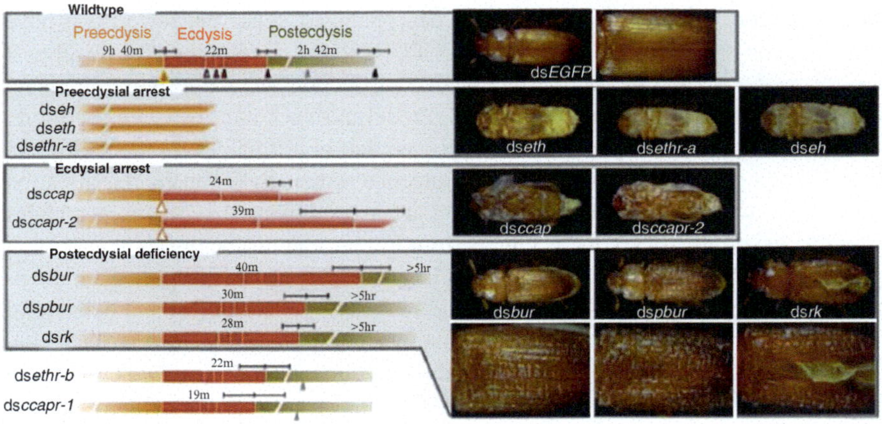

Fig. 4.2 Eclosion phenotypes of RNAi targeting four neuropeptides and their receptors in *T. castaneum* (Figure is reproduced with permission from Arakane et al. 2008)

In total, 74 Rhodopsin-like GPCRs, 19 Secretin receptor-like GPCRs, 11 Metabotropic glutamate receptor-like GPCRs, and seven Atypical GPCRs have been identified in *T. castaneum* (Bai et al. 2011).

To identify GPCRs that could be used as target sites for development of new insecticides we have performed a large-scale RNAi screening in the red flour beetle, *T. castaneum*. Double stranded RNA prepared using a 300–500 bp fragment of the gene coding for each of these 111 GPCRs was injected into 24 h old final instar larvae. Developmental abnormalities and mortality of dsRNA injected insects were recorded. The RNAi screen identified 25 GPCRs that may be useful as target sites for insecticide development, because knockdown in the expression of genes coding for these GPCRs caused more than 30% mortality. Among these, knockdown in the expression of eight GPCRs caused more than 90% mortality. As shown in Fig. 4.3, silencing of genes coding for TC012521/stan, TC009370/mthl, TC001872/Cirl, TC014055/fz and TC005545/smo caused severe mortality. Interestingly, knockdown in the expression of the gene coding for one of the GPCRs, dopamine-2 like receptor (TC007490) caused high lethality during the early larval stage. In *D. melanogaster*, dopamine-2 like receptor (D2R) is highly expressed in head and brain (http://www.flyatlas.org/) and D2R RNAi flies showed significantly decreased locomotor activity (Draper et al. 2007). Since knockdown in the expression of the gene coding for D2R caused severe problems and mortality during the early larval stage this receptor might be playing a critical role in the growth and development of beetle larvae perhaps by modulating neuronal development and locomotor activity. This first comprehensive RNAi screen on GPCRs provided some useful data and showed that it is possible to use high-throughput RNAi screens to identify target sites for insecticide development (Bai et al. 2011).

RNAi–aided reduction in the levels of some of the GPCRs showed severe phenotypes including arrest in development and failure to undergo ecdysis. For example,

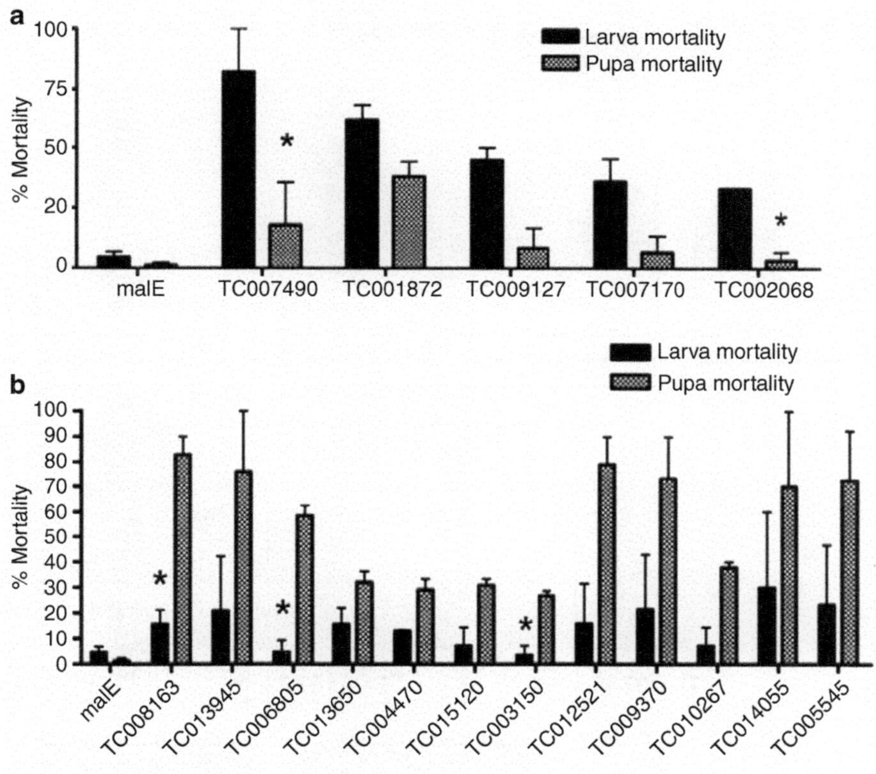

Fig. 4.3 Reduction in the levels of GPCRs by RNAi causes mortality during both larval and pupla stages. malE (control) or T. castaneum GPCR dsRNA was injected into 24 h old final instar larvae. Number of larvae and pupae died after each treatment were recorded. Mean ± SE of two independent experiments are shown (Figure is reproduced with permission from Bai et al. 2011)

knockdown in the expression of gene coding for bursicon receptor, Tcrk, caused problems in cuticle tanning, wing expansion as well as development and expansion of integumentary structures and adult eclosion (Bai and Palli 2010). Reduction in levels of TC007490/D2R blocked larval-pupal metamorphosis (Fig. 4.4e). Reduction in the levels of TC007490/D2R also caused problems with growth and only a small portion of tested insects fully grown and reached quiscent stage. TC001872/Cirl knockdown blocked larval-pupal metamorphosis (Fig. 4.4f). Interestingly, 60% of the TC001872/Cirl RNAi larvae dsRNA completed larval-pupal metamorphosis and died during the pupal stage. TC012521/stan RNAi insects died during pharate adult stage. Reduction in TC012521/stan levels caused problems in adult eclosion (Fig. 4.4g). Reduction in the levels of TC014055/fz and TC009370/mthl caused problems during both larval-pupal and pupal-adult ecdysis (Fig. 4.4i, j). Reduction in TC005545/smo levels led to death during the early pupal stages and the pupal wings were not fully extended in these insects (Fig. 4.4k) (Bai et al. 2011).

Fig. 4.4 Various Phenotypes observed after RNAi aided reduction the levels of GPCRs (**a**) control final instar larva; (**b**) insect in quiescent stage; (**c**) control pupa; (**d**) control adult; (**e**) TC007490/D2R RNAi phenotype; (**f**) TC001872/Cirl RNAi phenotype.; Accumulation of the ommatidia (*black arrow*) at quiescent stage is shown at the *upper-left panel* at higher magnification. (**g**) TC012521/stan RNAi Phenotype with wings attached to the ventral side of the abdomen (*black arrow*); (**h**) TC014055/fz RNAi Phenotype, the *black arrow* points to the split in the dorsal thoracic region; (**i**) PTC005545/smo RNAi Phenotype showing unextended pupal wings (*white arrow*) and (**j**) TC009370/mthl RNAi Phenotype showing improperly folded wings (*white arrow*) and unshed exuviae (*white arrow head*). Scale bar: 1.0 mm (Figure is reproduced with permission from Bai et al. 2011)

5 Summary and Conclusions

Within the next few years, whole genome sequence will not be limited for model insect species. Genome sequence information for more and more non-model insects, especially for those economically and ecologically important species, will become available, which will greatly speed up pesticide discovery. GPCR is one of the largest multi-gene families that play crucial roles in diverse developmental, physiological and behavior responses, which is why GPCRs have drawn the most attention in the pharmaceutical industry (Pierce et al. 2002; Gether 2000). Thus, about 40% of therapeutic drugs target human GPCRs (Filmore 2004). Although many insect GPCRs have been deorphanized, the knowledge gained from these studies has not yet been applied for discovery of new insecticides. Large-scale RNAi screen has been successfully used in dissecting signal transduction networks *in vitro* (Kiger et al. 2003; Nybakken et al. 2005) as well as *in vivo* (Cronin et al. 2009). RNAi screens have been used to identify 25 GPCRs that are critical for growth, development and survival of the red flour beetle (Bai et. al. 2011). In the future, many more studies using information available from insect genomes that are being sequenced will help to explore this class of molecules in the development of pest control agents.

Acknowledgements The research in Palli laboratory was supported by the National Science Foundation (IBN0421856), the National Institute of Health (GM070559-07), and the National Research Initiative of the USDA-NIFA (2011-67013-30143). This report is contribution number 11-08-075 from the Kentucky Agricultural Experimental Station.

References

Adams MD, Celniker SE, Holt RA, Evans CA, Gocayne JD, Amanatides PG, Scherer SE, Li PW, Hoskins RA, Galle RF, George RA, Lewis SE, Richards S, Ashburner M, Henderson SN, Sutton GG, Wortman JR, Yandell MD, Zhang Q, Chen LX, Brandon RC, Rogers YH, Blazej RG, Champe M, Pfeiffer BD, Wan KH, Doyle C, Baxter EG, Helt G, Nelson CR, Gabor GL, Abril JF, Agbayani A, An HJ, Andrews-Pfannkoch C, Baldwin D, Ballew RM, Basu A, Baxendale J, Bayraktaroglu L, Beasley EM, Beeson KY, Benos PV, Berman BP, Bhandari D, Bolshakov S, Borkova D, Botchan MR, Bouck J, Brokstein P, Brottier P, Burtis KC, Busam DA, Butler H, Cadieu E, Center A, Chandra I, Cherry JM, Cawley S, Dahlke C, Davenport LB, Davies P, de Pablos B, Delcher A, Deng Z, Mays AD, Dew I, Dietz SM, Dodson K, Doup LE, Downes M, Dugan-Rocha S, Dunkov BC, Dunn P, Durbin KJ, Evangelista CC, Ferraz C, Ferriera S, Fleischmann W, Fosler C, Gabrielian AE, Garg NS, Gelbart WM, Glasser K, Glodek A, Gong F, Gorrell JH, Gu Z, Guan P, Harris M, Harris NL, Harvey D, Heiman TJ, Hernandez JR, Houck J, Hostin D, Houston KA, Howland TJ, Wei MH, Ibegwam C, Jalali M, Kalush F, Karpen GH, Ke Z, Kennison JA, Ketchum KA, Kimmel BE, Kodira CD, Kraft C, Kravitz S, Kulp D, Lai Z, Lasko P, Lei Y, Levitsky AA, Li J, Li Z, Liang Y, Lin X, Liu X, Mattei B, McIntosh TC, McLeod MP, McPherson D, Merkulov G, Milshina NV, Mobarry C, Morris J, Moshrefi A, Mount SM, Moy M, Murphy B, Murphy L, Muzny DM, Nelson DL, Nelson DR, Nelson KA, Nixon K, Nusskern DR, Pacleb JM, Palazzolo M, Pittman GS, Pan S, Pollard J, Puri V, Reese MG, Reinert K, Remington K, Saunders RD, Scheeler F, Shen H, Shue BC, Siden-Kiamos I, Simpson M, Skupski MP, Smith T, Spier E, Spradling AC, Stapleton M, Strong R, Sun E, Svirskas R, Tector C, Turner R, Venter E, Wang AH, Wang X, Wang ZY,

Wassarman DA, Weinstock GM, Weissenbach J, Williams SM, Woodage T, Worley KC, Wu D, Yang S, Yao QA, Ye J, Yeh RF, Zaveri JS, Zhan M, Zhang G, Zhao Q, Zheng L, Zheng XH, Zhong FN, Zhong W, Zhou X, Zhu S, Zhu X, Smith HO, Gibbs RA, Myers EW, Rubin GM, Venter JC (2000) The genome sequence of Drosophila melanogaster. Science 287(5461): 2185–2195. doi:8392[pii]

Aikins MJ, Schooley DA, Begum K, Detheux M, Beeman RW, Park Y (2008) Vasopressin-like peptide and its receptor function in an indirect diuretic signaling pathway in the red flour beetle. Insect Biochem Mol Biol 38(7):740–748. doi:S0965-1748(08)00072-6[pii]10.1016/j.ibmb.2008.04.006

Arakane Y, Li B, Muthukrishnan S, Beeman RW, Kramer KJ, Park Y (2008) Functional analysis of four neuropeptides, EH, ETH, CCAP and bursicon, and their receptors in adult ecdysis behavior of the red flour beetle, Tribolium castaneum. Mech Dev 125(11–12):984–995. doi:S0925-4773(08)00137-8[pii]10.1016/j.mod.2008.09.002[doi]

Arakawa S, Gocayne JD, McCombie WR, Urquhart DA, Hall LM, Fraser CM, Venter JC (1990) Cloning, localization, and permanent expression of a Drosophila octopamine receptor. Neuron 4(3):343–354. doi:0896-6273(90)90047-J[pii]

Bai H, Palli SR (2010) Functional characterization of bursicon receptor and genome-wide analysis for identification of genes affected by bursicon receptor RNAi. Dev Biol 344(1):248–258. doi:S0012-1606(10)00289-7[pii]10.1016/j.ydbio.2010.05.003

Bai H, Zhu F, Shah K, Palli S (2011) Large-scale RNAi screen of G protein-coupled receptors involved in larval growth, molting and metamorphosis in the red flour beetle. BMC Genomics 12(1):388

Baker JD, Truman JW (2002) Mutations in the Drosophila glycoprotein hormone receptor, rickets, eliminate neuropeptide-induced tanning and selectively block a stereotyped behavioral program. J Exp Biol 205(Pt 17):2555–2565

Balfanz S, Strunker T, Frings S, Baumann A (2005) A family of octopamine [corrected] receptors that specifically induce cyclic AMP production or Ca2+ release in Drosophila melanogaster. J Neurochem 93(2):440–451. doi:JNC3034[pii]10.1111/j.1471-4159.2005.03034.x

Beggs KT, Hamilton IS, Kurshan PT, Mustard JA, Mercer AR (2005) Characterization of a D2-like dopamine receptor (AmDOP3) in honey bee, Apis mellifera. Insect Biochem Mol Biol 35(8):873–882. doi:S0965-1748(05)00093-7[pii]10.1016/j.ibmb.2005.03.005

Belmont M, Cazzamali G, Williamson M, Hauser F, Grimmelikhuijzen CJ (2006) Identification of four evolutionarily related G protein-coupled receptors from the malaria mosquito Anopheles gambiae. Biochem Biophys Res Commun 344(1):160–165. doi:S0006-291X(06)00670-X[pii]10.1016/j.bbrc.2006.03.117

Birgul N, Weise C, Kreienkamp HJ, Richter D (1999) Reverse physiology in drosophila: identification of a novel allatostatin-like neuropeptide and its cognate receptor structurally related to the mammalian somatostatin/galanin/opioid receptor family. EMBO J 18(21): 5892–5900. doi:10.1093/emboj/18.21.5892

Birse RT, Johnson EC, Taghert PH, Nassel DR (2006) Widely distributed Drosophila G-protein-coupled receptor (CG7887) is activated by endogenous tachykinin-related peptides. J Neurobiol 66(1):33–46. doi:10.1002/neu.20189

Blackburn MB, Jaffe H, Kochansky J, Raina AK (2001) Identification of four additional myoinhibitory peptides (MIPs) from the ventral nerve cord of Manduca sexta. Arch Insect Biochem Physiol 48(3):121–128. doi:10.1002/arch.1064[pii]10.1002/arch.1064

Blenau W (2005) Cellular actions of biogenic amines. Arch Insect Biochem Physiol 59(3):99–102. doi:10.1002/arch.20072

Blenau W, Baumann A (2001) Molecular and pharmacological properties of insect biogenic amine receptors: lessons from Drosophila melanogaster and Apis mellifera. Arch Insect Biochem Physiol 48(1):13–38. doi:10.1002/arch.1055[pii]10.1002/arch.1055

Blenau W, Erber J, Baumann A (1998) Characterization of a dopamine D1 receptor from Apis mellifera: cloning, functional expression, pharmacology, and mRNA localization in the brain. J Neurochem 70(1):15–23

Blenau W, Balfanz S, Baumann A (2000) Amtyr1: characterization of a gene from honeybee (Apis mellifera) brain encoding a functional tyramine receptor. J Neurochem 74(3):900–908

Bockaert J, PhilippePin J (1999) Molecular tinkering of G protein-coupled receptors: an evolutionary success. EMBO J 18(7):1723–1729

Bockaert J, Pin JP (1999) Molecular tinkering of G protein-coupled receptors: an evolutionary success. EMBO J 18(7):1723–1729. doi:10.1093/emboj/18.7.1723

Bonasio R, Zhang G, Ye C, Mutti NS, Fang X, Qin N, Donahue G, Yang P, Li Q, Li C, Zhang P, Huang Z, Berger SL, Reinberg D, Wang J, Liebig J (2010) Genomic comparison of the ants Camponotus floridanus and Harpegnathos saltator. Science 329(5995):1068–1071. doi:329/5995/1068[pii]10.1126/science.1192428

Brody T, Cravchik A (2000) Drosophila melanogaster G protein-coupled receptors. J Cell Biol 150(2):F83–F88

Broeck JV (2001) Insect G protein-coupled receptors and signal transduction. Arch Insect Biochem Physiol 48(1):1–12. doi:10.1002/arch.1054[pii]10.1002/arch.1054[doi]

Brown BE (1975) Proctolin: a peptide transmitter candidate in insects. Life Sci 17(8):1241–1252

Cazzamali G, Grimmelikhuijzen CJ (2002) Molecular cloning and functional expression of the first insect FMRFamide receptor. Proc Natl Acad Sci USA 99(19):12073–12078. doi:10.1073/pnas.192442799192442799[pii]

Cazzamali G, Saxild N, Grimmelikhuijzen C (2002) Molecular cloning and functional expression of a Drosophila corazonin receptor. Biochem Biophys Res Commun 298(1):31–36. doi:S0006291X02023987[pii]

Cazzamali G, Hauser F, Kobberup S, Williamson M, Grimmelikhuijzen CJ (2003) Molecular identification of a Drosophila G protein-coupled receptor specific for crustacean cardioactive peptide. Biochem Biophys Res Commun 303(1):146–152. doi:S0006291X03003024[pii]

Cazzamali G, Klaerke DA, Grimmelikhuijzen CJ (2005a) A new family of insect tyramine receptors. Biochem Biophys Res Commun 338(2):1189–1196. doi:S0006-291X(05)02321-1[pii]10.1016/j.bbrc.2005.10.058

Cazzamali G, Torp M, Hauser F, Williamson M, Grimmelikhuijzen CJ (2005b) The Drosophila gene CG9918 codes for a pyrokinin-1 receptor. Biochem Biophys Res Commun 335(1):14–19. doi:S0006-291X(05)01497-X[pii]10.1016/j.bbrc.2005.07.038

Chen W, Shields TS, Stork PJ, Cone RD (1995) A colorimetric assay for measuring activation of Gs- and Gq- coupled signaling pathways. Anal Biochem 226(2):349–354. doi:S0003269785712353[pii]

Chen J, Zhang Y, Shen P (2008) A protein kinase C activity localized to neuropeptide Y-like neurons mediates ethanol intoxication in Drosophila melanogaster. Neuroscience 156(1):42–47. doi:S0306-4522(08)01033-6[pii]10.1016/j.neuroscience.2008.07.008

Clark MC, Baro DJ (2007) Arthropod D-2 receptors positively couple with cAMP through the Gi/o protein family. Comp Biochem Physiol B Biochem Mol Biol 146(1):9–19. doi:10.1016/j.cbpb.2006.08.018

Corey JL, Quick MW, Davidson N, Lester HA, Guastella J (1994) A cocaine-sensitive Drosophila serotonin transporter: cloning, expression, and electrophysiological characterization. Proc Natl Acad Sci USA 91(3):1188–1192

Cronin SJ, Nehme NT, Limmer S, Liegeois S, Pospisilik JA, Schramek D, Leibbrandt A, de Simoes RM, Gruber S, Puc U, Ebersberger I, Zoranovic T, Neely GG, von Haeseler A, Ferrandon D, Penninger JM (2009) Genome-wide RNAi screen identifies genes involved in intestinal pathogenic bacterial infection. Science 325(5938):340–343. doi:1173164[pii]10.1126/science.1173164

Dai L, Dewey EM, Zitnan D, Luo CW, Honegger HW, Adams ME (2008) Identification, developmental expression, and functions of bursicon in the tobacco hawkmoth, Manduca sexta. J Comp Neurol 506(5):759–774. doi:10.1002/cne.21575

Downer KE, Haselton AT, Nachman RJ, Stoffolano JG Jr (2007) Insect satiety: sulfakinin localization and the effect of drosulfakinin on protein and carbohydrate ingestion in the blow fly, Phormia regina (Diptera: Calliphoridae). J Insect Physiol 53(1):106–112. doi:S0022-1910(06)00196-X[pii]10.1016/j.jinsphys.2006.10.013

Draper I, Kurshan PT, McBride E, Jackson FR, Kopin AS (2007) Locomotor activity is regulated by D2 like receptors in Drosophila: an anatomic and functional analysis. Dev Neurobiol 67(3):378–393. doi:10.1002/dneu.20355[doi]

Durocher Y, Perret S, Thibaudeau E, Gaumond MH, Kamen A, Stocco R, Abramovitz M (2000) A reporter gene assay for high-throughput screening of G-protein-coupled receptors stably or transiently expressed in HEK293 EBNA cells grown in suspension culture. Anal Biochem 284(2):316–326. doi:10.1006/abio.2000.4698S0003-2697(00)94698-0[pii]

Ebert PR, Rowland JE, Toma DP (1998) Isolation of seven unique biogenic amine receptor clones from the honey bee by library scanning. Insect Mol Biol 7(2):151–162. doi:10.1046/j.1365-2583.1998.72059.x

Egerod K, Reynisson E, Hauser F, Cazzamali G, Williamson M, Grimmelikhuijzen CJ (2003a) Molecular cloning and functional expression of the first two specific insect myosuppress in receptors. Proc Natl Acad Sci USA 100(17):9808–9813. doi:10.1073/pnas.16321971001632197100[pii]

Egerod K, Reynisson E, Hauser F, Williamson M, Cazzamali G, Grimmelikhuijzen CJ (2003b) Molecular identification of the first insect proctolin receptor. Biochem Biophys Res Commun 306(2):437–442. doi:S0006291X03009975[pii]

Feng G, Hannan F, Reale V, Hon YY, Kousky CT, Evans PD, Hall LM (1996) Cloning and functional characterization of a novel dopamine receptor from Drosophila melanogaster. J Neurosci 16(12):3925–3933

Filmore D (2004) It's a GPCR world. Mod Drug Discov 11(7):24–28

Fredriksson R, Schioth HB (2005) The repertoire of G-protein-coupled receptors in fully sequenced genomes. Mol Pharmacol 67(5):1414–1425. doi:mol.104.009001[pii]10.1124/mol.104.009001

Fredriksson R, Lagerstrom MC, Lundin LG, Schioth HB (2003) The G-protein-coupled receptors in the human genome form five main families. Phylogenetic analysis, paralogon groups, and fingerprints. Mol Pharmacol 63(6):1256–1272. doi:10.1124/mol.63.6.125663/6/1256[pii]

Gether U (2000) Uncovering molecular mechanisms involved in activation of G protein-coupled receptors. Endocr Rev 21(1):90–113

Gotzes F, Balfanz S, Baumann A (1994) Primary structure and functional characterization of a Drosophila dopamine receptor with high homology to human D1/5 receptors. Receptors Channels 2(2):131–141

Grimmelikhuijzen CJ, Cazzamali G, Williamson M, Hauser F (2007) The promise of insect genomics. Pest Manag Sci 63(5):413–416. doi:10.1002/ps.1352

Grohmann L, Blenau W, Erber J, Ebert PR, Strunker T, Baumann A (2003) Molecular and functional characterization of an octopamine receptor from honeybee (Apis mellifera) brain. J Neurochem 86(3):725–735. doi:1876[pii]

Hallem EA, Dahanukar A, Carlson JR (2006) Insect odor and taste receptors. Annu Rev Entomol 51:113–135. doi:10.1146/annurev.ento.51.051705.113646

Han KA, Millar NS, Grotewiel MS, Davis RL (1996) DAMB, a novel dopamine receptor expressed specifically in Drosophila mushroom bodies. Neuron 16(6):1127–1135. doi:S0896-6273(00)80139-7[pii]

Han KA, Millar NS, Davis RL (1998) A novel octopamine receptor with preferential expression in Drosophila mushroom bodies. J Neurosci 18(10):3650–3658

Hansen KK, Stafflinger E, Schneider M, Hauser F, Cazzamali G, Williamson M, Kollmann M, Schachtner J, Grimmelikhuijzen CJ (2010) Discovery of a novel insect neuropeptide signaling system closely related to the insect adipokinetic hormone and corazonin hormonal systems. J Biol Chem 285(14):10736–10747. doi:M109.045369[pii]10.1074/jbc.M109.045369

Hauser F, Nothacker HP, Grimmelikhuijzen CJ (1997) Molecular cloning, genomic organization, and developmental regulation of a novel receptor from Drosophila melanogaster structurally related to members of the thyroid-stimulating hormone, follicle-stimulating hormone, luteinizing hormone/choriogonadotropin receptor family from mammals. J Biol Chem 272(2):1002–1010

Hauser F, Cazzamali G, Williamson M, Blenau W, Grimmelikhuijzen CJ (2006) A review of neurohormone GPCRs present in the fruitfly Drosophila melanogaster and the honey bee Apis mellifera. Prog Neurobiol 80(1):1–19. doi:S0301-0082(06)00080-3[pii]10.1016/j.pneurobio.2006.07.005[doi]

Hauser F, Cazzamali G, Williamson M, Park Y, Li B, Tanaka Y, Predel R, Neupert S, Schachtner J, Verleyen P, Grimmelikhuijzen CJ (2008a) A genome-wide inventory of neurohormone GPCRs in the red flour beetle Tribolium castaneum. Front Neuroendocrinol 29(1):142–165. doi:S0091-3022(07)00058-1[pii]10.1016/j.yfrne.2007.10.003[doi]

Hauser F, Cazzamali G, Williamson M, Park Y, Li B, Tanaka Y, Predel R, Neupert S, Schachtner J, Verleyen P, Grimmelikhuijzen CJP (2008b) A genome-wide inventory of neurohormone GPCRs in the red flour beetle Tribolium castaneum. Front Neuroendocrinol 29(1):142–165. doi:10.1016/j.yfrne.2007.10.003

Hearn MG, Ren Y, McBride EW, Reveillaud I, Beinborn M, Kopin AS (2002) A Drosophila dopamine 2 like receptor: molecular characterization and identification of multiple alternatively spliced variants. Proc Natl Acad Sci USA 99(22):14554–14559. doi:10.1073/pnas.202498299202498299[pii]

Hector CE, Bretz CA, Zhao Y, Johnson EC (2009) Functional differences between two CRF-related diuretic hormone receptors in Drosophila. J Exp Biol 212(19):3142–3147. doi:212/19/3142[pii]10.1242/jeb.033175

Hill CA, Fox AN, Pitts RJ, Kent LB, Tan PL, Chrystal MA, Cravchik A, Collins FH, Robertson HM, Zwiebel LJ (2002) G protein-coupled receptors in Anopheles gambiae. Science 298(5591):176–178. doi:10.1126/science.1076196[doi]298/5591/176[pii]

Holman GM, Cook BJ, Nachman RJ (1986) Primary structure and synthesis of a blocked myotropic neuropeptide isolated from the cockroach, Leucophaea maderae. Comp Biochem Physiol C 85(1):219–224

Horn F, Bettler E, Oliveira L, Campagne F, Cohen FE, Vriend G (2003) GPCRDB information system for G protein-coupled receptors. Nucleic Acids Res 31(1):294–297

Huang J, Zhang Y, Li M, Wang S, Liu W, Couble P, Zhao G, Huang Y (2007) RNA interference-mediated silencing of the bursicon gene induces defects in wing expansion of silkworm. FEBS Lett 581(4):697–701. doi:S0014-5793(07)00072-5[pii]10.1016/j.febslet.2007.01.034

Huang J, Ohta H, Inoue N, Takao H, Kita T, Ozoe F, Ozoe Y (2009) Molecular cloning and pharmacological characterization of a Bombyx mori tyramine receptor selectively coupled to intracellular calcium mobilization. Insect Biochem Mol Biol 39(11):842–849. doi:S0965-1748(09)00148-9[pii]10.1016/j.ibmb.2009.10.001

Hummon AB, Richmond TA, Verleyen P, Baggerman G, Huybrechts J, Ewing MA, Vierstraete E, Rodriguez-Zas SL, Schoofs L, Robinson GE, Sweedler JV (2006) From the genome to the proteome: uncovering peptides in the Apis brain. Science 314(5799):647–649. doi:314/5799/647[pii]10.1126/science.1124128

Hyun S, Lee Y, Hong ST, Bang S, Paik D, Kang J, Shin J, Lee J, Jeon K, Hwang S, Bae E, Kim J (2005) Drosophila GPCR Hanisa receptor for the circadian clock neuropeptide PDF. Neuron 48(2):267–278. doi:S0896-6273(05)00700-2[pii]10.1016/j.neuron.2005.08.025

Ignell R, Root CM, Birse RT, Wang JW, Nassel DR, Winther AM (2009) Presynaptic peptidergic modulation of olfactory receptor neurons in Drosophila. Proc Natl Acad Sci USA 106(31):13070–13075. doi:0813004106[pii]10.1073/pnas.0813004106

Iversen A, Cazzamali G, Williamson M, Hauser F, Grimmelikhuijzen CJ (2002a) Molecular cloning and functional expression of a Drosophila receptor for the neuropeptides capa-1 and-2. Biochem Biophys Res Commun 299(4):628–633. doi:S0006291X02027092[pii]

Iversen A, Cazzamali G, Williamson M, Hauser F, Grimmelikhuijzen CJ (2002b) Molecular identification of the first insect ecdysis triggering hormone receptors. Biochem Biophys Res Commun 299(5):924–931. doi:S0006291X02027985[pii]

Johnson EC, Bohn LM, Barak LS, Birse RT, Nassel DR, Caron MG, Taghert PH (2003a) Identification of Drosophila neuropeptide receptors by G protein-coupled receptors-beta-arrestin 2 interactions. J Biol Chem 278(52):52172–52178. doi:10.1074/jbc.M306756200M306756200[pii]

Johnson EC, Garczynski SF, Park D, Crim JW, Nassel DR, Taghert PH (2003b) Identification and characterization of a G protein-coupled receptor for the neuropeptide proctolin in Drosophila melanogaster. Proc Natl Acad Sci USA 100(10):6198–6203. doi:10.1073/pnas.1030108100 1030108100[pii]

Johnson EC, Bohn LM, Taghert PH (2004) Drosophila CG8422 encodes a functional diuretic hormone receptor. J Exp Biol 207(Pt 5):743–748

Johnson EC, Shafer OT, Trigg JS, Park J, Schooley DA, Dow JA, Taghert PH (2005) A novel diuretic hormone receptor in Drosophila: evidence for conservation of CGRP signaling. J Exp Biol 208(Pt 7):1239–1246. doi:208/7/1239[pii]10.1242/jeb.01529

Jorgensen LM, Hauser F, Cazzamali G, Williamson M, Grimmelikhuijzen CJ (2006) Molecular identification of the first SIFamide receptor. Biochem Biophys Res Commun 340(2):696–701. doi:S0006-291X(05)02812-3[pii]10.1016/j.bbrc.2005.12.062

Kiger AA, Baum B, Jones S, Jones MR, Coulson A, Echeverri C, Perrimon N (2003) A functional genomic analysis of cell morphology using RNA interference. J Biol 2(4):27. doi:10.1186/1475-4924-2-27

Kim YJ, Bartalska K, Audsley N, Yamanaka N, Yapici N, Lee JY, Kim YC, Markovic M, Isaac E, Tanaka Y, Dickson BJ (2010) MIPs are ancestral ligands for the sex peptide receptor. Proc Natl Acad Sci USA 107(14):6520–6525. doi:0914764107[pii]10.1073/pnas.0914764107

Kirkness EF, Haas BJ, Sun W, Braig HR, Perotti MA, Clark JM, Lee SH, Robertson HM, Kennedy RC, Elhaik E, Gerlach D, Kriventseva EV, Elsik CG, Graur D, Hill CA, Veenstra JA, Walenz B, Tubio JM, Ribeiro JM, Rozas J, Johnston JS, Reese JT, Popadic A, Tojo M, Raoult D, Reed DL, Tomoyasu Y, Kraus E, Mittapalli O, Margam VM, Li HM, Meyer JM, Johnson RM, Romero-Severson J, Vanzee JP, Alvarez-Ponce D, Vieira FG, Aguade M, Guirao-Rico S, Anzola JM, Yoon KS, Strycharz JP, Unger MF, Christley S, Lobo NF, Seufferheld MJ, Wang N, Dasch GA, Struchiner CJ, Madey G, Hannick LI, Bidwell S, Joardar V, Caler E, Shao R, Barker SC, Cameron S, Bruggner RV, Regier A, Johnson J, Viswanathan L, Utterback TR, Sutton GG, Lawson D, Waterhouse RM, Venter JC, Strausberg RL, Berenbaum MR, Collins FH, Zdobnov EM, Pittendrigh BR (2010) Genome sequences of the human body louse and its primary endosymbiont provide insights into the permanent parasitic life style. Proc Natl Acad Sci USA 107(27):12168–12173. doi:1003379107[pii]10.1073/pnas.1003379107

Klose MK, Dason JS, Atwood HL, Boulianne GL, Mercier AJ (2010) Peptide-induced modulation of synaptic transmission and escape response in Drosophila requires two G-protein-coupled receptors. J Neurosci 30(44):14724–14734. doi:30/44/14724[pii]10.1523/JNEUROSCI.3612-10.2010

Knight PJ, Pfeifer TA, Grigliatti TA (2003) A functional assay for G-protein-coupled receptors using stably transformed insect tissue culture cell lines. Anal Biochem 320(1):88–103. doi:S0003269703003543[pii]

Kolakowski LF Jr (1994) GCRDB: a G-protein-coupled receptor database. Receptors Channels 2(1):1–7

Kreienkamp HJ, Larusson HJ, Witte I, Roeder T, Birgul N, Honck HH, Harder S, Ellinghausen G, Buck F, Richter D (2002) Functional annotation of two orphan G-protein-coupled receptors, Drostar1 and −2, from Drosophila melanogaster and their ligands by reverse pharmacology. J Biol Chem 277(42):39937–39943. doi:10.1074/jbc.M206931200M206931200[pii]

Kubiak TM, Larsen MJ, Burton KJ, Bannow CA, Martin RA, Zantello MR, Lowery DE (2002) Cloning and functional expression of the first Drosophila melanogaster sulfakinin receptor DSK-R1. Biochem Biophys Res Commun 291(2):313–320. doi:10.1006/bbrc.2002.6459S0006291X02964594[pii]

Lagerstrom MC, Schioth HB (2008) Structural diversity of G protein-coupled receptors and significance for drug discovery. Nat Rev Drug Discov 7(4):339–357. doi:nrd2518[pii]10.1038/nrd2518

Larsen MJ, Burton KJ, Zantello MR, Smith VG, Lowery DL, Kubiak TM (2001) Type A allatostatins from Drosophila melanogaster and Diplotera puncata activate two Drosophila allatostatin receptors, DAR-1 and DAR-2, expressed in CHO cells. Biochem Biophys Res Commun 286(5):895–901. doi:10.1006/bbrc.2001.5476S0006-291X(01)95476-2[pii]

Lear BC, Zhang L, Allada R (2009) The neuropeptide PDF acts directly on evening pacemaker neurons to regulate multiple features of circadian behavior. PLoS Biol 7(7):e1000154. doi:10.1371/journal.pbio.1000154

Lee KS, You KH, Choo JK, Han YM, Yu K (2004) Drosophila short neuropeptide F regulates food intake and body size. J Biol Chem 279(49):50781–50789. doi:10.1074/jbc.M407842200M407842200[pii]

Lenz C, Sondergaard L, Grimmelikhuijzen CJ (2000a) Molecular cloning and genomic organization of a novel receptor from Drosophila melanogaster structurally related to mammalian galanin receptors. Biochem Biophys Res Commun 269(1):91–96. doi:10.1006/bbrc.2000.2251S0006-291X(00)92251-4[pii]

Lenz C, Williamson M, Grimmelikhuijzen CJ (2000b) Molecular cloning and genomic organization of an allatostatin preprohormone from Drosophila melanogaster. Biochem Biophys Res Commun 273(3):1126–1131. doi:10.1006/bbrc.2000.3062S0006-291X(00)93062-6[pii]

Lenz C, Williamson M, Hansen GN, Grimmelikhuijzen CJ (2001) Identification of four Drosophila allatostatins as the cognate ligands for the Drosophila orphan receptor DAR-2. Biochem Biophys Res Commun 286(5):1117–1122. doi:10.1006/bbrc.2001.5475S0006-291X(01)95475-0[pii] -&gt

Lingo PR, Zhao Z, Shen P (2007) Co-regulation of cold-resistant food acquisition by insulin-and neuropeptide Y-like systems in Drosophila melanogaster. Neuroscience 148(2):371–374. doi:S0306-4522(07)00789-0[pii]10.1016/j.neuroscience.2007.06.010

Loveall BJ, Deitcher DL (2010) The essential role of bursicon during Drosophila development. BMC Dev Biol 10:92. doi:1471-213X-10-92[pii]10.1186/1471-213X-10-92

Lu H-L, Kersch CN, Taneja-Bageshwar S, Pietrantonio PV (2011) A calcium bioluminescence assay for functional analysis of mosquito (Aedes aegypti) and tick (Rhipicephalus microplus) G protein-coupled receptors. J Vis Exp (50):e2732

Luo CW, Dewey EM, Sudo S, Ewer J, Hsu SY, Honegger HW, Hsueh AJ (2005) Bursicon, the insect cuticle-hardening hormone, is a heterodimeric cystine knot protein that activates G protein coupled receptor LGR2. Proc Natl Acad Sci USA 102(8):2820–2825. doi:0409916102[pii]10.1073/pnas.0409916102

Maestro JL, Aguilar R, Pascual N, Valero ML, Piulachs MD, Andreu D, Navarro I, Belles X (2001) Screening of antifeedant activity in brain extracts led to the identification of sulfakinin as a satiety promoter in the German cockroach. Are arthropod sulfakinins homologous to vertebrate gastrins-cholecystokinins? Eur J Biochem 268(22):5824–5830. doi:2527[pii]

Maqueira B, Chatwin H, Evans PD (2005) Identification and characterization of a novel family of Drosophila beta-adrenergic-like octopamine G-protein coupled receptors. J Neurochem 94(2):547–560. doi:JNC3251[pii]10.1111/j.1471-4159.2005.03251.x

Meeusen T, Mertens I, Clynen E, Baggerman G, Nichols R, Nachman RJ, Huybrechts R, DeLoof A, Schoofs L (2002) Identification in Drosophila melanogaster of the invertebrate G protein-coupled FMRFamide receptor. Proc Natl Acad Sci USA 99(24):15363–15368. doi:10.1073/pnas.252339599252339599[pii]

Mendive FM, Van Loy T, Claeysen S, Poels J, Williamson M, Hauser F, Grimmelikhuijzen CJ, Vassart G, Vanden Broeck J (2005) Drosophila molting neurohormone bursicon is a heterodimer and the natural agonist of the orphan receptor DLGR2. FEBS Lett 579(10):2171–2176. doi:S0014-5793(05)00324-8[pii]10.1016/j.febslet.2005.03.006

Mertens I, Vandingenen A, Johnson EC, Shafer OT, Li W, Trigg JS, De Loof A, Schoofs L, Taghert PH (2005) PDF receptor signaling in Drosophila contributes to both circadian and geotactic behaviors. Neuron 48(2):213–219. doi:S0896-6273(05)00776-2[pii]10.1016/j.neuron.2005.09.009

Meyering-Vos M, Muller A (2007) RNA interference suggests sulfakinins as satiety effectors in the cricket Gryllus bimaculatus. J Insect Physiol 53(8):840–848. doi:S0022-1910(07)00112-6[pii]10.1016/j.jinsphys.2007.04.003

Mitri C, Parmentier M-L, Pin J-P, Bockaert Jl, Grau Y (2004) Divergent evolution in metabotropic glutamate receptors. J Biol Chem 279(10):9313–9320. doi:10.1074/jbc.M310878200

Mitri C, Soustelle L, Framery Brn, Bockaert Jl, Parmentier M-L, Grau Y (2009) Plant insecticide L-Canavanine repels *Drosophila* via the insect orphan GPCR DmX. PLoS Biol 7(6):e1000147

Mitsumasu K, Ohta H, Tsuchihara K, Asaoka K, Ozoe Y, Niimi T, Yamashita O, Yaginuma T (2008) Molecular cloning and characterization of cDNAs encoding dopamine receptor-1 and -2 from brain-suboesophageal ganglion of the silkworm, Bombyx mori. Insect Mol Biol 17(2):185–195. doi:10.1111/j.1365-2583.2008.00792.x

Nachman RJ, Teal PE, Ujvary I (2001) Comparative topical pheromonotropic activity of insect pyrokinin/PBAN amphiphilic analogs incorporating different fatty and/or cholic acid components. Peptides 22(2):279–285. doi:S0196-9781(00)00380-6[pii]

Nassel DR (2002) Neuropeptides in the nervous system of Drosophila and other insects: multiple roles as neuromodulators and neurohormones. Prog Neurobiol 68(1):1–84. doi:S0301008202000576[pii]

Nassel DR, Winther AM (2010) Drosophila neuropeptides in regulation of physiology and behavior. Prog Neurobiol 92(1):42–104. doi:S0301-0082(10)00100-0[pii]10.1016/j.pneurobio.2010.04.010

Nene V, Wortman JR, Lawson D, Haas B, Kodira C, Tu ZJ, Loftus B, Xi Z, Megy K, Grabherr M, Ren Q, Zdobnov EM, Lobo NF, Campbell KS, Brown SE, Bonaldo MF, Zhu J, Sinkins SP, Hogenkamp DG, Amedeo P, Arensburger P, Atkinson PW, Bidwell S, Biedler J, Birney E, Bruggner RV, Costas J, Coy MR, Crabtree J, Crawford M, Debruyn B, Decaprio D, Eiglmeier K, Eisenstadt E, El-Dorry H, Gelbart WM, Gomes SL, Hammond M, Hannick LI, Hogan JR, Holmes MH, Jaffe D, Johnston JS, Kennedy RC, Koo H, Kravitz S, Kriventseva EV, Kulp D, Labutti K, Lee E, Li S, Lovin DD, Mao C, Mauceli E, Menck CF, Miller JR, Montgomery P, Mori A, Nascimento AL, Naveira HF, Nusbaum C, O'Leary S, Orvis J, Pertea M, Quesneville H, Reidenbach KR, Rogers YH, Roth CW, Schneider JR, Schatz M, Shumway M, Stanke M, Stinson EO, Tubio JM, Van zee JP, Verjovski-Almeida S, Werner D, White O, Wyder S, Zeng Q, Zhao Q, Zhao Y, Hill CA, Raikhel AS, Soares MB, Knudson DL, Lee NH, Galagan J, Salzberg SL, Paulsen IT, Dimopoulos G, Collins FH, Birren B, Fraser-Liggett CM, Severson DW (2007) Genome sequence of Aedes aegypti, a major arbovirus vector. Science 316(5832):1718–1723. doi:1138878[pii]10.1126/science.1138878

Nichols R, Egle JP, Langan NR, Palmer GC (2008) The different effects of structurally related sulfakinins on Drosophila melanogaster odor preference and locomotion suggest involvement of distinct mechanisms. Peptides 29(12):2128–2135. doi:S0196-9781(08)00359-8[pii]10.1016/j.peptides.2008.08.010

Nybakken K, Vokes SA, Lin TY, McMahon AP, Perrimon N (2005) A genome-wide RNA interference screen in Drosophila melanogaster cells for new components of the Hh signaling pathway. Nat Genet 37(12):1323–1332. doi:ng1682[pii]10.1038/ng1682

Ohta H, Tsuchihara K, Mitsumasu K, Yaginuma T, Ozoe Y, Asaoka K (2009) Comparative pharmacology of two D1-like dopamine receptors cloned from the silkworm Bombyx mori. Insect Biochem Mol Biol 39(5–6):342–347. doi:10.1016/j.ibmb.2009.01.011

Olsen SS, Cazzamali G, Williamson M, Grimmelikhuijzen CJ, Hauser F (2007) Identification of one capa and two pyrokinin receptors from the malaria mosquito Anopheles gambiae. Biochem Biophys Res Commun 362(2):245–251. doi:S0006-291X(07)01438-6[pii]10.1016/j.bbrc.2007.06.190

Onai T, FitzGerald MG, Arakawa S, Gocayne JD, Urquhart DA, Hall LM, Fraser CM, McCombie WR, Venter JC (1989) Cloning, sequence analysis and chromosome localization of a Drosophila muscarinic acetylcholine receptor. FEBS Lett 255(2):219–225. doi:0014-5793(89)81095-6[pii]

Ono H, Yoshikawa H (2004) Identification of amine receptors from a swallowtail butterfly, Papilio xuthus L.: cloning and mRNA localization in foreleg chemosensory organ for recognition of host plants. Biochem Mol Biol 34(12):1247–1256. doi:10.1016/j.ibmb.2004.08.009

Orchard I, Belanger JH, Lange AB (1989) Proctolin: a review with emphasis on insects. J Neurobiol 20(5):470–496. doi:10.1002/neu.480200515

Pan L, Broadie KS (2007) Drosophila fragile X mental retardation protein and metabotropic glutamate receptor A convergently regulate the synaptic ratio of ionotropic glutamate receptor subclasses. J Neurosci 27(45):12378–12389. doi:10.1523/jneurosci.2970-07.2007

Park Y, Adams ME (2005) Insect G protein-coupled receptors: recent discoveries and implications. In: Lawrence IG, Kostas I, Sarjeet SG, Lawrence IG, Kostas I, Sarjeet SG (eds) Comprehensive molecular insect science. Elsevier, Amsterdam, pp 143–171

Park WJ, Liu J, Adler PN (1994) Frizzled gene expression and development of tissue polarity in the Drosophila wing. Dev Genet 15(4):383–389. doi:10.1002/dvg.1020150410[doi]

Park Y, Kim YJ, Adams ME (2002) Identification of G protein-coupled receptors for Drosophila PRXamide peptides, CCAP, corazonin, and AKH supports a theory of ligand-receptor coevolution. Proc Natl Acad Sci USA 99(17):11423–11428. doi:10.1073/pnas.162276199[doi]162276199[pii]

Park Y, Kim YJ, Dupriez V, Adams ME (2003) Two subtypes of ecdysis-triggering hormone receptor in Drosophila melanogaster. J Biol Chem 278(20):17710–17715. doi:10.1074/jbc.M301119200M301119200[pii]

Perez DM (2005) From plants to man: the GPCR "tree of life". Mol Pharmacol 67(5):1383–1384. doi:mol.105.011890[pii]10.1124/mol.105.011890[doi]

Pierce KL, Premont RT, Lefkowitz RJ (2002) Seven-transmembrane receptors. Nat Rev Mol Cell Biol 3(9):639–650. doi:10.1038/nrm908[doi]nrm908[pii]

Pietrantonio PV, Jagge C, McDowell C (2001) Cloning and expression analysis of a 5HT7-like serotonin receptor cDNA from mosquito Aedes aegypti female excretory and respiratory systems. Insect Mol Biol 10(4):357–369. doi:10.1046/j.0962-1075.2001.00274.x

Poels J, Birse RT, Nachman RJ, Fichna J, Janecka A, Vanden Broeck J, Nassel DR (2009) Characterization and distribution of NKD, a receptor for Drosophila tachykinin-related peptide 6. Peptides 30(3):545–556. doi:S0196-9781(08)00449-X[pii]10.1016/j.peptides.2008.10.012

Pollock VP, McGettigan J, Cabrero P, Maudlin IM, Dow JA, Davies SA, Pollock VP, McGettigan J, Cabrero P, Maudlin IM, Dow JA, Davies SA (2004) Conservation of capa peptide-induced nitric oxide signalling in Diptera. J Exp Biol 207(Pt23):4135–4145. doi:207/23/4135[pii]10.1242/jeb.01255

Predel R, Nachman RJ (2001) Efficacy of native FXPRLamides (pyrokinins) and synthetic analogs on visceral muscles of the American cockroach. J Insect Physiol 47(3):287–293. doi:S002219100000113X[pii]

Price MD, Merte J, Nichols R, Koladich PM, Tobe SS, Bendena WG (2002) Drosophila melanogaster flatline encodes a myotropin orthologue to Manduca sexta allatostatin. Peptides 23(4):787–794. doi:S0196978101006490[pii]

Rao KR, Mohrherr CJ, Riehm JP, Zahnow CA, Norton S, Johnson L, Tarr GE (1987) Primary structure of an analog of crustacean pigment-dispersing hormone from the lubber grasshopper Romalea microptera. J Biol Chem 262(6):2672–2675

Richards S, Gibbs RA, Weinstock GM, Brown SJ, Denell R, Beeman RW, Gibbs R, Bucher G, Friedrich M, Grimmelikhuijzen CJ, Klingler M, Lorenzen M, Roth S, Schroder R, Tautz D, Zdobnov EM, Muzny D, Attaway T, Bell S, Buhay CJ, Chandrabose MN, Chavez D, Clerk-Blankenburg KP, Cree A, Dao M, Davis C, Chacko J, Dinh H, Dugan-Rocha S, Fowler G, Garner TT, Garnes J, Gnirke A, Hawes A, Hernandez J, Hines S, Holder M, Hume J, Jhangiani SN, Joshi V, Khan ZM, Jackson L, Kovar C, Kowis A, Lee S, Lewis LR, Margolis J, Morgan M, Nazareth LV, Nguyen N, Okwuonu G, Parker D, Ruiz SJ, Santibanez J, Savard J, Scherer SE, Schneider B, Sodergren E, Vattahil S, Villasana D, White CS, Wright R, Park Y, Lord J, Oppert B, Brown S, Wang L, Weinstock G, Liu Y, Worley K, Elsik CG, Reese JT, Elhaik E, Landan G, Graur D, Arensburger P, Atkinson P, Beidler J, Demuth JP, Drury DW, Du YZ, Fujiwara H, Maselli V, Osanai M, Robertson HM, Tu Z, Wang JJ, Wang S, Song H, Zhang L, Werner D, Stanke M, Morgenstern B, Solovyev V, Kosarev P, Brown G, Chen HC, Ermolaeva O, Hlavina W, Kapustin Y, Kiryutin B, Kitts P, Maglott D, Pruitt K, Sapojnikov V, Souvorov A, Mackey AJ, Waterhouse RM, Wyder S, Kriventseva EV, Kadowaki T, Bork P, Aranda M, Bao R, Beermann A, Berns N, Bolognesi R, Bonneton F, Bopp D, Butts T, Chaumot A, Denell RE, Ferrier DE, Gordon CM, Jindra M, Lan Q, Lattorff HM, Laudet V, von Levetsow C, Liu Z, Lutz R, Lynch JA, da Fonseca RN, Posnien N, Reuter R, Schinko JB, Schmitt C, Schoppmeier M, Shippy TD, Simonnet F, Marques-Souza H, Tomoyasu Y, Trauner J, Vander Zee M, Vervoort M, Wittkopp N, Wimmer EA, Yang X, Jones AK, Sattelle DB, Ebert PR, Nelson D, Scott JG, Muthukrishnan S, Kramer KJ, Arakane Y, Zhu Q, Hogenkamp D, Dixit R, Jiang H, Zou Z, Marshall J, Elpidina E, Vinokurov K, Oppert C, Evans J, Lu Z, Zhao P, Sumathipala N, Altincicek B, Vilcinskas A, Williams M, Hultmark D, Hetru C, Hauser F, Cazzamali G, Williamson M, Li B, Tanaka Y, Predel R, Neupert S, Schachtner J, Verleyen P, Raible F, Walden KK, Angeli S, Foret S, Schuetz S, Maleszka R, Miller SC, Grossmann D (2008) The genome of the model beetle and pest Tribolium castaneum. Nature 452(7190):949–955. doi:nature06784[pii]10.1038/nature06784

Robinson GE, Hackett KJ, Purcell-Miramontes M, Brown SJ, Evans JD, Goldsmith MR, Lawson D, Okamuro J, Robertson HM, Schneider DJ (2011) Creating a buzz about insect genomes. Science 331(6023):1386. doi:331/6023/1386[pii]10.1126/science.331.6023.1386

Roeder T (2005) Tyramine and octopamine: ruling behavior and metabolism. Annu Rev Entomol 50:447–477. doi:10.1146/annurev.ento.50.071803.130404

Roller L, Yamanaka N, Watanabe K, Daubnerova I, Zitnan D, Kataoka H, Tanaka Y (2008) The unique evolution of neuropeptide genes in the silkworm Bombyx mori. Insect Biochem Mol Biol 38(12):1147–1157

Rosenkilde C, Cazzamali G, Williamson M, Hauser F, Sondergaard L, De Lotto R, Grimmelikhuijzen CJ (2003) Molecular cloning, functional expression, and gene silencing of two Drosophila receptors for the Drosophila neuropeptide pyrokinin-2. Biochem Biophys Res Commun 309(2):485–494. doi:S0006291X03016139[pii]

Rotte C, Krach C, Balfanz S, Baumann A, Walz B, Blenau W (2009) Molecular characterization and localization of the first tyramine receptor of the American cockroach (Periplaneta americana). Neuroscience 162(4):1120–1133. doi:S0306-4522(09)00980-4[pii]10.1016/j.neuroscience.2009.05.066

Saudou F, Hen R (1994) 5-Hydroxytryptamine receptor subtypes in vertebrates and invertebrates. Neurochem Int 25(6):503–532

Saudou F, Amlaiky N, Plassat JL, Borrelli E, Hen R (1990) Cloning and characterization of a Drosophila tyramine receptor. EMBO J 9(11):3611–3617

Saudou F, Boschert U, Amlaiky N, Plassat JL, Hen R (1992) A family of Drosophila serotonin receptors with distinct intracellular signalling properties and expression patterns. EMBO J 11(1):7–17

Schlenstedt J, Balfanz S, Baumann A, Blenau W (2006) Am5-HT7: molecular and pharmacological characterization of the first serotonin receptor of the honeybee (Apis mellifera). J Neurochem 98(6):1985–1998. doi:JNC4012[pii]10.1111/j.1471-4159.2006.04012.x

Schoofs L, Holman GM, Hayes TK, Nachman RJ, De Loof A (1990) Locustatachykinin I and II, two novel insect neuropeptides with homology to peptides of the vertebrate tachykinin family. FEBS Lett 261(2):397–401. doi:0014-5793(90)80601-E[pii]

Schoofs L, Holman GM, Proost P, Van Damme J, Hayes TK, De Loof A (1992) Locustakinin, a novel myotropic peptide from Locusta migratoria, isolation, primary structure and synthesis. Regul Pept 37(1):49–57

Schulte G, Bryja V (2007) The Frizzled family of unconventional G-protein-coupled receptors. Trends Pharmacol Sci 28(10):518–525. doi:S0165-6147(07)00208-8[pii]10.1016/j.tips.2007.09.001[doi]

Sellami A, Agricola HJ, Veenstra JA (2011) Neuroendocrine cells in Drosophila melanogaster producing GPA2/GPB5, a hormone with homology to LH, FSH and TSH. Gen Comp Endocrinol 170(3):582–588. doi:S0016-6480(10)00404-1[pii]10.1016/j.ygcen.2010.11.015

Shapiro RA, Wakimoto BT, Subers EM, Nathanson NM (1989) Characterization and functional expression in mammalian cells of genomic and cDNA clones encoding a Drosophila muscarinic acetylcholine receptor. Proc Natl Acad Sci USA 86(22):9039–9043

Soehler S, Neupert S, Predel R, Stengl M (2008) Examination of the role of FMRFamide-related peptides in the circadian clock of the cockroach Leucophaea maderae. Cell Tissue Res 332(2):257–269. doi:10.1007/s00441-008-0585-9

Srivastava DP, Yu EJ, Kennedy K, Chatwin H, Reale V, Hamon M, Smith T, Evans PD (2005) Rapid, nongenomic responses to ecdysteroids and catecholamines mediated by a novel Drosophila G-protein-coupled receptor. J Neurosci 25(26):6145–6155. doi:25/26/6145[pii]10.1523/JNEUROSCI.1005-05.2005

Stables J, Green A, Marshall F, Fraser N, Knight E, Sautel M, Milligan G, Lee M, Rees S (1997) A bioluminescent assay for agonist activity at potentially any G-protein-coupled receptor. Anal Biochem 252(1):115–126. doi:S0003-2697(97)92308-3[pii]10.1006/abio.1997.2308

Stafflinger E, Hansen KK, Hauser F, Schneider M, Cazzamali G, Williamson M, Grimmelikhuijzen CJ (2008) Cloning and identification of an oxytocin/vasopressin-like receptor and its ligand from insects. Proc Natl Acad Sci USA 105(9):3262–3267. doi:0710897105[pii]10.1073/pnas.0710897105

Starratt AN, Brown BE (1975) Structure of the pentapeptide proctolin, a proposed neurotransmitter in insects. Life Sci 17(8):1253–1256

Staubli F, Jorgensen TJ, Cazzamali G, Williamson M, Lenz C, Sondergaard L, Roepstorff P, Grimmelikhuijzen CJ (2002) Molecular identification of the insect adipokinetic hormone receptors. Proc Natl Acad Sci USA 99(6):3446–3451. doi:10.1073/pnas.05255649999/6/3446[pii]

Sudo S, Kuwabara Y, Park JI, Hsu SY, Hsueh AJ (2005) Heterodimeric fly glycoprotein hormone-alpha2 (GPA2) and glycoprotein hormone-beta5 (GPB5) activate fly leucine-rich repeat-containing G protein-coupled receptor-1 (DLGR1) and stimulation of human thyrotropin receptors by chimeric fly GPA2 and human GPB5. Endocrinology 146(8):3596–3604. doi:en.2005-0317[pii]10.1210/en.2005-0317

Sugamori KS, Demchyshyn LL, McConkey F, Forte MA, Niznik HB (1995) A primordial dopamine D1 like adenylyl cyclase-linked receptor from Drosophila melanogaster displaying poor affinity for benzazepines. FEBS Lett 362(2):131–138. doi:0014-5793(95)00224-W[pii]

Terhzaz S, Rosay P, Goodwin SF, Veenstra JA (2007) The neuropeptide SIFamide modulates sexual behavior in Drosophila. Biochem Biophys Res Commun 352(2):305–310. doi:S0006-291X(06)02461-2[pii]10.1016/j.bbrc.2006.11.030

Thamm M, Balfanz S, Scheiner R, Baumann A, Blenau W (2010) Characterization of the 5-HT1A receptor of the honeybee (Apis mellifera) and involvement of serotonin in phototactic behavior. Cell Mol Life Sci 67(14):2467–2479. doi:10.1007/s00018-010-0350-6

The HoneyBee Genome Sequencing Consortium (2006) Insights into social insects from the genome of the honeybee Apis mellifera. Nature 443(7114):931–949

The International Aphid Genomics Consortium (2010) Genome sequence of the pea aphid Acyrthosiphon pisum. PLoS Biol 8(2):e1000313. doi:10.1371/journal.pbio.1000313

The International Silkworm Genome Consortium (2008) The genome of a lepidopteran model insect, the silkworm Bombyx mori. Insect Biochem Mol Biol 38(12):1036–1045. doi:S0965-1748(08)00209-9[pii]10.1016/j.ibmb.2008.11.004

Touhara K, Vosshall LB (2009) Sensing odorants and pheromones with chemosensory receptors. Annu Rev Physiol 71:307–332. doi:10.1146/annurev.physiol.010908.163209

Truman JW (2005) Hormonal control of insect ecdysis: endocrine cascades for coordinating behavior with physiology. Vitam Horm 73:1–30. doi:S0083-6729(05)73001-6[pii]10.1016/S0083-6729(05)73001-6[doi]

Veenstra JA (1989) Isolation and structure of corazonin, a cardioactive peptide from the American cockroach. FEBS Lett 250(2):231–234. doi:0014-5793(89)80727-6[pii]

Veenstra JA (2009) Does corazonin signal nutritional stress in insects? Insect Biochem Mol Biol 39(11):755–762. doi:S0965-1748(09)00138-6[pii]10.1016/j.ibmb.2009.09.008

Verlinden H, Vleugels R, Marchal E, Badisco L, Tobback J, Pfluger HJ, Blenau W, Vanden Broeck J (2010) The cloning, phylogenetic relationship and distribution pattern of two new putative GPCR-type octopamine receptors in the desert locust (Schistocerca gregaria). J Insect Physiol 56(8):868–875. doi:S0022-1910(10)00068-5[pii]10.1016/j.jinsphys.2010.03.003

Von Nickisch-Rosenegk E, Krieger J, Kubick S, Laage R, Strobel J, Strotmann J, Breer H (1996) Cloning of biogenic amine receptors from moths (Bombyx mori and Heliothis virescens). Insect Biochem Mol Biol 26(8–9):817–827

Wei Z, Baggerman G, Nachman RJ, Goldsworthy G, Verhaert P, De Loof A, Schoofs L (2000) Sulfakinins reduce food intake in the desert locust, Schistocerca gregaria. J Insect Physiol 46(9):1259–1265. doi:S0022191000000469[pii]

Werren JH, Richards S, Desjardins CA, Niehuis O, Gadau J, Colbourne JK, Beukeboom LW, Desplan C, Elsik CG, Grimmelikhuijzen CJ, Kitts P, Lynch JA, Murphy T, Oliveira DC, Smith CD, Van de Zande L, Worley KC, Zdobnov EM, Aerts M, Albert S, Anaya VH, Anzola JM, Barchuk AR, Behura SK, Bera AN, Berenbaum MR, Bertossa RC, Bitondi MM, Bordenstein SR, Bork P, Bornberg-Bauer E, Brunain M, Cazzamali G, Chaboub L, Chacko J, Chavez D, Childers CP, Choi JH, Clark ME, Claudianos C, Clinton RA, Cree AG, Cristino AS, Dang PM, Darby AC, de Graaf DC, Devreese B, Dinh HH, Edwards R, Elango N, Elhaik E, Ermolaeva O, Evans JD, Foret S, Fowler GR, Gerlach D, Gibson JD, Gilbert DG, Graur D, Grunder S, Hagen DE, Han Y, Hauser F, Hultmark D, Hunter HCt, Hurst GD, Jhangian SN, Jiang H, Johnson RM, Jones AK, Junier T, Kadowaki T, Kamping A, Kapustin Y, Kechavarzi B, Kim J, Kiryutin B,

Koevoets T, Kovar CL, Kriventseva EV, Kucharski R, Lee H, Lee SL, Lees K, Lewis LR, Loehlin DW, Logsdon JM Jr, Lopez JA, Lozado RJ, Maglott D, Maleszka R, Mayampurath A, Mazur DJ, McClure MA, Moore AD, Morgan MB, Muller J, Munoz-Torres MC, Muzny DM, Nazareth LV, Neupert S, Nguyen NB, Nunes FM, Oakeshott JG, Okwuonu GO, Pannebakker BA, Pejaver VR, Peng Z, Pratt SC, Predel R, Pu LL, Ranson H, Raychoudhury R, Rechtsteiner A, Reese JT, Reid JG, Riddle M, Robertson HM, Romero-Severson J, Rosenberg M, Sackton TB, Sattelle DB, Schluns H, Schmitt T, Schneider M, Schuler A, Schurko AM, Shuker DM, Simoes ZL, Sinha S, Smith Z, Solovyev V, Souvorov A, Springauf A, Stafflinger E, Stage DE, Stanke M, Tanaka Y, Telschow A, Trent C, Vattathil S, Verhulst EC, Viljakainen L, Wanner KW, Waterhouse RM, Whitfield JB, Wilkes TE, Williamson M, Willis JH, Wolschin F, Wyder S, Yamada T, Yi SV, Zecher CN, Zhang L, Gibbs RA (2010) Functional and evolutionary insights from the genomes of three parasitoid Nasonia species. Science 327(5963):343–348. doi:327/5963/343[pii]10.1126/science.1178028

Wicher D, Derst C, Gautier H, Lapied B, Heinemann SH, Agricola HJ (2007) The satiety signaling neuropeptide perisulfakinin inhibits the activity of central neurons promoting general activity. Front Cell Neurosci 1:3. doi:10.3389/neuro.03.003.2007

Winther AM, Acebes A, Ferrus A (2006) Tachykinin-related peptides modulate odor perception and locomotor activity in Drosophila. Mol Cell Neurosci 31(3):399–406. doi:S1044-7431(05)00253-8[pii]10.1016/j.mcn.2005.10.010

Witz P, Amlaiky N, Plassat JL, Maroteaux L, Borrelli E, Hen R (1990) Cloning and characterization of a Drosophila serotonin receptor that activates adenylate cyclase. Proc Natl Acad Sci USA 87(22):8940–8944

Woodhead AP, Stay B, Seidel SL, Khan MA, Tobe SS (1989) Primary structure of four allatostatins: neuropeptide inhibitors of juvenile hormone synthesis. Proc Natl Acad Sci USA 86(15): 5997–6001

Wu Q, Wen T, Lee G, Park JH, Cai HN, Shen P (2003) Developmental control of foraging and social behavior by the Drosophila neuropeptide Y-like system. Neuron 39(1):147–161. doi:S0896627303003969[pii]

Wu Q, Zhao Z, Shen P (2005) Regulation of aversion to noxious food by Drosophila neuropeptide Y-and insulin-like systems. Nat Neurosci 8(10):1350–1355. doi:nn1540[pii]10.1038/nn1540

Yapici N, Kim YJ, Ribeiro C, Dickson BJ (2008) A receptor that mediates the post-mating switch in Drosophila reproductive behaviour. Nature 451(7174):33–37. doi:nature06483[pii]10.1038/nature06483

Chapter 5
Bursicon as a Potential Target for Insect Control

Shengzhang Dong and Qisheng Song

1 Introduction

Bursicon is an insect heterodimeric neuropeptide hormone that is secreted from the central nervous system (CNS) into the hemolymph to regulate various aspects of cuticle tanning (sclerotization and melanization) and wing expansion in a diverse group of insects (Luo et al. 2005; Honegger et al. 2008). Insects are encased in a semi-rigid exoskeleton, which provides support for muscle attachment and protection of internal organs from physical damage and microbial invasion, but also limits insect growth. Insects must periodically shed their old exoskeletons and form new ones (i.e. molting) through a stereotyped sequence of ecdysis behaviors to accommodate growth or change in shape during postembryonic development. The newly formed cuticle is usually soft and flexible, which permits body growth but lacks support and protective function. Insect must quickly harden its newly formed exoskeleton after each ecdysis to resume its support and protective functions. During the last metamorphic molt, the cuticle tanning process is regulated by a neuropeptide called bursicon.

Bursicon was first discovered five decades ago as a blood-borne tanning hormone in the blowflies *Calliphora erythrocephala* that regulates cuticle tanning in newly emerged adults (Cottrell 1962; Fraenkel and Hsiao 1965). Despite over four decade's effort, the molecular identity of bursicon was not elucidated until 2005. Now we know that the functional bursicon consists of two cystine knot proteins, named as bursicon (burs) or bursicon α (burs α), and partner of bursicon (pburs) or bursicon β (burs β) (Luo et al. 2005; Mendive et al. 2005). The sequences of two bursicon monomers are highly conserved among insects, and this conservation is extended to other arthropods, and even to echinoderms (Luo et al. 2005). Besides being

S. Dong • Q. Song (✉)
Division of Plant Sciences-Entomology, University of Missouri,
Columbia, MO 65211, USA
e-mail: dongs@missouri.edu; songq@missouri.edu

I. Ishaaya et al. (eds.), *Advanced Technologies for Managing Insect Pests*,
DOI 10.1007/978-94-007-4497-4_5,

involved in cuticle tanning, bursicon exerts its effects on various aspects of the wing expansion and maturation process (Baker and Truman 2002; Dewey et al. 2004). In *Drosophila*, bursicon acts through a specific *Drosophila* leucine-rich repeats-containing G-protein-coupled receptor 2 (DLGR2) (Luo et al. 2005; Mendive et al. 2005), encoded by the rickets gene (Baker and Truman 2002). The recent characterization of bursicon's molecular nature and discovery of its receptor intensified interest in bursicon studies, especially in elucidating bursicon signaling components downstream of the receptor, its mode of action, and regulation of its release.

Insect neuropeptides and their receptors regulate virtually all aspects of insect life, including metabolism, homeostasis, development, reproduction, molting and behavior (Altstein and Nässel 2010; Bendena 2010; Mercier et al. 2007; Scherkenbeck and Zdobinsky 2009; Smith and Rybczynski 2012). They would be excellent candidates for development of new methods for pest control because they show insect target specificity and environmental compatibility. Although they are impractical for the direct use as insect control agents due to their instability in the environment and poor solubility in organic solutions, recent advances in peptide drug delivery systems in invertebrate and transgenic techniques in crops make this possibility a reality (Scherkenbeck and Zdobinsky 2009; Teal et al. 1999). As an important neuropeptide hormone during insect development, bursicon, which only affects insect stages in the process of molting (Baker and Truman 2002; Honegger et al. 2008), would appear to be an ideal target for designing a green pesticide with high safety to environment and other non-target organisms. Recent studies show that in addition to the classic roles of bursicon heterodimer in mediating cuticle tanning and wing expansion, bursicon could also form homodimers to mediate prophylactic innate immunity in *Drosophila* via the immudeficiency (IMD) pathway (An et al. 2012). In this chapter, we will review recent advances in bursicon research and development of neuropeptides for pest control, and then discuss the possibilities of bursicon as a target for pest control.

2 Structure of Bursicon

2.1 Discovery

In 1962, evidence from two parallel studies using a ligated blowfly assay showed that cuticular tanning in insects was regulated by a hormone (Cottrell 1962; Fraenkel and Hsiao 1962). When the blowfly, *Calliphora erythocephala* adults were ligated between the head and thorax immediately after emergence, neck ligation prevented the cuticle of thoracic and abdominal area from being tanned. To demonstrate the probable hormonal nature of the "endogenous tanning agent", a test solution containing hemolymph from a newly emerged fly adult that had an already darkened cuticle was injected into the neck ligated and un-tanned recipient fly. The cuticle of the recipient animal tanned due to hemolymph transfusion. Similarly, when a test solution containing CNS homogenate from a newly emerged fly was injected into a ligated fly, the untanned cuticle in the recipient fly was quickly tanned. Based on this finding, researchers determined that the tanning results from a factor released into the hemolymph from the fly CNS, and named the factor 'bursicon' (derived from the Greek word 'bursikos', meaning 'related to tanning') (Fraenkel and Hsiao 1965). Now, bursicon's presence has been established in almost all insect orders (Honegger et al. 2008; Song 2012).

2.2 *Molecular Structure of Bursicon*

After characterization as a hormone, various experiments were performed to determine the molecular nature of bursicon. In the first two decades after its discovery, bursicon was considered to be a single protein hormone with an estimated molecular size around 30–60 kDa, depending on methods and insect species that were used (Fraenkel et al. 1966; Fraenkel 1975; Seligman and Doy 1973; Taghert and Truman 1982).

The first decade of this century witnessed great progress in the study of bursicon, during which a putative spot for bursicon was separated on 2-dimensional gels and was micro-sequenced, resulting in the amino acid sequence of five short peptides (Honegger et al. 2002). By comparing these partial sequences with the *Drosophila* genome sequence published in 2000 (Adams et al. 2000), a *Drosophila* gene *CG13419* was independently identified by three of the five partial sequences as a candidate bursicon gene. Reverse genetic approaches demonstrated that *CG13419* was a bursicon gene since mutations in this gene impaired cuticle tanning and wing expansion in *Drosophila* adults (Dewey et al. 2004). The mature protein without an N-terminal signal sequence was predicted to contain 141 amino acids with an approximate molecular weight of 15 kDa, which was half molecular weight of the detected bursicon from gel elution and 2D gel electrophoresis. These findings plus some other early experiments suggested that a bioactive bursicon might be a homodimer (Honegger et al. 2004). However, the recombinant protein of *CG13419* shows no tanning activity in the ligated fly assay despite forming a homodimer (Luo et al. 2005).

The protein encoded by *CG13419* belongs to a class of cystine knot proteins (CKP) that contains a consensus framework produced by six cysteines. CKP in vertebrates could form either homodimers or heterodimers (Vitt et al. 2001), which shed light on the potential sequence for another bursicon gene. The breakthrough studies came simultaneously in 2005 from two independent research groups (Luo et al. 2005; Mendive et al. 2005). Using the previously identified bursicon sequence, another *Drosophila* gene *CG15284*, which codes the only other CKP in the *Drosophila* genome, was identified. Thus, CG13419 was named as burs α and CG15284 as burs β. Both studies showed that burs α and burs β heterodimers were capable of activating cuticle tanning and wing expansion through a G-protein coupled receptor (GPCR) DLGR2, encoded by the rickets gene in *Drosophila*. Burs α and burs β can also form homodimers in vivo, but their functions remained unknown (Luo et al. 2005; Honegger et al. 2008).

2.3 Localization and Expression of Bursicon

The localization of bursicon was initially determined by assaying the activity of extracts from various parts of the CNS and hemolymph samples at different developmental stages via the "ligated fly assay". Bursicon activity was demonstrated in almost all parts of the CNS including brain, corpora cardiaca (CC) and corpora allata (CA) complex, and thoracic and abdominal ganglia depending upon insect species analyzed (Ewer and Reynolds 2002). Further studies demonstrated that bursicon subunits are only expressed in a set of neurons in the subesophageal, thoracic and abdominal ganglia in *Drosophila* (Dewey et al. 2004; Luo et al. 2005) and several other insect species (Dai et al. 2008; Wang et al. 2008), and there is a close association between bursicon immunoreactive (IR) and custacean cardioactive peptide (CCAP) neurons (Luan et al. 2006; Peabody et al. 2008). CCAP-expressing neurons (N_{CCAP}) consist of two functionally distinct groups, one releases bursicon into the hemolymph and the other regulates its release. The first group, named N_{CCAP}-c929, includes 14 bursicon-expressing neurons of the abdominal ganglion, and suppression of activity within this group blocks bursicon release into the hemolymph together with tanning and wing expansion. The second group, named N_{CCAP}-R, consists of N_{CCAP} neurons outside the c929. The neurotransmission and protein kinase A (PKA) activity are required in N_{CCAP}-R to regulate bursicon secretion from N_{CCAP}-c929. N_{CCAP} forms an interacting neuronal network responsible for the regulation and release of bursicon (Luan et al. 2006). However, the detailed mechanisms that regulate the release of bursicon are not fully understood yet.

Recently, several studies showed that bursicon transcripts were not only expressed during adult molting, but also found during the larval and pupae stages (Loveall and Deitcher 2010; Mendive et al. 2005; Luo et al. 2005; Wang et al. 2008). Luo et al. (2005) demonstrated that *burs α* and *burs β* are both expressed during all stages of *Drosophila* development, including early third-instar larvae, late third-instar larvae (wandering stage), white pupae, prepupae, pupae, pharate adults and <2days old adults. But, bursicon expression at the larval stage in *Drosophila* is limited to type III boutons of the neuromuscular junction (NMJ) (Loveall and Deitcher 2010). In *M. domestica*, both *burs α* and *burs β* were transcribed as early as in the first instar larvae, reached the maximum level in pharate adults, and declined sharply after adult emergence (Fig. 5.1). In *Bombyx mori*, *burs α* transcript was first detected in larval and had two peaks in pupae (Huang et al. 2007). Based on these data, it is likely that bursicon is released contemporaneously with ecdysis other than eclosion, and bursicon signaling is essential to insect growth and development. However, the bursicon activities in insects were only demonstrated during adult eclosion. Therefore, it is unclear that whether bursicon activities are diverse in other stages of development.

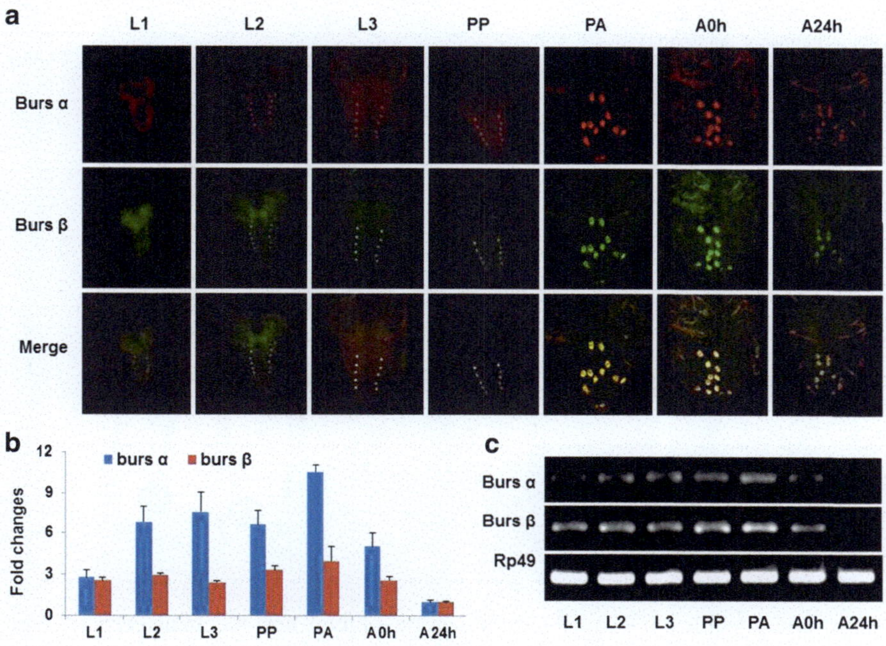

Fig. 5.1 Expression patterns of bursicon α and β transcripts in the CNS of housefly *Musca domestica*. (**a**) Fluorescence in situ hybridization (FISH) of *burs* α and β transcripts in the CNS. The CNSs were dissected from the indicated developmental stages, fixed and hybridized with antisense *burs* α (*red*) and *burs* β (*green*) probes. The CNS samples were visualized under a confocal microscope and photographed. *Yellow* in the merged pictures indicates the CNSs with the co-localized *burs* α and β signals. (**b**) Relative quantification PCR analysis of *burs* α and β transcripts in the CNS. (**c**) RT-PCR analysis of *burs* α and β transcripts in the CNS. *L1* first instar larvae, *L2* second instar larvae, *L3* third instar larvae, *PP* prepupae, *PA* pharate adults, *A0h* 0-h adults, *A24h* 24-h adults. The data represent the mean ± SE of three biological replicates (Modified from Wang et al. 2008)

2.4 Evolution of Insect Bursicon

Both insect bursicon subunits belong to the vertebrate cystine knot protein (CKP) family and contain 11 cysteine residues in conserved positions (Fig. 5.2). The cystine knot family-members contain two additional cysteines that form a third disulfide bond that penetrates the ring structure, thus forming a cystine knot with ten amino acids, of which six are cysteine residues. The intrusion of the additional disulfide bond through the cystine ring confines the amino acid residue between the second and third cysteine to a glycine, as any other amino acid at this position would imply severe steric hindrance for the formation of the knot (Vitt et al. 2001). The CKP family contains the glycoprotein hormones, transforming growth factor-b, platelet-derived growth factor subfamily members, mucins, norrie disease protein, von Willebrand factor (vWF), bone morphogenetic protein antagonists (BMP), and slit-like proteins. The glycoprotein hormones are characterized by a

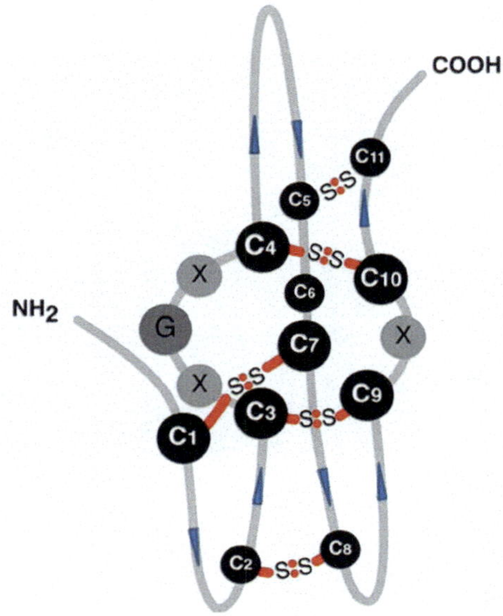

Fig. 5.2 Predicted cystine knot structure of bursicon α and bursicon β (Adapted from Luo et al. 2005)

strong and specific noncovalent dimerization of two non-identical carbohydrate (CHO)-containing subunits, designated alpha (α) and beta (β). The glycoprotein hormones are a family of structurally related hormones that include follitropin (FSH), Iutropin (LH), thyrotropin (TSH) and chorionic gonadotropin (CG) (Hearn and Gomme 2000). Only the intact α and β heterodimer is able to bind specifically to the hormone receptor; free α- or β-subunit is biologically inactive (Pierce and Parsons 1981). The growth factors (neurotrophins), nerve growth factor (NGF), transforming growth factor-beta (TGF- β), and platelet-derived growth factor (PDGF) exist as hetero- or homo- dimers (Hearn and Gomme 2000).

By exploring the present powerful bioinformatics and molecular tools, cDNAs corresponding to homologues of both *burs α* and *burs β* have been identified in dozens of insect and other arthropod species (Honegger et al. 2008; Song 2012). A sequence alignment based on amino acids from 12 insect and other arthropod species showed that insect *burs α* and *burs β* had high sequence identities in insects (Fig. 5.3). *Burs α* shared 53 conserved amino acids, and among which 11 are cysteines. *Burs β* shared 37 conserved amino acids, among which 11 are also cysteines except human body louse, *Pediculus humanus corporis*, which has eight cysteines. Phylogenetic trees based on amino acid sequences also showed that both *burs α* and *burs β* are very conserved among insect and other arthropod species, which is confidently clustered into a clad with human 10-membered cystine knot subfamily, but *burs α* and *burs β* showed the different evolutionary relationships with human cysteine knot proteins (Fig. 5.4). *Burs α* seems to be more close to BMP-antagonist, while *burs β* is more close to glycoprotein hormones. It was suggested that they may be evolved from different ancestors.

Besides the sequence conservation that occurs in insect bursicon, their functions are also much conserved among insects, and bursicon from different insects showed

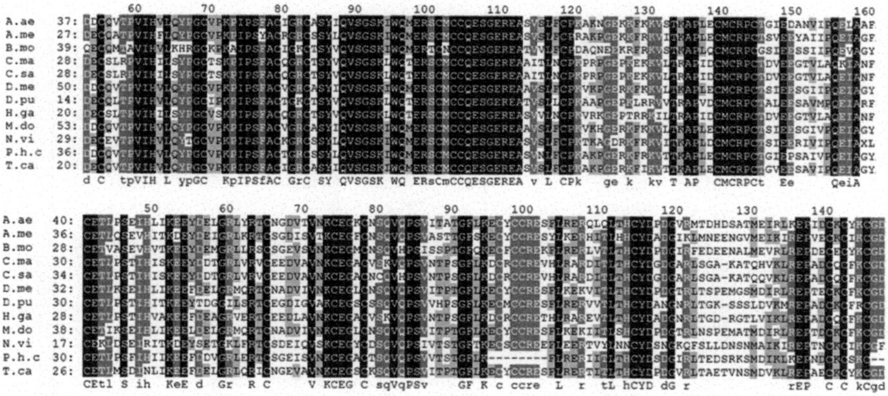

Fig. 5.3 Alignments of bursicon α (*upper*) and β (*lower*) amino acids in 12 insect and other arthropod species. A.ae: *Aedes aegypti* (Accession Nos. P85316 for burs α and EAT34015 for burs β); A.me: *Apis mellifera* (NP_001091704 for burs α and NP_001035352 for burs β); B.mo: *Bombyx mori* (NP_001091845 for burs α and NP_001037289 for burs β); C.ma: *Carcinus maenas* (ABX55995 for burs α and ABX55997 for burs β); C.sa: *Callinectes spidus* (ACG50067 for burs α and ACG50066 for burs β); D.me: *Drosophila melanogaster* (AAF55915 for burs α and AAF53396 for burs β); D.pu: *Daphnia pulex* (EFX87749.1 for burs α and EFX87546.1 for burs β); H.ga: *Homarus gammarus* (ADI86242 for burs α and ADI86243 for burs β); M.do: *Musca domestica* (ABO20870 for burs α and ABO20869 for burs β); N.vi: *Nasonia vitripennis* (NP_001155852 for burs α and XP_001601162 for burs β); P.h.c: *Pediculus humanus corporis* (XP_002430782 for burs α and XP_002430781 for burs β); T.ca: *Tribolium castaneum* (ABA03053 for burs α and ABA40403 for burs β)

a cross-species activity. For example, the extracts from CNS or hemolymph of several dipteran species including, *Sarcophaga bullata* (Baker and Truman 2002; Fogal and Fraenkel 1969), *Phormia regina* (Fraenkel and Hsiao 1965), and *Lucilia spp.* (Seligman and Doy 1973) had cross-species tanning activities in the neck-ligated fly assay. The bioactive cuticle tanning factor present in the CNS and hemolymph of newly emerged adults has been identified through molecular cloning and confirmed in the neck-ligated fly assay using r-bursicon heterodimer (An et al. 2008; Luo et al. 2005; Mendive et al. 2005). The cross-species activity of bursicon has also been demonstrated using r-bursicon heterodimer. For example, *Drosophila r*-bursicon heterodimer expressed in a mammalian cell line exhibited a strong tanning activity in the housefly, *Musca domestica* (An et al. 2009). Surprisingly, lobster nervous system homogenates were reported to have bursicon activity in the ligated fly bioassay (Kostron et al. 1995) and this observation was supported by the presence of pburs-like transcript in the cDNA database of the loster *Homarus americanus* (Van Loy et al. 2007), illustrating the remarkable conservation of bursicon in invertebrate species. In addition, bursicon shows a high activity even in lower concentration. Injection with 0.5 μl recombinant heterodimer bursicon (about 60 ng) or CNS extract of newly emerged flies (0.5 CNS equivalent/fly) can induce strong cuticle tanning activity in the neck-ligated fly assay (An et al. 2008; Luo et al. 2005; Mendive et al. 2005).

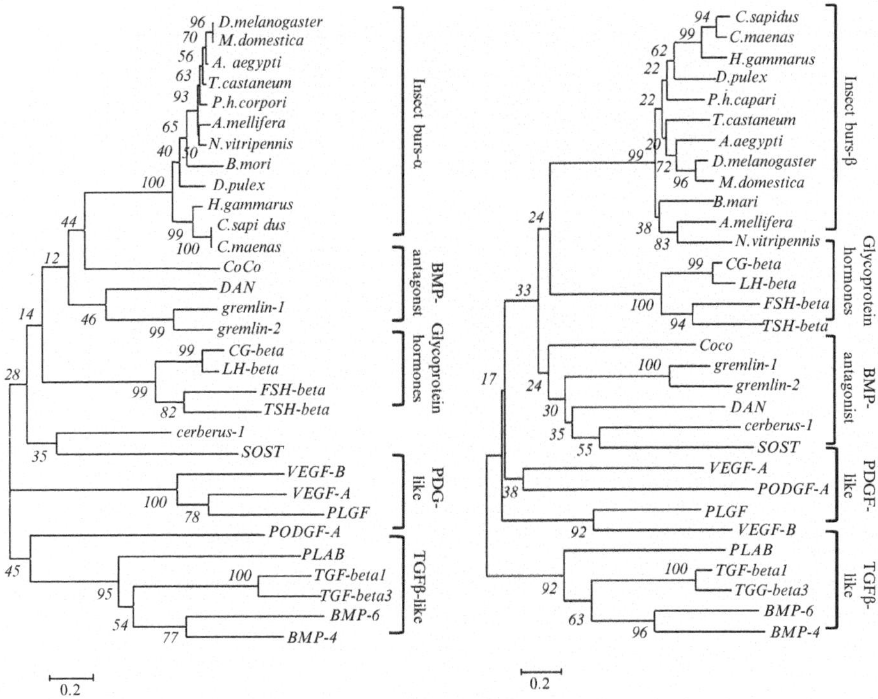

Fig. 5.4 Phylogenetic relationships between bursicon α (*left*) or β (*right*) and human 10-membered cystine knot subfamily. The phylogenetic trees (neighbor-joining) were constructed using the MEGA version 3.1 on the amino acids. Numbers in the nodes correspond to bootstrap values in 1,000 replicates. The sequences of insect bursicon α and β used for phylogenetic analysis were the same as those in Fig. 5.3. gremlin1 (Accession No. EAW92267); gremlin2 (AAH46632); DAN(BAA92265): differential screening-selected gene aberrative in neuroblastoma; SOST(AAQ88990): sclerostin; Cerberus-1: cerberus-related 1 (AAK92484); Coco (BAC82440); PLGF(P49763): placenta growth factor; PDGF-A(AAA60045): platelet-derived growth factor A; VEGF-A(P15692): vascular endothelial growth factor A; VEGF-B(P49765): vascular endothelial growth factor B; CG-beta (P01233): choriogonadotropin subunit beta; LH-beta (P01229): luteinizing hormone subunit beta; FSH-beta (P01225): follicle-stimulating hormone beta subunit; TSH-beta (P01222): thyroid-stimulating hormone subunit beta; BMP-4 (BAA06410): bone morphogenetic protein 4; BMP-6 (NP_001709); PLAB (AAB88913): placental bone morphogenic protein; TGF beta1 (P01137): transforming growth factor beta-1; TGF beta3 (P10600)

3 Role of Bursicon on Insect Development

3.1 Bursicon Receptor and Signaling Pathway

In recent years, a family of leucine-rich repeat-containing G protein-coupled receptors (LGRs) was identified, which constitutes a unique cluster of transmembrane proteins sharing a large leucine-rich extracellular domain for hormone binding

(Nishi et al. 2000). Genes in this family are conserved in invertebrates and vertebrates. Based on overall sequence similarity and the architecture of the ectodomain, i.e., the number of leucine-rich repeats and the presence or absence of a low density lipoprotein motif, LGRs can be classified into three subgroups: subgroup A, vertebrate glycoprotein hormone receptors, comprising vertebrate receptors for cystine knot-forming gonadotropins [follicle stimulating hormone (FSH), luteinizing hormone (LH), choriogonadotropin (CG) and thyroid stimulating hormone (TSH)] (Grossmann et al. 1997; Themmen and Huhtaniemi 2000); subgroup B, *D. melanogaster* LGR2 (DLGR2) (Eriksen et al. 2000) and vertebrate orphan receptors LGR4, 5, 6 (Hsu et al. 2002); and subgroup C, receptors for mammalian insulin/relaxin-related proteins, harboring a N-terminal low density lipoprotein (LDL) motif (Hsu et al. 2002; Kumagai et al. 2002).

In *Drosophila*, there are four different LGRs i.e. DLGR1, DLGR2, DLGR3 and DLGR4. These receptors are closely related to the mammalian glycoprotein hormone (LH/CG, FSH and TSH) receptors and are characterized by having a large, horseshoe-shaped extracellular N-terminal domain containing 9–18 leucine-rich repeats (Hauser et al. 2006). DLGR1 or CG7665 was identified and could be activated by a heterodimer, composed of the two *Drosophila* homologues of the human glycoprotein subunits glycoprotein hormone-alpha 2 (GPA2) and glycoprotein hormone-beta 5 (GPB5) (Sudo et al. 2005). Genetic analyses in *D. melanogaster* revealed that DLGR2, encoded by the *rickets* (*rk*) gene, was the receptor for the bursicon heterodimer mediating the cuticle tanning and wing expansion processes. Mutations of *rickets* or *burs α* have caused defects in cuticle tanning and wing expansion after adult eclosion (Baker and Truman 2002; Dewey et al. 2004). It has been demonstrated recently that the recombinant bursicon (*r*-bursicon) heterodimer can also bind to and activate DLGR2, which in turn leads to dose-dependent intracellular increase in adenylyl cyclase activity and cAMP production in the mammalian 293T cells and COS-7 cells that over-express DLGR2 (Luo et al. 2005; Mendive et al. 2005). A competitive binding assay also demonstrated that bursicon heterodimer binds to DLGR2 with high specificity (Luo et al. 2005). Another interesting phenomenon is that DLGR2 is highly selective for the bursicon heterodimer, and that neither burs α nor burs β homodimer proteins could induce the activation of DLGR2 and downstream signaling pathway (Honegger et al. 2008). Thus DLGR2 has been well recognized as the bursicon heterodimer receptor. However, burs α or burs β can also form homodimers (Luo et al. 2005), suggesting that another receptor for burs α or burs β homodimer exist in *Drosophila*.

Upon activation by bursicon, DLGR2 has been shown to activate the cAMP/PKA signaling pathway, which is required for transduction of the hormonal signal that induces wing epidermal cell death, eventually leading to wing maturation and cuticle tanning after eclosion (Fig. 5.5) (Kimura et al. 2004; Luo et al. 2005; Mendive et al. 2005). Davis et al. (2007) recently investigated the role of CCAP neurons, bursicon, and *rickets* in the tanning process. They showed that the levels of epidermal dopa decarboxylase (DDC) and epidermal tyrosine hydroxylase (TH encoded by the gene pale, *ple*) transcripts do not change at eclosion. By contrast, the levels of the epidermal TH enzyme fall prior to eclosion then increase sharply immediately

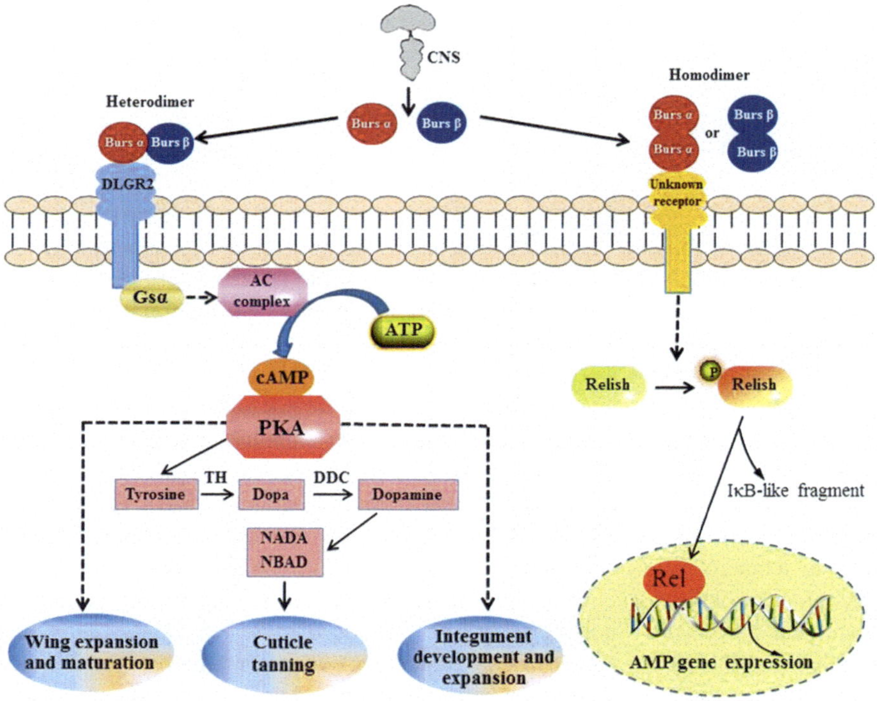

Fig. 5.5 Hypothesized bursicon signaling pathways in *Drosophila*, showing the known interactions between bursicon heterodimer and its receptor, DLGR2, for cuticle tanning and wing expansion and between bursicon homodimer and its unknown receptor leading to the transcription of AMPs. *TH* tyrosine hydroxylase, *DDC* dopa decarboxylase, *AMP* antimicrobial peptides, *NADA* N-acetyldopamine, *NBAD* N-β-alanyldopamine

following eclosion. Flies mutants for *burs α* (Dewey et al. 2004) and *rk* (Baker and Truman 2002) show defects in tanning and are defective in TH activation (Davis et al. 2007); whereas these mutant flies show relatively normal increases in epidermal TH expression after eclosion, they fail to phosphorylate TH, the active form of TH. When cAMP, the second messenger for bursicon, was injected, TH was highly phosphorylated. However, TH was not phosphorylated in the presence of PKA inhibitor. Thus, these results show that bursicon binds to DLGR2 receptor, stimulates cAMP production, and activates cAMP-dependent PKA to phosphorylate TH. The activated TH then converts tyrosine to DOPA, the precursor of tanning agents. This is the first report to provide convincing data that bursicon plays a role in cuticle tanning by acting at a strategic point in the biochemistry and at a strategic location in the insects (Davis et al. 2007). Despite these reports, our knowledge of signaling components involved in the bursicon-stimulated cuticle tanning process remains rudimentary. Besides *Drosophila*, bursicon receptor was also studied in the red flour beetle, *Tribolium castaneum*. RNA interference (RNAi) and microarray results showed that the bursicon receptor gene (*Tcrk*) is required for cuticle tanning and wing expansion in *T. castaneum*, as well as development and expansion of integumentary structures and adult eclosion (Arakane et al. 2008; Bai and Palli 2010).

3.2 Role in Cuticle Tanning

Cuticular tanning, a process by which cuticles are stabilized by incorporation of phenolic compounds, has been studied for many years. Tanning of the newly formed, soft cuticle in insects includes two separate processes, termed sclerotization (hardening) and melanization (or darkening) (Honegger et al. 2008). Before eclosion, the newly formed insect cuticle consists of a thin layer of hydrophobic, waxy, chitin-free epicuticle and a thick layer of protein- and chitin-rich procuticle (Locke 2001). After eclosion, insect cuticular tanning occurs immediately, which hardens and darkens insect cuticle, and provides enough protection from desiccation (or from imbibing excessive water in the case of aquatic insects) and pathogens. Figure 5.6 shows the essential role of recombinant bursicon heterodimer and CNS homogenates (containing bursicon) in inducing cuticle tanning in the ligated *Drosophila* adults. Tanning is a chemical process by which certain regions of the insect cuticle undergo an irreversible change from a soft layer into a stiffer and harder structure. Sclerotization mainly occurs in the exocuticle and epicuticle, which are the two outer most layers in insect cuticle. Sclerotized cuticle has decreased deformability, extractability of matrix proteins, and increased resistance towards dissolvants and digestive enzymes (Andersen 2005).

Although the exact metabolic pathway involved in the insect cuticle tanning process is not yet elucidated clearly, especially regarding the precise regional and temporal regulation of the various steps in the process, it is now widely accepted that cuticle tanning mostly involves post-translational modifications that lead to covalent cross-linking between cuticular proteins and catecholamines (Andersen 2005, 2010, 2012). In *Drosophila*, dopamine is considered the central molecule for both sclerotization and melanization. It is produced from the amino acid tyrosine, which is first hydroxylated to 3,4-dihydroxyphenylalanine (DOPA) by TH, which is encoded by the pale (*ple*) locus (Neckameyer and White 1993). The conversion of DOPA to dopamine is catalyzed by DDC (Livingstone and Tempel 1983), encoded by *Ddc* (Hirsh and Davidson 1981). Seligman et al. (1969) hypothesized that the conversion of tyrosine to DOPA is regulated by bursicon and the conversion process is suggested in hemocytes (Mills and Whitehead 1970). But the conversion of tyrosine to DOPA in hemocytes has largely been refuted (Reynolds 1983). It is most likely that bursicon may act on DLGR2 receptors in the epidermal walls to stimulate the hydroxylation of tyrosine to DOPA since the epidermis is the main tissue involved in either the production of the metabolites necessary for tanning or as the cell layer through which the dopamine metabolites must cross to reach the extracellular space (Moussian 2010). The direct evidence of the involvement of bursicon in the conversion of tyrosine to DOPA in epidermis is from the report by Davis et al. (2007). In their

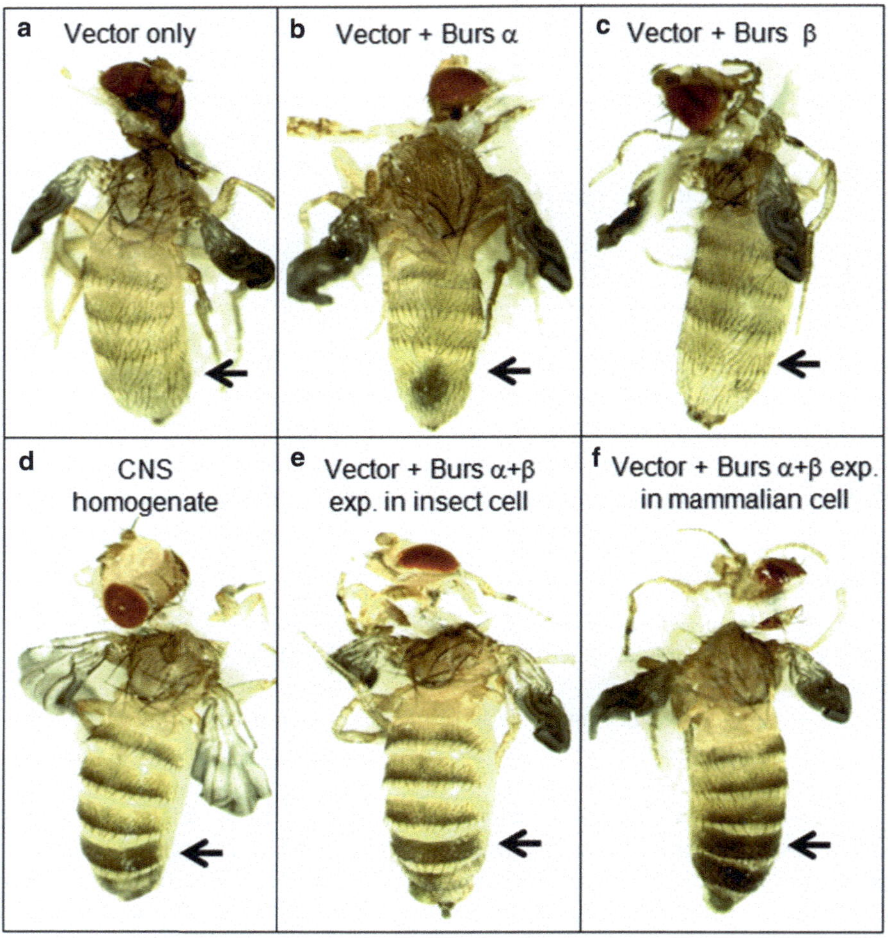

Fig. 5.6 Functional assay of the r-bursicon heterodimer in ligated flies. Newly emerged flies were ligated between the head and thorax at emergence. These flies with unsclerotized cuticle at 1 h after neck-ligation were injected with 0.5 μl of cell culture transfected with blank pcDNA 3.1 vector as a sham control (**a**) or with the purified r-burs α (**b**) or r-burs β (**c**) or r-burs heterodimer expressed in insect High Five™ cells (**e**) and in mammalian HEK293 cells (**f**). The CNS homogenate (0.5 CNS equivalent/fly) from newly emerged flies was used as a positive control (**d**). The *arrow indicates* the area with the untanned cuticle (*light color*) in control (**a**–**c**) and the tanned cuticle (*darkened*) in the animals injected with r-bursicon heterodimer (**d**–**f**) (Modified from An et al. 2008)

investigation, bursicon has been clearly shown to stimulate phosphorylation of TH via PKA, thus activates TH to convert tyrosine into the precursor of tanning agent DOPA in epidermis. The cuticle tanning process is an enzymes-mediated metabolic process, and the enzymes participated in the process include, but not limit to, diphenoloxidases (DPO), laccases, peroxidase, TH and DDC (Moussian 2010). Dopamine is also converted to insoluble melanin via 5,6-dihydroxyindole. Melanin can be linked to granular proteins or may be distributed throughout the cuticular matrix, giving the cuticle a dark color. Melanin probably also forms bonds with cuticular proteins, thereby contributing to cuticular strength (Andersen 2005).

The roles of bursicon signaling in insect cuticle tanning were also demonstrated in several other insects. In the red flour beetle *T. castaneum*, injection of *Tcrk* dsRNA during 1-day old final instar larvae stages delayed and reduced cuticle tanning in *T. castaneum*, which suggested that bursicon receptor is required for complete cuticle tanning during the pupal stage (Bai and Palli 2010). However, injection of dsRNA for bursicon and *Tcrk* in *T. castaneum* during the pharate adult stages did not affect normal cuticle tanning (Arakane et al. 2008), indicating that cuticle tanning initiate before pupariation. A recent study showed that a decrease in laccase 2 (*TcLac2*) mRNA levels were detected in *Tcrk* RNAi bettles, suggesting that bursicon receptor may influence pupal cuticle tanning by regulating the expression of TcLac2 gene in *T. castaneum* since laccase 2 (one of the tyrosinases in *T. castaneum*) and TH are two key enzymes that control cuticle sclerotization processes in *T. castaneum* (Arakane et al. 2005; Gorman and Arakane 2010). In the tobacco hornworm, *Manduca sexta*, recombinant bursicon caused the color of wing cuticles turning from white to yellow or brown, suggesting that bursicon causes wing tanning (Dai et al. 2008). In the silkworm, *B. mori*, however, no distinct tanning phenotype was observed when bursicon gene expression was knocked down by injection of *burs* α dsRNA (Huang et al. 2007).

3.3 Role in Wing Expansion

Bursicon's role in wing expansion and maturation and its impact on its structural changes has been documented especially well in *Drosophila* by using mutants and transgenic flies to manipulate the bursicon and cell signaling pathways. Recent research showed that mutation in the bursicon genes or injection of bursicon dsRNA caused the failure of initiation of the behavioral program for wing expansion and resulted in defects in wing expansion in the fruit fly *D. melanogaster* (Dewey et al. 2004), the silkworm *B. mori* (Huang et al. 2007) and the red flour beetle, *T. castaneum* (Arakane et al. 2008), which indicates that bursicon is the hormone required to initiate this process in insects. The wings of newly emerged fly adults are folded and immature, but when newly eclosed flies are neck-ligated, which is believed to block the release of bursicon into hemolymph, the deficiency of bursicon by the ligation caused the failures in wing expansion. It was also shown that mutation of the rickets gene, which encodes the DLGR2 receptor as a bursicon heterodimer receptor, inhibited wing expansion in *Drosophila* (Kimura et al. 2004). In *T. castaneum*, pupae developed from the larvae injected with the bursicon receptor (*Tcrk*) dsRNA showed shortened wings and elongated abdomens (Bai and Palli 2010), and injection of *Tcrk* dsRNA into pharate pupae caused a wrinkled elytra phenotype (Arakane et al. 2008). It was hypothesized that reduction in the expression of *Tcrk* blocked bursicon signaling pathway in *T. castaneum*.

The insect wing expansion process is another physiological event occurring after insect ecdysis and accompanying cuticle sclerotization. The expansion and maturation of insect wings goes through several stages, which are accompanied by transitions in the wing's underlying cellular architecture (Kiger et al. 2007). First, upon larval pupariation the imaginal wing discs evaginate. Subsequent epithelial cell expansion, without further cell proliferation, causes the wings to become compactly folded within the confines of the pupal case, prior to secretion of the adult wing cuticle. Then, upon eclosion the wing cuticle is pale and pliable, and soon an increase in blood pressure forces the wings to expand. Within approximately an hour the dorsal and ventral cuticular panels of each wing have expanded and bonded. Subsequent tanning over a period of several hours forms a strong, flexible flight organ.

Although it is not clear how wing expansion in insects is controlled, it has been demonstrated that wing development and expansion is strongly associated with programmed cell death and removal of cell debris from the wing tissue (Kiger et al. 2007; Kimura et al. 2004; Natzle et al. 2008). Shortly after eclosion, hemolymph pressure forces the expansion of the folded wing blade. This is caused by ingestion of air and tonic contraction of abdominal muscles (Baker and Truman 2002). The timing of wing expansion after eclosion in *Drosophila* and the observation that *rk* and *burs* mutants cannot expand their wings provide strong evidence for bursicon's function in the wing expansion motor program (Baker and Truman 2002; Dewey et al. 2004). The mature wing is formed by the fusion of the ventral and dorsal layers of cuticle, which are secreted by the underlying epidermis. At the same time, the dorsal and ventral epithelial cell layers delaminate from the outside cuticle, undergo a process called epithelial-mesenchymal transition (EMT), and then exit the wing accompanied by the initiation of a programmed cell death under the control of a yet to be identified signaling pathway (Kiger et al. 2007; Natzle et al. 2008). It has been shown that stimulation of components downstream of bursicon, such as a membrane permeant analog of cAMP, or ectopic expression of constitutively active forms of G proteins or PKA, induced precocious cell death; and conversely, cell death was inhibited in wing clones lacking G protein or PKA function. Therefore, activation of the cAMP/PKA signaling pathway is likely required for transduction of the hormonal signal that induces wing epidermal cell death after eclosion (Kimura et al. 2004).

3.4 Novel Function in Innate Immunity

The function of bursicon heterodimer has been demonstrated in several insects to regulate cuticle tanning, wing expansion and integument development/expansion in pupal-adult stages via a specific receptor DLGR2 (Honegger et al. 2008; Song 2012). Recent studies reveal several lines of evidences suggesting that bursicon may have novel functions beyond cuticle tanning and wing expansion. First, besides forming heterodimer, burs α and burs β can also form homodimers both in vivo and in vitro (Luo et al. 2005; Honegger et al. 2008); Second, *burs* α and *burs* β are expressed in all stages from larval to adult in *Drosophila* (Luo et al. 2005; Honegger et al. 2008) and *M. domestica* (Fig. 5.1); Third, a DNA microarray analysis in the neck-ligated flies using r-bursicon revealed that the expression of 87 genes were up or down regulated by r-bursicon, and seven of them are the immune-related genes (An et al. 2008). These led to the hypothesis that bursicon is involved in regulating innate immune gene expression. When the neck-ligated *Drosophila* adults or the 24 h-old files (without ligation), which displayed low levels of bursicon transcripts and AMP genes, were injected with the r-burs α-α and r-burs β-β homodimers, the transcript levels of the representative AMP genes were up-regulated (An et al. 2012), demonstrating a role for bursicon homodimers in mediating AMP gene transcription *in vivo* and *in vitro*. Most importantly, the homodimer-induced gene expression was translated into operational AMPs that reduce bacterial populations when the supernatant from the homodimer injected flies were incubated with Gram negative bacterial *Escherichia coli* while this was not the case in the control flies injected with the blank vector transfected sample (An et al. 2012). The homodimers influenced AMP gene expression via the IMD intracellular pathway by activating the NF-κB transcription factor Relish (An et al. 2012). Up-regulation of the AMP genes is a normal occurrence at the time of molting. Our results reveal a molecular mechanism of CNS-regulated prophylactic innate immunity that operates during vulnerable molting periods and support our hypothesis that the CNS influences innate immunity via secretion of a neurohormone and thus expands the biological roles of bursicon beyond cuticle tanning and wing expansion (Fig. 5.5).

4 Neuropeptides for Controlling Pests

4.1 Application of Neuropeptides on Controlling Pests

Neuropeptides comprise the most diverse group of intercellular signaling molecules in animals and regulate vital physiological processes as hormones, neuromodulators or neurotransmitters. Most neuropeptides are released from specialized neurosecretory cells of the insect CNS. Neuropeptides are diverse in structure, localization and function. Their only common feature is that they are all synthesized as peptide precursor proteins, also called prepropolypeptides (preprohormones), which then undergo post-translational processing (cleavage) in the endoplasmic reticulum-Golgi network and electron-dense secretory granules (Nässel 2002; Scherkenbeck and Zdobinsky 2009). Most neuropeptides interact with GPCRs, thereby generating an intracellular response (Altstein and Nässel 2010; Bendena 2010). By means of traditional biochemical and molecular techniques combined with bioinformatics technology, a large number of neuropeptides and hormonal peptides have been identified from a variety of insects and their putative functions tested in different bioassays. In *Drosophila*, about 119 neuropeptide encoding genes and 200 encoding peptide and protein GPCRs have been predicted from the genome (Hauser et al. 2006;

Hewes and Taghert 2001; Liu et al. 2006), and 46 neuropeptides derived from 19 of these precursors could be biochemically characterized, including adipokinetic hormones (AKHs) and hypertrehalosaemic peptides, allatostatins, bursicon, ecdysis triggering hormone (ETH), crustacean cardioactive peptide (CCAP), diuretic peptide hormones (DHs), FMRFamide-related peptides (FaRPs), ion transport peptide (ITP), leucokinins, neuropeptide F, prothoracicotropic hormone (PTTH), tachykinin-related peptides, and seven encoding insulin-like peptides (Nässel 2002).

Insect neuropeptides regulate many physiological and behavioral processes during development, such as regulation of carbohydrate and lipid metabolism (AKH and insulin-like peptides), activation of ovary maturation (neuroparsins and insulin-like peptides), triggering of ecdysis or eclosion behavior (PTTH, allatotropin, CCAP and ETH), regulation of water balance and feeding behavior (ITP, DHs and neuropeptide F), modulation of visceral muscle contractions (FaRPs, pyrokinins, tachykinins and allatostatins) (Altstein and Nässel 2010; Bendena 2010; Mercier et al. 2007; Scherkenbeck and Zdobinsky 2009). A variety of neuropeptide actions are clearly known in insects. AKHs, was first identified from the locust *Locusta migratoria* (Stone et al. 1976) and produced in the neurosecretory cells of the glandular lobe of the corpora cardiac (CC), functioned as mobilization of lipids from fat body to hemolymph in locusts, increase of blood hemolymph trehalose levels in several insects, stimulation of heart beat frequency in cockroaches, inhibition of protein synthesis in locust and cricket, inhibition of fatty acid and RNA synthesis in locust fat body (Orchard 1987; Van der Horst et al. 2001). Insects CCAP was first identified in the locust *L. migratoria* and was mainly present in the brain and in the ventral nerve cord as well as in the endocrine cells of the midgut. CCAP plays a pivotal role in the induction of AKH release from the corpora cardiac in the locust *L. migratoria* (Veelaert et al. 1997) and ecdysis motor activity and adult eclosion in *M. sexta* and *Drosophila* and is part of a hormonal cascade including ecdysone, eclosion hormone and ETH (Gammie and Truman 1997, 1999; Zitnan and Adams 2000). Insect DHs are commonly produced by a subset of cells in the protocerebral median neurosecretory cell (MNC) group of the brain and neurosecretory cells in the abdominal ganglia. DHs composed of three main families: the corticotropin-releasing factor (CRF)-related peptides, calcitonin (CT)-like peptides and the insect kinins (Coast et al. 2002). DHs induce cyclic AMP production in Malpighian tubules and assist in the formation of urine, as well as regulating water and iron balance in the insect. PTTH, secreted by the corpus allata and then released into the hemolymph, acts on the prothoracic gland (PG) by activating its tyrosine kinase receptor Torso (Rewitz et al. 2009). PTTH induces PG to secrete the molting hormone precursor ecdysone which is then converted to the active form, 20 hydroxyecdysone (20E) in peripheral tissues and stimulates the molting process (Mizoguchi et al. 2001).

Insect neuropeptide receptors, which bind and are activated by the closely related peptide hormones, are also considered along with neuropeptides as possible targets for development of insecticides. Insect neuropeptide receptors can be classified into two major groups: single transmembrane receptors and heptahelical (7TM) or GPCRs. GPCRs comprised the majority of insect peptide and protein hormone receptors. In *Drosophila*, about 200 genes coding GPCRs have been identified from the genome, and are classified into four families: rhodopsin-like (family A), secretin receptor-like (family B), metabotropic glutamate receptor-like (family C) and atypical receptors (family D) (Brody and Cravchik 2000; Hauser et al. 2006). The neuropeptide-receptor signaling pathway play critical roles in processes as diverse as development, ecdysis, energy metabolism, water and salt balances, behavior and reproduction (Van Hiel et al. 2010).

Neuropeptides and their receptors are involved in the control of almost all key functions in insects and thus would be considered to be targets for the development of a novel generation of selective, non-neurotoxic insecticides and appear to be ideal candidates for pest management strategies. However, insect neuropeptides and their receptors have a number of characteristics that make them rather unsuitable candidates for insect control agents (Scherkenbeck and Zdobinsky 2009; Teal et al. 1999). First, peptides are unstable in the environment, and suffer rapid degradation in the hemolymph and digestive system of insects. Secondly, peptides show poor solubility in organic solutions and do not penetrate insect cuticle, which contains surface layers composed of an apolar lipid matrix that inhibits penetration of polar compounds like peptides. Teal et al. (1999) reported that development of the synthetic analogues, mimetics, agonists or antagonists, capable of penetrating the insect cuticle, may be effective tools in combating insect pests in an environmentally more-sound manner than with conventional pesticides.

4.2 Potential Application of Bursicon for Controlling Pests

Bursicon is one of the neuropeptides regulating insect development, and its molecular structure, functions and signaling pathway have been described above, and most of them are well studied. Recently, several reports suggested that bursicon would be used as a potential target site for the design of novel, safe and selective compounds to control pests. Firstly, the *bursicon* mutant flies are not able to expand their wings after eclosion and show a prolonged retention of the elongate abdomen shape characteristic of a newly eclosed fly (Dewey et al. 2004). When treated with *burs* α dsRNA in pupae, most of the silkworm *B. mori* failed to initiate the behavioral program for wing expansion (Huang et al. 2007). Weak postecdysis activity, wrinkled elytra and deficiency in proper folding of the hindwing were observed in the red flour beetle, *T. castaneum* after treated with either *burs* α dsRNA or *burs* β dsRNA (Arakane et al. 2008). It seemed that bursicon-signaling pathway for wing expansion is common in insects. Therefore, bursicon could be used for constructing a special agent affecting insect wing expansion, especially for the pests relying on migration and far distance fly. Secondly, the known bursicon bioactivity seems to be restricted to very short periods of time during the insect's lifespan. Bursicon activities are demonstrated only during pupae-adult molting period in *Drosophila* (Baker and Truman 2002), and hemolymph from 1 day-old adults showed no activities for wing expansion and cuticle tanning (Baker and Truman 2002; Fraenkel and Hsiao 1965). Therefore, bursicon may be considered as a specifically targeted agent for novel pesticides. Third, bursicon are well conserved in insects. *Burs* α and *burs* β showed high sequence identity among 12 insect species (Fig. 5.3). Besides the *bursicon* sequence conservation, bursicon also showed cross activities among insects. Such as ganglia extracts from *P. americana* can stimulate cuticle tanning in *S. bullata* (Luo et al. 2005).

DLGR2, the receptor of bursicon, which is encoded by *rickets* and belongs to GPCRs, also appears to be a possible target for development of insect pest control agents (Maule et al. 2002; Van Hiel et al. 2010). As the first Type B LGR to be fully characterized in insects, DLGR2 acts as a bursicon heterodimer receptor and regulates wing expansion and cuticle tanning after adult eclosion (Luo et al. 2005; Mendive et al. 2005). The *rickets* mutant flies fail to expand their wings and thorax and delay the onset of tanning (sclerotization and melanization) (Baker and Truman 2002). When treated with *rickets* dsRNA, the beetle *T. castaneum* showed postecdysis defects, including weak postecdysis activity, wrinkled elytra and a failure to retract the hindwing (Arakane et al. 2008; Bai and Palli 2010). Therefore, a substance that interferes with the DLGR2 would be impossible to kill indiscriminately the entire population, but would act as a regulator agent, for example during an epidemic expansion of a pest (Van Hiel et al. 2010).

Like other neuropeptides and their receptors, bursicon and its receptor also have a number of characteristics that make them unsuitable for direct use as insect control agents (Scherkenbeck and Zdobinsky 2009; Teal et al. 1999), but recent advances in molecular biology make it possible to reconsider bursicon and its receptor as potential candidates for pest control. RNA interference (RNAi) has been widely used as an effective tool for study of gene function by down-regulating gene expression via double-stranded RNA (dsRNA) treatment. Injection of dsRNA for a specific gene is the most widely used method to deliver dsRNA molecules into insect tissues, but it is impractical for insect pest control (Amdam et al. 2003; Cruz et al. 2006; Dong and Friedrich 2005). Indirect feeding of dsRNA through dsRNA spraying appears to be effective in some insects (Araujo et al. 2006; Wang et al. 2011; Turner et al. 2006), but it costs a lot to synthesize and spray dsRNA. Therefore, both injection and feeding are most likely unsuitable for controlling field pests. Baum et al. (2007) and Mao et al. (2007) developed a new strategy called the plant-mediated herbivorous insect RNAi: suppressing a critical insect gene(s) by feeding insects with plant tissues engineered to produce a specific insect dsRNA. When transgenic corn plants were engineered to express dsRNAs direct against V-type ATPase A gene, an insect specific gene, and these plants showed a significant reduction of feeding damage by western corn rootworm *Diabrotica virgifera virgifera* in a growth chamber assay, suggesting that the RNAi system can be exploited to control insect pests via expression of an insect-specific-gene dsRNA in plants (Baum et al. 2007). Similarly, when tobacco and *Arabidopsis* plants were engineered to produce dsRNAs directed against a cytochrome P450 gene, *CYP6AE14*, from the cotton bollworm *Helicoverpa armigera,* the bollworm larvae fed with plant material of both transgenic species, showed significant reduction of the endogenous *CYP6AE14* transcript and increased sensitivity to the cotton metabolite gossypol when transferred to artificial diets. The insect RNAi effect triggered by ingestion of transgenic dsRNA-producing plants is gene-specific and could be used to protect crop plants against insect damage (Price and Gatehouse 2008). Besides lepidopteran and coleopteran plant pests (Baum et al. 2007; Mao et al. 2007), the plant-mediated herbivorous insect RNAi was also applied to a phloem-sucking hemipteran insect, rice planthopper, *Nilaparvata lugens* (Zha et al. 2011). However, the challenge from using RNAi to produce insect-pest-resistant plants is that which insect gene should be targeted and which dsRNA delivery system is required for delivering target genes outside of midgut epithelium cells (Price and Gatehouse 2008). Bursicon and its receptor showed the high potential for insect control, and dsRNA treatment can dramatically decrease the bursicon activities and disrupt insect development. It was suggested that bursicon or its receptor could be used as a potential target gene for developing transgenic RNAi plants. However, a suitable delivery system that could bring bursicon or its receptor dsRNA through the midgut barrier to its target sites in CNS (bursicon synthesis site) or epithelial cells (bursicon receptor site) needs to be identified before the RNAi approach could be effectively used in transgenic crops for pest control.

5 Conclusion

Neuropeptides and their receptors are major players in insect whole life and are considered to be a top research subject as potential targets for the production of selective and non-neurotoxic control agents to combat insect pests. Recent advances in understanding of signaling mechanism of insect neuropeptides as well as development of peptide drugs and agents in invertebrate provide new insights into generation of neuropeptide agonists/antagonist and conversion into nonpeptidergic small molecule (NPSM). Compared to other neuropeptides, bursicon and its receptor are well studied in *Drosophila* and other insects in recent years, and signaling pathway for controlling cuticle tanning and wing expansion is not fully but mainly defined. Like all the other neuropeptides, due to poor bioavailability, pharmacokinetics and short half-life the peptides themselves, the bursicon-receptor system cannot be used for controlling insect pests directly. However, with an increase in research into the discovery of non-peptide small molecules that functions as mimics for neuropeptides and the development of RNAi transgenic plants, scientists will conquer this difficulty and turn the bursicon controlling pests to a real success in the near future.

References

Adams MD, Celniker SE, Holts RA et al (2000) The genome sequence of *Drosophila melanogaster*. Science 287:2185–2195

Altstein M, Nässel DR (2010) Neuropeptide signaling in insects. Adv Exp Med Biol 692:155–165

Amdam GV, Simões ZL, Guidugli KR, Norberg K, Omholt SW (2003) Disruption of vitellogenin gene function in adult honeybees by intra-abdominal injection of double-stranded RNA. BMC Biotechnol 3:1

An S, Wang S, Gilbert L, Beerntsen B, Ellersieck M, Song Q (2008) Global identification of bursicon-regulated genes in *Drosophila melanogaster*. BMC Genomics 9:424

An S, Wang S, Song Q (2009) Identification of a novel bursicon-regulated transcriptional regulator, md13379, in the house fly *Musca domestica*. Arch Insect Biochem Physiol 70:106–121

An S, Dong S, Wang Q, Li S, Gilbert LI, Stanley D, Song Q (2012) Insect neuropeptide bursicon homodimers induce innate immune and stress genes during molting by activating the NF-κB transcription factor Relish. PLoS One 7(3):e34510

Andersen SO (2005) Cuticular sclerotization and tanning. In: Gilbert LI, Iatrou K, Gill SS (eds) Comprehensive molecular insect science, vol 4. Elsevier/Pergamon, Amsterdam, pp 145–170

Andersen SO (2010) Insect cuticular sclerotization: a review. Insect Biochem Mol Biol 40: 166–178

Andersen SO (2012) Cuticular sclerotization and tanning. In: Gilbert LI (ed) Insect molecular biology and biochemistry. Academic, San Diego, pp 67–192

Arakane Y, Muthukrishnan S, Beeman RW, Kanost MR, Kramer KJ (2005) Laccase 2 is the phenoloxidase gene required for beetle cuticle tanning. Proc Natl Acad Sci 102:11337–11342

Arakane Y, Li B, Muthukrishnan S, Beeman RW, Kramer KJ, Park Y (2008) Functional analysis of four neuropeptides, EH, ETH, CCAP and bursicon, and their receptors in adult ecdysis behavior of the red flour beetle, *Tribolium castaneum*. Mech Dev 125:984–995

Araujo RN, Santos A, Pinto FS, Gontijo NF, Lehane MJ, Pereira MH (2006) RNA interference of the salivary gland nitrophorin 2 in the triatomine bug *Rhodnius prolixus* (Hemiptera: Reduviidae) by dsRNA ingestion or injection. Insect Biochem Mol Biol 36:683–693

Bai H, Palli SR (2010) Functional characterization of bursicon receptor and genome-wide analysis for identification of genes affected by bursicon receptor RNAi. Dev Biol 344:248–258

Baker J, Truman JW (2002) Mutations in the *Drosophila* glycoprotein hormone receptor, rickets, eliminate neuropeptide-induced tanning and selectively block a stereotyped behavioral program. J Exp Biol 205:2555–2565

Baum JA, Bogaert T, Clinton W, Heck GR, Feldmann P, Ilagan O, Johnson S, Plaetinck G, Munyikwa T, Pleau M, Vaughn T, Roberts J (2007) Control of coleopteran insect pests through RNA interference. Nat Biotechnol 25:1322–1326

Bendena WG (2010) Neuropeptide physiology in insects. Adv Exp Med Biol 692:166–691

Brody T, Cravchik A (2000) *Drosophila melanogaster* G protein coupled receptors. J Cell Biol 150:F83–F88

Coast GM, Orchard I, Phillips JE, Schooley DA (2002) Insect diuretic and antidiuretic hormones. Adv Insect Physiol 29:279–409

Cottrell CB (1962) The imaginal ecdysis of blowflies. The control of cuticular hardening and darkening. J Exp Biol 39:395–411

Cruz J, Mané-Padrós D, Bellés X, Martín D (2006) Functions of the ecdysone receptor isoform-A in the hemimetabolous insect *Blattella germanica* revealed by systemic RNAi in vivo. Dev Biol 297:158–171

Dai L, Dewey EM, Zitnan D, Luo CW, Honegger HW, Adams ME (2008) Identification, developmental expression, and functions of bursicon in the tobacco hawkmoth, *Manduca sexta*. J Comp Neurol 506:759–774

Davis MM, O'Keefe SL, Primrose DA, Hodgetts RB (2007) A neuropeptide hormone cascade controls the precise onset of post-eclosion cuticular tanning in *Drosophila melanogaster*. Development 134:4395–4404

Dewey EM, McNabb SL, Ewer J, Kuo GR, Takanishi CL, Truman JW, Honegger HW (2004) Identification of the gene encoding bursicon, an insect neuropeptide responsible for cuticle sclerotization and wing spreading. Curr Biol 14:1208–1213

Dong Y, Friedrich M (2005) Nymphal RNAi: systemic RNAi mediated gene knockdown in juvenile grasshopper. BMC Biotechnol 5:25

Eriksen KK, Hauser F, Schiott M, Pedersen KM, Sondergaard L, Grimmelikhuijzen CJ (2000) Molecular cloning, genomic organization, developmental regulation, and a knock-out mutant of a novel leu-rich repeats-containing G protein-coupled receptor (DLGR-2) from *Drosophila melanogaster*. Genome Res 10:924–938

Ewer J, Reynolds S (2002) Neuropeptide control of molting in insects. In: Pfaff DW, Arnold AP, Fahrbach SE, Etgen AM, Rubin RT (eds) Hormones, brain and behavior. Academic, San Diego

Fogal W, Fraenkel G (1969) The role of bursicon in melanization and endocuticle formation in the adult flesh fly, *Sarcophaga bullata*. J Insect Physiol 15:1235–1247

Fraenkel G (1975) Interactions between ecdysone, bursicon, and other endocrines during puparium formation and adult emergence in flies. Am Zool 15(Suppl 1):29–48

Fraenkel G, Hsiao C (1962) Hormonal and nervous control of tanning in the fly. Science 138:27–29

Fraenkel G, Hsiao C (1965) Bursicon, a hormone which mediates tanning of the cuticle in the adult fly and other insects. J Insect Physiol 11:513–556

Fraenkel G, Hsiao C, Seligman M (1966) Properties of bursicon: an insect protein hormone that controls cuticular tanning. Science 151:91–93

Gammie SC, Truman JW (1997) Neuropeptide hierarchies and the activation of sequential motor behaviors in the hawkmoth *Manduca sexta*. J Neurosci 17:4389–4397

Gammie SC, Truman JW (1999) Eclosion hormone provides a link between ecdysis-triggering hormone and crustacean cardioactive peptide in the neuroendocrine cascade that controls ecdysis behavior. J Exp Biol 202:343–352

Gorman MJ, Arakane Y (2010) Tyrosine hydroxylase is required for cuticle sclerotization and pigmentation in *Tribolium castaneum*. Insect Biochem Mol Biol 40:267–273

Grossmann M, Weintraub BD, Szkudlinski MW (1997) Novel insights into the molecular mechanisms of human thyrotropin action: structural, physiological, and therapeutic implications for the glycoprotein hormone family. Endocr Rev 18:476–501

Hauser F, Williamson M, Cazzamali G, Grimmelikhuijzen CJ (2006) Identifying neuropeptide and protein hormone receptors in *Drosophila melanogaster* by exploiting genomic data. Brief Funct Genomic Proteomic 4:321–330

Hearn MT, Gomme PT (2000) Molecular architecture and biorecognition processes of the cystine knot protein superfamily: part I. The glycoprotein hormones. J Mol Recognit 13:223–278

Hewes RS, Taghert PH (2001) Neuropeptides and neuropeptide receptors in the *Drosophila melanogaster* genome. Genome Res 11:1126–1142

Hirsh J, Davidson N (1981) Isolation and characterization of the dopa decarboxylase gene of *Drosophila melanogaster*. Mol Cell Biol 1:475–485

Honegger HW, Market D, Pierce LA, Dewey EM, Kostron B, Wilson M, Choi D, Klukas KA, Mesce KA (2002) Cellular localization of bursicon using antisera against partial peptide sequences of this insect cuticle-sclerotizing neurohormone. J Comp Neurol 452:163–177

Honegger HW, Dewey EM, Kostron B (2004) From bioassays to *Drosophila* genetics: strategies for characterizing an essential insect neurohormone, bursicon. Symp Biol Hung 55:91–102

Honegger HW, Dewey EM, Ewer J (2008) Bursicon, the tanning hormone of insects: recent advances following the discovery of its molecular identity. J Comp Physiol 194:989–1005

Hsu SY, Nakabayashi K, Nishi S, Kumagai J, Kudo M, Sherwood OD, Hsueh AJ (2002) Activation of orphan receptors by the hormone relaxin. Science 295:671–674

Huang J, Zhang Y, Li M, Wang S, Liu W, Couble P, Zhao G, Huang Y (2007) RNA interference-mediated silencing of the bursicon gene induces defects in wing expansion of silkworm. FEBS Lett 581:697–701

Kiger JA, Natzle JE, Kimbrell DA, Paddy MR, Kleinhesselink K, Gree MM (2007) Tissue remodeling during maturation of the *Drosophila* wing. Dev Biol 301:178–191

Kimura K, Kodama A, Hayasaka Y, Ohta T (2004) Activation of the cAMP/PKA signaling pathway is required for post-ecdysial cell death in wing epidermal cells of *Drosophila melanogaster*. Development 131:1597–1606

Kostron B, Marquardt K, Kaltenhauser U, Honegger HW (1995) Bursicon, the cuticular sclerotizing hormone – comparison of its molecular mass in different insects. J Insect Physiol 41:1045–1053

Kumagai J, Hsu SY, Matsumi H, Roh JS, Fu P, Wade JD, Bathgate RA, Hsueh AJ (2002) INSL3/Leydig insulin-like peptide activates the LGR8 receptor important in testis descent. J Biol Chem 277:31283–31286

Liu F, Baggerman G, D'Hertog W, Verleyen P, Schoofs L, Wets G (2006) In silico identification of new secretory peptide genes in *Drosophila melanogaster*. Mol Cell Proteomics 5:510–522

Livingstone MS, Tempel BL (1983) Genetic dissection of monoamine transmitter synthesis in *Drosophila*. Nature 303:67–70
Locke M (2001) The Wigglesworth Lecture: insects for studying fundamental problems in biology. J Insect Physiol 47:495–507
Loveall BJ, Deitcher DL (2010) The essential role of bursicon during *Drosophila* development. BMC Dev 10:92–108
Luan H, Lemon WC, Peabody NC, Pohl JB, Zelensky PK, Wang D, Nitabach MN, Holmes TC, White BH (2006) Functional dissection of a neuronal network required for cuticle tanning and wing expansion in *Drosophila*. J Neurosci 26:573–584
Luo CW, Dewey EM, Sudo S, Ewer J, Hsu SY, Honegger HW, Hsueh AJ (2005) Bursicon, the insect cuticle-hardening hormone, is a heterodimeric cystine knot protein that activates G protein-coupled receptor LGR2. Proc Natl Acad Sci 102:2820–2825
Mao YB, Cai WJ, Wang JW, Hong GJ, Tao XY, Wang LJ, Huang YP, Chen XY (2007) Silencing a cotton bollworm P450 monooxygenase gene by plant-mediated RNAi impairs larval tolerance of gossypol. Nat Biotechnol 25:1307–1313
Maule AG, Mousley A, Marks NJ, Day TA, Thompson DP, Geary TG, Halton DW (2002) Neuropeptide signaling systems – potential drug targets for parasite and pest control. Curr Top Med Chem 2:733–758
Mendive FM, Van Loy T, Claeysen S, Poels J, Williamson M, Hauser F, Grimmelikhuijzen CJ, Vassart G, Vanden Broeck J (2005) *Drosophila* molting neurohormone bursicon is a heterodimer and the natural agonist of the orphan receptor DLGR2. FEBS Lett 579:2171–2176
Mercier J, Doucet D, Retnakaran A (2007) Molecular physiology of crustacean and insect neuropeptides. J Pestic Sci 32:345–359
Mills RR, Whitehead DL (1970) Hormonal control of tanning in the American cockroach: changes in blood cell permeability during ecdysis. J Insect Physiol 16:331–340
Mizoguchi A, Ohashi Y, Hosoda K, Ishibashi J, Kataoka H (2001) Developmental profile of the changes in the prothoracicotropic hormone titer in hemolymph of the silkworm *Bombyx mori*: correlation with ecdysteroid secretion. Insect Biochem Mol Biol 31:349–358
Moussian B (2010) Recent advances in understanding mechanisms of insect cuticle differentiation. Insect Biochem Mol Biol 40:363–375
Nässel DR (2002) Neuropeptides in the nervous system of *Drosophila* and other insects: multiple roles as neuromodulators and neurohormones. Prog Neurobiol 68:1–84
Natzle JE, Kiger JA Jr, Green MM (2008) Bursicon signaling mutations separate epithelial-mesenchymal transition from programmed cell death during *Drosophila melanogaster* wing maturation. Genetics 180:885–893
Neckameyer WS, White K (1993) *Drosophila* tyrosine hydroxylase is encoded by the *pale* locus. J Neurogenet 8:189–199
Nishi S, Hsu SY, Zell K, Hsueh AJ (2000) Characterization of two fly LGR (leucine-rich repeat-containing, G protein-coupled receptor) proteins homologous to vertebrate glycoprotein hormone receptors: constitutive activation of wild-type fly LGR1 but not LGR2 in transfected mammalian cells. Endocrinology 141:4081–4090
Orchard I (1987) Adipokinetic hormone: an update. J Insect Physiol 33:451–463
Peabody NC, Diao F, Luan H, Wang H, Dewey EM, Honegger HW, White BH (2008) Bursicon functions within the *Drosophila* central nervous system to modulate wing expansion behavior, hormone secretion, and cell death. J Neurosci 28:14379–14391
Pierce JG, Parsons TF (1981) Glycoprotein hormones: structure and function. Annu Rev Endocrinol 50:465–495
Price DR, Gatehouse JA (2008) RNAi-mediated crop protection against insects. Trends Biotechnol 26:393–400
Rewitz KF, Yamanaka M, Gilbert LI, O'Connor MB (2009) The insect neuropeptide PTTH activates receptor tyrosine kinase Torso to initiate metamorphosis. Science 326:1403–1405
Reynolds SE (1983) Bursicon. In: Downer RGH, Laufer H (eds) Endocrinology of insects. Alan R. Liss. Inc., New York

Scherkenbeck J, Zdobinsky T (2009) Insect neuropeptides: structures, chemical modifications and potential for insect control. Bioorg Med Chem 17:4071–4084

Seligman M, Doy FA (1973) Hormonal regulation of disaggregation of cellular fragments in the haemolymph of *Lusilia cuprina*. J Insect Physiol 19:125–135

Seligman IM, Friedman S, Fraenkel G (1969) Bursicon mediation of tyrosine hydroxylation during tanning in their adult cuticle in the fly *Sarcophaga bullata*. J Insect Physiol 15:553–562

Smith W, Rybczynski R (2012) Prothoracicotropic hormone. In: Gilbert LI (ed) Insect endocrinology. Academic, San Diego, pp 1–62

Song Q (2012) Bursicon, a neuropeptide hormone that controls cuticle tanning and wing expansion. In: Gilbert LI (ed) Insect endocrinology. Academic, San Diego, pp 93–105

Stone JV, Mordue W, Betley KE, Morris HR (1976) Structure of locust adipokinetic hormone, a neurohormone that regulates lipid utilization during flight. Nature 265:207–221

Sudo S, Kuwabara Y, Park JI, Hsu SY, Hsueh AJ (2005) Heterodimeric fly glycoprotein hormone-alpha2 (GPA2) and glycoprotein hormone-beta5 (GPB5) activate fly leucine-rich repeat-containing G protein-coupled receptor-1 (DLGR1) and stimulation of human thyrotropin receptors by chimeric fly GPA2 and human GPB5. Endocrinology 146:3596–3604

Taghert PH, Truman JW (1982) The distribution and molecular characteristics of the tanning hormone, bursicon, in the tobacco hornworm *Manduca sexta*. J Exp Biol 98:373–383

Teal PE, Meredith JA, Nachman RJ (1999) Development of amphiphylic mimics of insect neuropeptides for pest control. Ann N Y Acad Sci 897:348–360

Themmen APN, Huhtaniemi IT (2000) Mutations of gonadotropins and gonadotropin receptors: elucidating the physiology and pathophysiology of pituitary–gonadal function. Endocr Rev 21:551–583

Turner CT, Davy MW, MacDiarmid RM, Plummer KM, Birch NP, Newcomb RD (2006) RNA interference in the light brown apple moth, *Epiphyas postvittana* (Walker) induced by double-stranded RNA feeding. Insect Mol Biol 15:383–391

Van der Horst DJ, Van Marrewijk WJA, Diederen JHB (2001) Adipokinetic hormones of insects: release, signal transduction and responses. Int Rev Cytol 211:179–240

Van Hiel MB, Van Loy T, Poels J, Vandersmissen HP, Verlinden H, Badisco L, Vanden Broeck J (2010) Neuropeptide receptors as possible targets for development of insect pest control agents. Adv Exp Med Biol 692:211–226

Van Loy T, Van Hiel MB, Vandermissen HP, Poels J, Mendive F, Vassart G, Vanden Broeck J (2007) Evolutionary conservation of bursicon in the animal kingdom. Gen Comp Endocrinol 153:59–63

Veelaert D, Passier P, Devreese B, Vanden Broeck J, Van Beeumen J, Vullings H, Diederen J, Schoofs L, De Loof A (1997) Isolation and characterization of an adipokinetic hormone release-inducing factor in locusts: the crustacean cardioactive peptide. Endocrinology 138:138–142

Vitt UA, Hsu SY, Hsueh AJ (2001) Evolution and classification of cystine knot-containing hormones and related extracellular signal molecules. Mol Endocrinol 15:681–694

Wang S, An S, Song Q (2008) Transcriptional expression of bursicon and novel bursicon-regulated genes in the house fly *Musca domestica*. Arch Insect Biochem Physiol 68:100–112

Wang Y, Zhang H, Li H, Miao X (2011) Second-generation sequencing supply an effective way to screen RNAi targets in large scale for potential application in pest insect control. PLoS One 6(4):e18644

Zha W, Peng X, Chen R, Du B, Zhu L, He G (2011) Knockdown of midgut genes by dsRNA-transgenic plant-mediated RNA interference in the hemipteran insect *Nilaparvata lugens*. PLoS One 6(5):e20504

Zitnan D, Adams ME (2000) Excitatory and inhibitory roles of central ganglia in initiation of the insect ecdysis behavioural sequence. J Exp Biol 203:1329–1340

Chapter 6
Cell-Based Screening Systems for Insecticides

Guy Smagghe and Luc Swevers

1 Introduction

In the middle of the previous century, more than 50 years ago, insect cell culture began with the successful establishment of the first continuous insect cell lines by Dr. Grace in Australia (Grace 1962) and by Dr. Gao in China (Gaw 1958), and this happened independently from each other. To date the field has grown to the extent that over 500 insect cell lines have been established from many insect species representing numerous insect orders and from several different tissue sources (reviewed by Lynn 2007; Smagghe et al. 2009). These cell lines are used in different fields as research tools in studies on signaling mechanisms and to test hypotheses about gene expression and evidently also in screening programs designed to discover new insecticide chemistries and assess their insecticide potential. This screening research is leading to the development of high-throughput screening (HTS) technologies that are essential in the search for new insect control agents, and furthermore, to accelerate their development (Chap. 7 by Nakagawa and Harada 2012, this book). A few insect cell lines are also used in routine industrial processes to produce (target) proteins of biomedical significance. Hence, both primary cell cultures and established lines can be used in basic biological studies meant to reveal how insect cells work.

G. Smagghe (✉)
Department of Crop Protection, Faculty of Bioscience Engineering,
Ghent University, 9000 Ghent, Belgium
e-mail: guy.smagghe@ugent.be

L. Swevers
Insect Molecular Genetics and Biotechnology, Institute of Biosciences & Applications,
National Centre for Scientific Research "Demokritos", Aghia Paraskevi, Athens, Greece
e-mail: swevers@bio.demokritos.gr

I. Ishaaya et al. (eds.), *Advanced Technologies for Managing Insect Pests*,
DOI 10.1007/978-94-007-4497-4_6, © Springer Science+Business Media Dordrecht 2012

To date an average of ~10 billion USD is spent per year for synthetic insecticides to control pest insects of importance in agriculture and human health (Beckmann and Haack 2003). The problems associated with the classical neurotoxic insecticide groups, including insect resistance and environmental concerns, have encouraged the development of more insect-specific insecticide screening procedures and bioassays. The suitability of such strategy is demonstrated by various insect growth regulators (IGRs) that impair the endocrine regulation of growth, development (e.g. molting/metamorphosis) and reproduction (Dhadialla et al. 2005). Successes include ecdysteroid receptor (EcR) agonists or "molting accelerating compounds" (MACs) (e.g. tebufenozide, methoxyfenozide), juvenile hormone analogues (JHAs) (e.g. methoprene, fenoxycarb and pyriproxyfen), and chitin synthesis inhibitors (e.g. benzoylphenylureas, BPUs). Additionally, there are newer synthetic insecticides that act specifically on insect nervous systems, energy metabolism (e.g. pyridalyl) and muscle (neural) targets (e.g. neonicotinoids, spinosyns, avermectins, flubendiamide and chlorantraniliprole). There is also a surge of interest in biological control agents, led by the insecticidal *Bacillus thuringiensis* (*Bt*) endotoxins. In our view, discovering and deploying these modern anti-insect chemistries, whenever from nature or synthetic sources, underscore the potential power of the approach. Nonetheless, this is limited by contemporary screening procedures.

At early screening stages, there is an increasing interest in the development of *in vitro* methods to replace conventional animal toxicity tests. The ultimate goal is to achieve an alternative system that allows for rapid testing of candidate compounds, formulations and finished products and enables the accurate prediction of toxic efficacy at the whole animal level. There are certain key requirements in developing an alternative cell-based testing procedure. These include dependable intra- and inter-laboratory reproducibility, high predictive power for correct toxicity assessment decisions, relevance to the type of compounds to be tested, and low cost/benefit ratio. Because of the very large inventory of natural and synthetic chemicals potentially useful in insect pest management programs, the ideal systems would be amenable to HTS technology. HTS is widely used in the private sector drug discovery programs in which thousands of potential pharmaceuticals are screened for desired biological activity on a daily basis. Established insect cell lines, joined with HTS procedures, will contribute to rapid screening of many materials and accelerate the discovery of novel environmentally-safe control agents. Tests employing cell cultures can be readily automated. Cell-based assays can enable the discovery of new modes of action for insecticide candidates (Harada et al. 2011). This review chapter is designed to present some significant recent examples and advances, using that approach with EcR reporter systems as a paradigm, and to offer a vision of the future of cell-based screening strategies.

2 Paradigm: Screening Systems for Molting Hormone Analogs (or "Molting Accelerating Compounds", MACs) Discovery

2.1 *Introduction*

The concept of interfering with insect endocrine systems as a selective mechanism to control pest insects was proposed by Williams (1969), which paved the way for the design of "third generation pesticides". The major insect hormones, the terpenoid juvenile hormones (JH) and the steroid molting hormones or ecdysteroids were first targets.

The molecular mechanism of action of the insect molting hormone 20-hydroxyecdysone (20E) was elucidated during the last decade of the twentieth century and followed the cloning of the ecdysone receptor (EcR) of *Drosophila* (Koelle et al. 1991). It was shown that EcR forms a heterodimer complex with the Ultraspiracle (USP)/retinoid-X-receptor (RXR), and upon ligand binding, it induces the expression of target genes with characteristic binding elements (*i.e.* the ecdysone response elements (EcREs)), in their regulatory regions (Yao et al. 1993; Thomas et al. 1993; reviewed by Henrich 2005; Fahrbach et al. 2012). During its action *in vivo*, 20E induces the expression of a conserved regulatory cascade consisting of early and late genes (Thummel 1997; Riddiford et al. 2000; King-Jones and Thummel 2005); early genes typically encode transcription factors (such as nuclear receptors (E75 and HR3), zinc-finger proteins (Broad-Complex) and ETS-domain containing transcription factors (E74)) that forward and amplify the ecdysone response, while late genes generally comprise 'realizator' genes that establish the phenotypical changes (Thummel 1997). It is noted that, besides its contribution to the understanding of insect molting and metamorphosis, the elucidation of the molecular mechanism of the primary ecdysone response, *i.e.* the mechanism by which 20E activates the EcR/USP(RXR) complex at target sites, also provided the necessary tools for the development of sensitive and specific screening systems for the 20E analogs, with important applications for control of insect pests.

2.2 *Cell-Based Screening Assays*

Early assays to evaluate ecdysteroid activity were based on the displacement of binding of a radioactive ecdysteroid (^{3}H- or ^{125}I-labeled ponasterone A; PonA) to dipteran and lepidopteran (Cherbas et al. 1988; Nakagawa et al. 2000, 2002a, b, 2005) and coleopteran cell extracts (Harada et al. 2011) or imaginal disc extracts (Terentiou et al. 1993; Smagghe and Degheele 1995; Smagghe et al. 1996). More recently, after the cloning of complete ORFs of EcR and USP, *in vitro* transcription/translation systems were used to evaluate binding affinity and specificity for a particular species (Minakuchi et al. 2003, 2004, 2007; Ogura et al. 2005a; review by Nakagawa 2005; Nakagawa and Henrich 2009). Other assays employed organ cultures to evaluate molting hormone activity such as those based on the evagination of imaginal discs (Smagghe et al. 2000) and the promotion of chitin synthesis by cuticle fragments (Nakagawa et al. 1998; review by Nakagawa 2005). Insect cell lines generally respond to ecdysteroids and non-steroidal ecdysone agonists by cessation of cell proliferation and formation of filamentous extensions (Wing 1988; Sohi et al. 1995; Siaussat et al. 2007; Mosallanejad et al. 2008) and this property has indeed been exploited for development of screening systems for ecdysone analogs (e.g. the B_{II} bioassay; Dinan et al. 2001; Harmatha et al. 2002).

Because of the design of ecdysone agonists, earlier assays based on displacement of PonA binding or *in vitro* culture of tissues were elaborate, time-consuming, and not readily amenable to high-throughput format. Assays based on inhibition of cell proliferation could be scaled up easily but could be considered as lacking specificity since many other compounds that do not exhibit ecdysteroid activity would be expected to interfere with the assay (Soin et al. 2010b).

With the advent of EcRE-based reporter systems, highly specific and sensitive cell-based screening systems could be developed for HTS of ecdysone agonists. The reporter plasmids typically consist of a basal reporter cassette (basal promoter followed by GFP or luciferase ORF and polyadenylation sequence) preceded by multiple copies of a high-affinity EcRE (for instance, seven copies of the EcRE identified in the *hsp27* promoter of *Drosophila*, see Swevers et al. 2004) (Fig. 6.1). When used in insect cell lines, ectopic expression of EcR and USP is not considered necessary, since sufficient amounts of endogenous EcR or USP are present to guarantee a robust response (Soin et al. 2010b). Because EcR/USP complexes from virtually all insect species can bind to the same EcRE (Henrich 2005), all that is needed for the development of a cell-based screening system for ecdysone analogs is therefore a tissue culture cell line that can be transfected efficiently with the reporter construct. Transfectable cell lines from different insect species therefore correspond to different species-specific screening systems for ecdysone analogs and their availability will allow the screening for ecdysone analogs that are specific for particular insect orders or (theoretically) individual species.

As shown in Fig. 6.1, cell-based screening systems were developed that are specific for lepidopteran insects with the use of pEcRE/b.act/GFP or pEcRE/b.act/Luc reporter plasmids (based respectively on GFP and luciferase reporters). Good examples are cell lines that have been derived from lepidopteran insects (Bm5 and Sl2 cells derived from *Bombyx mori* and *Spodoptera littoralis*, respectively; Swevers et al. 2004; Soin et al. 2010a), dipteran insects (S2 cell line; derived from *Drosophila melanogaster*; Soin et al. 2010b), and coleopteran insects (Ag3C cell line; derived from *Anthonomus grandis*; Soin et al. 2009). With the exception of Bm5 cells, EcRE-based reporter constructs were introduced by transfection and their presence in the cells in episomal form was only transient, therefore necessitating new transfections to be carried out for each new assay. For Bm5 cells, (semi-)clonal permanently transformed cell lines (Bm5/EcRE-GFP) are available that have incorporated EcRE/b.act/GFP

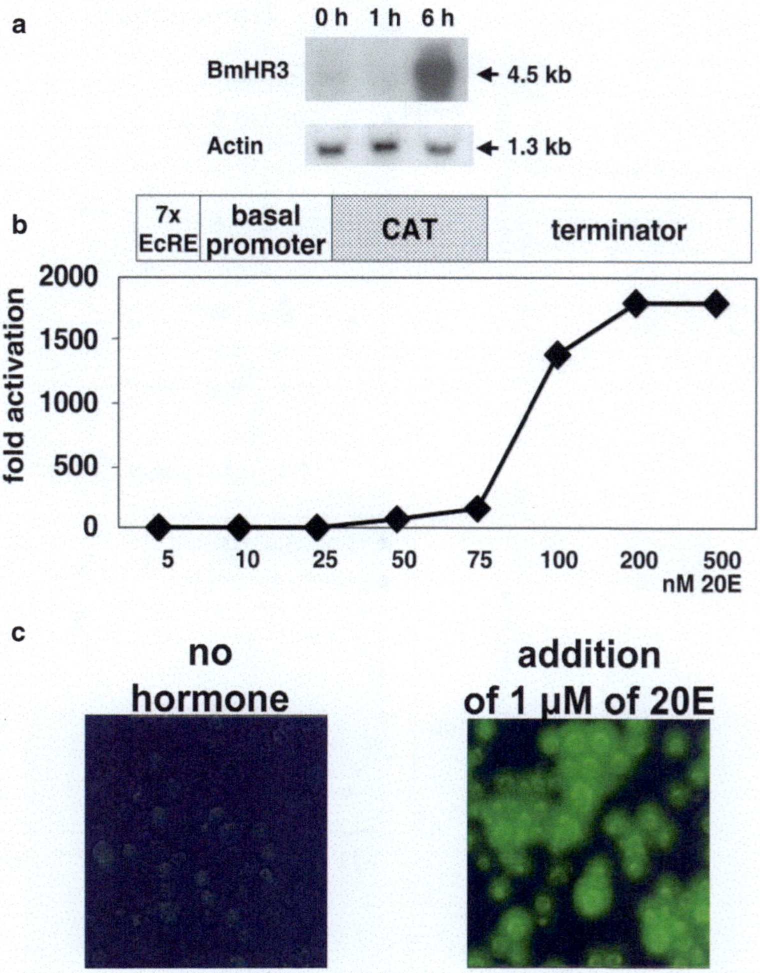

Fig. 6.1 Assessment of the primary response to ecdysteroids (20-hydroxyecdysone, 20E) in silk moth-derived Bm5 cells, transformed with the pBmbA/ERE.gfp construct for induction of *green fluorescence*. (**a**) Northern blot analysis of 20E primary response gene BmHR3 expression at intervals of administration of 1 μM of 20E. Actin hybridizations were carried out as control. Molecular weight of hybridizing mRNAs is shown at the *right*. (**b**) Induction of CAT activity from reporter construct pBmbA/ERE.cat after treatment with different concentrations of 20E (5–500 nM). Expression levels of induced relative to non-induced cells are indicated. Drawing of the reporter construct is at the *top*. (**c**) Observation of induction of *green fluorescence* by 20E by fluorescence microscopy. Untreated and treated transformed Bm5 cells are shown at *left* and *right*, respectively (×40) (Redrafted from Swevers et al. 2004)

cassettes in their genomes; upon addition of ecdysone agonist (tebufenozide), virtually all cells of this cell line will exhibit bright fluorescence (Swevers et al. 2004).

Then, as reported by Soin et al. (2010a), screening of compounds with ecdysone agonist activity using the *Spodoptera*-based Sl2 cell line after transfection and the

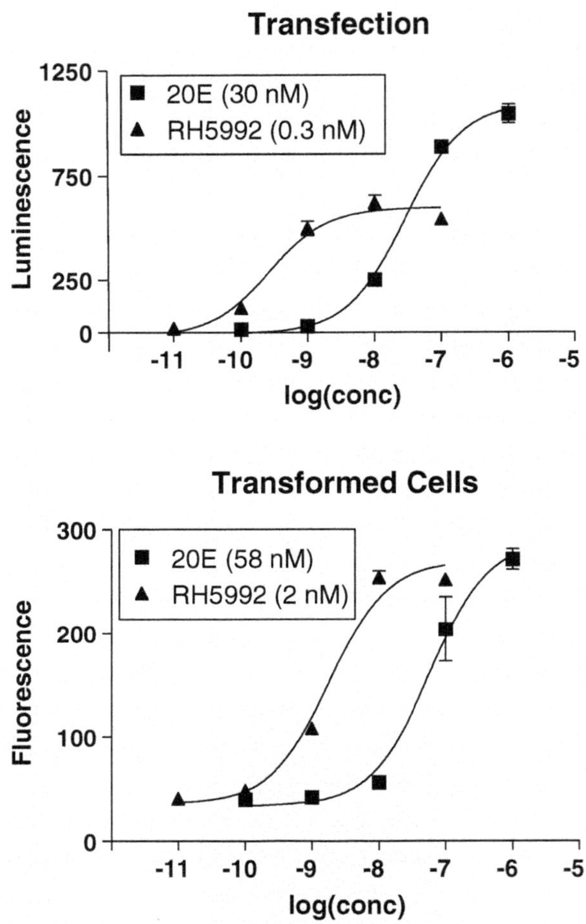

Fig. 6.2 Induction of ERE-reporter in transiently transfected Bm5 cells (luciferase, *top*) and permanently transformed Bm5/ERE-GFP cell line (GFP, *bottom*). Note that the inducibility of tebufenozide in transfected cells was at ~50% of the response by 20E, while in the transformed cell line similar inducibilities were observed for both compounds. At *right* are indicated the EC_{50} values

transformed *Bombyx*-derived Bm5 cell line gave virtual identical results with respect to the identification of active compounds. This indicated that reporter cassettes both in episomal form and after genomic integration can be used for screening. However, it was noted that differences with respect to the inducibility of the reporter can exist between the two forms in which the reporter cassette exists in the cells (Fig. 6.2). When Bm5 cells are transfected with the pEcRE/b.act/Luc reporter, the inducibility by the natural hormone 20E was approximately twofold higher than by the non-steroidal ecdysone agonist RH-5992. This inducibility was also referred to as the 'efficacy' of the compounds (Soin et al. 2010b). This difference was not observed in the transformed Bm5/EcRE-GFP cell line (with the reporter cassette permanently incorporated in the genome) which showed similar inducibility by both 20E and RH-5992 (Fig. 6.2). In both cases, however, very similar median effective concentrations (EC_{50}s) were obtained for both compounds (also referred to as the 'potency' of the compounds; Soin et al. 2010b). As noted before, the difference is likely caused by the status of the DNA of the reporter plasmid in the nuclei of the cells (permanent transgene versus transient episomal form) and invites further investigation. Deeper investigation into the causes of this difference may lead to new insights in the activation mechanism of gene transcription by EcR/USP, but for practical purposes, it is clear that both types of screening systems can be effectively used for the discovery of new compounds with ecdysone mimetic activity.

Today, screens are being carried out in an increasing higher throughput format, and for this reason, it is essential that measurements can be carried out on as small numbers of cells as possible because this material can easily be limiting. The Bm5/EcRE-GFP cell line has been successfully used in a droplet-based microfluidic system to obtain dose–response relationship measurements for the natural hormone 20E (Baret et al. 2010). As little as ~7,500 cells could be used per concentration to allow precise measurement of the EC_{50} of the hormone, indicating significant downscaling of the amount of cells necessary to obtain useful measurements.

An essential condition for the success of the approach of cell lines derived from different insect species is the requirement of functional uptake of DNA during transfection protocols. This condition is not always satisfied, as illustrated by the Se4 (derived from the lepidopteran *Spodoptera exigua*; Goodman et al. 2001) and the coleopteran BCIRL-Lepd-SL1 (derived from the Colorado potato beetle *Leptinotarsa decemlineata*; Long et al. 2002) and BRL-AG-3A (derived from the cotton boll weevil *Anthonomus grandis*; Stiles et al. 1992) cell lines, which do not express reporter genes following transfection (Swevers et al. 2008; Soin et al. 2009). To explore alternative means of gene transduction in these cell lines, recombinant baculoviruses were generated that have incorporated the EcRE/b.act/GFP cassette (Swevers et al. 2008). The baculoviruses used are BmNPV (*Bombyx mori* nuclear polyhedrosis virus) with a very limited host range that can only productively infect *B. mori* and closely related species (Maeda et al. 1993). Efficient transduction of the EcRE-based reporter was observed after infection of Se4 and BRL-AG-3A cells, which allowed evaluation of the activity of 20E and ecdysone agonists (Swevers et al. 2008; Soin et al. 2009). However, it was also observed that the recombinant BmNPV virus could not achieve gene transduction in the coleopteran BCIRL-Lepd-SL1 cells, indicating that this method does not work for all cell lines. However, it can be anticipated that, through careful screening, it would be possible to identify cell lines from all major insect orders that can be transduced with EcRE-based reporter constructs efficiently, either through liposome-based methods or through virus transduction, to allow screening for ecdysteroid activities.

2.3 Differences Among Cell Lines Derived from Insect Species of Different Orders

The availability of transfectable cell lines that belong to species of different insect orders has allowed the direct comparison of the efficiency by which synthetic nonsteroidal ecdysone agonists activate the EcR/USP complex in cells derived from different insect orders. This has been investigated in most detail by comparison of

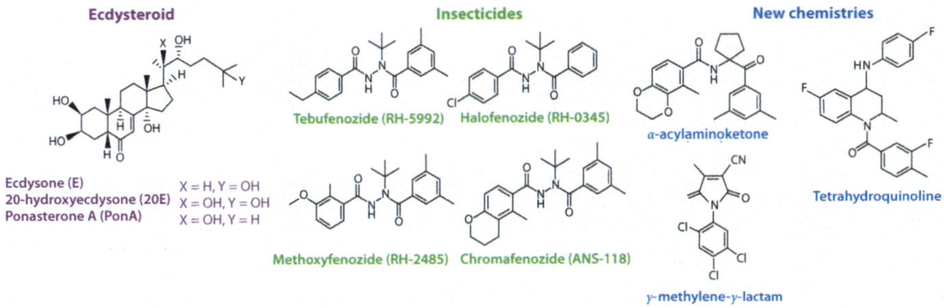

Fig. 6.3 Chemical structures of the molting hormones or ecdysteroids as ecdysone (E), 20-hydroxyecdysone (20E), ponasterone A (PonA), and stable nonsteroidal EcR ligands used as insecticides in agriculture/forestry together with a series of new chemistries: the DAH-based tebufenozide, methoxyfenozide and halofenozide (RH-5992, RH-2485 and RH-0345, respectively; all developed by Rohm and Haas, USA), chromafenozide (ANS-118, CM-001; jointly by Nippon Kayaku and Sankyo, Japan), tetrahydroquinoline (THQ) (by FMC, USA), α-acylaminoketone (AAK) (by Intrexon, USA), and γ-methylene-γ-lactams (by CSIRO, Australia)

the activity of the EcRE-based reporters in lepidopteran Bm5 and dipteran S2 cells (Soin et al. 2010b).

In this study, a library of non-steroidal ecdysone agonists containing different mother structures with diacylhydrazine (DAH) and other related analogues such as α-acylaminoketone (AAK) and tetrahydroquinoline (THQ) (Smith et al. 2003; Tice et al. 2003a, b; Palli et al. 2005b) (Fig. 6.3) was tested for activation of the EcRE-based reporter in S2 cells and the activities were compared with activities obtained in lepidopteran Bm5 cells (Wheelock et al. 2006; Soin et al. 2010a, b). While high activities for compounds with DAH or AAK structure were obtained in Bm5 cells, it was observed that in S2 cells none of the tested compound had a higher activity than 20E. In all cases, compounds were more active by 10- to 1,000-fold in Bm5 cells than in S2 cells, thus categorizing compounds with DAH or AAK structure specific for lepidopteran receptors. THQ compounds were generally much less active in Bm5 cells and actually two THQ compounds were identified that were capable to activate EcR/USP specifically in S2 cells (Soin et al. 2010b). However, also in these cases, activity was rather low (EC_{50} > 10 μM) and chemical structures need considerable optimization before any possible practical use. THQ compounds also display considerable cellular toxicity (Soin et al. 2010a) and the development of safer analogues is necessary.

When commercial DAH analogues were tested on transfected BRL-AG-3C cells, a coleopteran cell line derived from *A. grandis*, similar conclusions to tests done on S2 cells could be drawn. DAH analogues are much weaker with respect to EcR/USP activation in coleopteran cells than in lepidopteran cells (Soin et al. 2009). However, this finding is also somewhat paradoxical, since DAH analogues such as halofenozide (RH-0345) are actually used successfully to combat some coleopteran pests (Dhadialla et al. 1998). This illustrates that high activity in cell-based systems does not guarantee *per se* that a compound is successful in the field to control pests (see also below). Although significant ecdysteroid activity is essential, other parameters such as efficient uptake and resistance to metabolism and excretion also play a major role (see also below).

That it is possible to isolate specific ligands to particular groups of receptors is illustrated by mutation studies of the ligand-binding domain of the *Choristoneura fumiferana* (Lepidoptera) EcR (Kumar et al. 2004). It was found that a single amino acid change lead to discrimination between ecdysteroids and DAH analogs on one hand and THQ analogs on the other hand. This study can be interpreted as "reverse screening" of mutated receptors to respond specifically to ligands with different chemical structures.

2.4 Alternative Reporter Assays

When cell lines for a particular species are not available, one can try to overexpress its EcR and USP proteins in heterologous cell lines. If these cell lines are derived from insects, the problem arises of possible interference from endogenous receptors which could complicate the interpretation of results. Several strategies have been used to circumvent this problem.

A transformed *Drosophila* Kc cell line is available, L57-3-11, that expresses very low levels of endogenous DmEcR due to "parahomologous" gene targeting at the gene locus (Cherbas and Cherbas 1997). This cell line has been used to transfect heterologous EcR expression and EcRE-based reporter constructs to functionally characterize insect and even crustacean receptors (Swevers et al. 1996; Kumar et al. 2002; Verhaegen et al. 2010). Another strategy that has been explored is based on the knock-down of the endogenous receptor through RNAi (Hannan et al. 2009). Although these approaches can give good results, they still need to be watched with caution because of residual expression of endogenous receptors.

Another approach is based on expression constructs in which the EcR (full-length or hinge-ligand-binding domain-F region) is fused with the DNA-binding domain of the Gal4 transcription factor from yeast. These expression constructs are combined with expression constructs of USP/RXR that are fused with a transactivation domain (Gal4 or VP16) (or *vice versa* in two-hybrid format; Palli et al. 2003). When combined with reporter constructs containing copies of the UAS Gal4 binding site, activation at high levels is observed following addition of ecdysone agonists (Palli et al. 2003). This system was used to create orthogonal gene switch applications in plants and mammals (Palli et al. 2005a). Because the Gal4 DNA binding site comprises a unique sequence recognized specifically by the Gal4 factor, there is no interference at the level of DNA binding by endogenous EcR and USP/RXR in the insect cells. However, it can be assumed that heterologous dimers are formed between Gal4 fusions and endogenous receptors and that this process can interfere with the specificity of the assay.

2.5 Mammalian Cells

Mammalian cells have also been used to evaluate the primary ecdysone response after transfection of EcRE-based reporters and EcR and USP expression constructs (Palli et al. 2005a; Beatty et al. 2009). From the beginning, it was clear that activity of compounds could differ significantly between insect and mammalian cells in EcRE-based reporter assays. In the first published experiments using *Drosophila* EcR, it was observed that 20E was barely active, while significant activity was obtained with muristerone A (MurA) (Christopherson et al. 1992). The ecdysone response system was successfully applied as an inducible expression system in mammalian cells and transgenic mice, but this was only achieved after considerable engineering of DNA-binding (C-domain) and ligand-binding (E-domain) domains and co-expression of RXR (No et al. 1996; Suhr et al. 1998; see also review by Palli et al. 2005a).

Evaluation of the ecdysone response in mammalian cell lines has been used several times to gain insights in the activity of compounds against particular receptor isoforms or to investigate the cross-talk between JH and 20E at the EcR/USP complex (Henrich et al. 2003; Beatty et al. 2006). Significantly, it was observed that DAHs trigger a stronger response with beetle (*Leptinotarsa*) than with *Drosophila* EcR/USP, which is consistent with the use of those compounds to control coleopteran pests (Beatty et al. 2009).

2.6 Chemistry of Analogs

A more detailed description of ecdysteroids and non-steroidal anologs that were used in cell-based screening systems is presented in the Chap. 7 written by Nakagawa and Harada (this book). For use in insect pest control, ecdysteroids are not considered suitable compounds because of their high polarity, low stability and high synthesis costs (Dinan 1995). Screens have focused on the isolation of non-steroidal analogs that are sufficiently apolar to penetrate the insect's cuticle, are relatively resistant to metabolic enzymes and display high affinity for the ecdysone receptor complex. The group of chemicals that has been investigated mostly and for which commercial compounds were isolated are DAHs (Dhadialla et al. 1998, 2005; Sawada et al. 2003). Commercial DAH compounds, i.e. RH-2485 or methoxyfenozide, RH-0345 or halofenozide, and RH-5992 or tebufenozide from Rohm & Haas Co, Spring House, PA; now Dow AgroSciences, Indianapolis, IN; and ANS-118 or chromafenozide from Sankyo Agro Co. Ltd (presently Mitsui Agro Inc., Japan) and Nippon-Kayaku Co. Ltd (both Tokyo, Japan) (Dhadialla et al. 1998; Sawada et al. 2003), are mainly used to control lepidopteran pests although halofenozide formulations also exist against soil-dwelling coleopterans (Dhadialla et al. 1998, 2005). Other non-steroidal analogs include AAKs (which resemble much DAHs; Tice et al. 2003a, b) and THQs (Smith et al. 2003; Palli et al. 2005b) (Fig. 6.3). As discussed above, some compounds of the latter group may display higher affinity towards EcR/USP from dipteran insects (Soin et al. 2010b). Other structures with ecdysone mimetic activity have shown up in the literature such as 3,5-di-*tert*-butyl-4-hydroxy-*N*-isobutyl-benzamide (Mikitani 1996; Nakagawa 2005) but a systematic search of their derivatives has not been reported. Recently, a new class of γ-methylene γ-lactam ecdysone agonists that bind ecdysone receptors from *Bovicola ovis* (Phthiraptera) and *Lucilia cuprina* (Diptera) has been described (Birru et al. 2010) but their activity in cell-based reporter systems has not been tested. In this case, purified ligand-binding regions from the EcR and USP proteins of the insect pests were used in high-throughput binding assays based on displacement of fluorescein-inokosterone A conjugate (a fluorescent ecdysteroid analog; Graham et al. 2007).

Recently, non-steroidal compounds that specifically bind to EcRs of Lepidoptera and Coleoptera cells were found by virtual screening, although it is unknown whether these compounds are ecdysone agonists or antagonists (Harada et al. 2011; Chap. 7 by Nakagawa and Harada 2012, this book). In addition, since the activity of these compounds in terms of IC_{50} in binding assays is over 10 μM, the structural optimization of these compounds is required. The IC_{50} value of the most potent compound, *i.e.* PonA, is at 1–10 nM range.

2.7 Agreement with Structural Studies

A large library of DAH compounds was used to screen for ecdysteroid activity in Bm5/EcRE-GFP cells at different concentrations and the measured EC_{50} values were used to construct a three-dimensional quantitative structure-activity relationship (3-D QSAR) model that describes the ecdysone agonist activities of the DAH analogs (Wheelock et al. 2006). When comparative molecular field analysis (CoMFA) was used to visualize the steric and electrostatic potential fields that were favorable and unfavorable for biological activity, a high agreement was found between the CoMFA models and structural models of the ligand-binding pocket of *B. mori* EcR that were based on the crystal structure of *H. virescens* EcR (Wheelock et al. 2006). These studies indicate that the ligand-binding pocket of EcR represents the target of the EcRE-based reporter assays. A high degree of correlation of activities of compounds ($r^2 = 0.81–0.89$) was also found with other (low-throughput) assays that are based on Sf9 cells and *Chilo suppressalis* integument cultures (Nakagawa et al. 1998, 2000; Ogura et al. 2005a, b).

2.8 Validation: Larval Toxicity Assays

As insecticides, ecdysone agonists act by inducing a premature lethal molt. Because apolar DAHs can penetrate the cuticle efficiently, they are not readily metabolized, and therefore persist for a long time in the body, the insect becomes trapped in the molting process and slowly dies of starvation and desiccation (Retnakaran et al.

1995, 1997, 2003). Although high ecdysteroid activity is a requisite for a compound to have insecticidal activity, other parameters such as penetrability of the cuticle, slow catabolism and high persistence in insect tissues therefore also play a major role. Because the cell-based screening system only evaluates high ecdysteroid activity, it therefore does not guarantee *per se* that the compounds have high insecticidal activity in the field. It was indeed shown that from a series of new DAH compounds that were selected in the *in vitro* screening system, only one proved effective in larval toxicity assays against *Spodoptera littoralis* larvae (Soin et al. 2010a). Many more compounds were toxic against *Bombyx mori* larvae which confirmed the high sensitivity of this domesticated species against this chemical class of insecticides (Nakagawa et al. 1989a, b). Because differences between *Bombyx* and *Spodoptera* were minimal in cell-based assays, the larval toxicity assays therefore demonstrated that the species specificity of some compounds is not based on differences in the activation of the ecdysone receptor but rather on unidentified *in vivo* parameters such as permeability of the cuticle, uptake/excretion by the gut or metabolic detoxification (Soin et al. 2010a).

As noted before, four DAHs have been developed as commercial insecticides, primarily against lepidopteran insects. Because the commercial DAHs all display excellent *in vivo* parameters (high penetrability, low metabolism, low excretion), the specificity for lepidopteran insects is likely caused by its high efficiency to activate the EcR/USP complex (as shown by the cell-based reporter assays; Soin et al. 2009, 2010b). One exception so far is the observation that RH-0345 can be used to control soil-dwelling coleopterans despite the rather low activity of this compound in reporter assays with coleopteran cell lines (Soin et al. 2009). In this case the *in vivo* parameters could compensate for the lower activity against EcR/USP. It is also observed that commercial DAHs have toxic effects against larvae of mosquitoes and flies and that this toxicity is mediated through the induction of a premature molt (Smagghe et al. 2002; Beckage et al. 2004; Boudjelida et al. 2005). However, high concentrations are needed (EC_{50} of ~20 μg/L to ~2 mg/L) which indicates that the use of DAHs to control mosquito and fly larvae in their aquatic environment may not be very practical. In case of dipteran pests, the high concentrations that are needed in the field also reflect the high doses needed to activate the ecdysone hormone receptor-complex in the cell lines (Soin et al. 2010b).

2.9 The Search for Antagonists

The most straightforward use of the cell-based EcRE reporter assays is the screening for compounds with ecdysone agonist activity. Strong agonist activity is also the property on which the insecticidal activity is based, i.e. the capacity to induce a premature, lethal molt. On the other hand, some studies have actively searched for antagonists of the EcR/USP complex, i.e. compounds that prevent the activation of EcR/USP (Soin et al. 2010a). In these assays, cells were pre-incubated with candidate compounds (usually DAHs or compounds with ecdysteroid structure that did not display agonist activity in previous assays), which was then followed by the addition of ecdysone agonist. If the ecdysone agonist failed to activate the EcRE-based reporter, antagonist activity was suggested. In this approach, however, compounds with general toxicity could also inhibit activation of the reporter and appear as false positives. Thus, these types of assays should always be backed up by determination of the general toxicity of the compounds. Using this double approach, however, we were never able to identify with certainty 'true' ecdysone antagonist candidates (own unpublished results). Inhibition of EcRE-based reporter activity was always correlated to some extent by general toxicity of compounds (Soin et al. 2010b).

A better approach to search for new ecdysone antagonists would be based on assays in which the reporter is activated upon addition of the potential antagonist. According to general models of transcriptional regulation by nuclear hormone receptors, binding of antagonists can result in the recruitment of co-repressors to the ligand-binding domain (Jackson et al. 1997; Wagner et al. 1998). A Gal4 two-hybrid assay therefore can be devised in which addition of antagonist induces interaction between EcR and an insect nuclear co-repressor as SMRTER (Tsai et al. 1999).

It is noted that EcR antagonists so far have not been reported. Whether this is caused by the experimental approach to look for antagonists or reflects an intrinsic property of EcR/USP remains to be seen. Availability of ecdysone antagonists would be very valuable for developmental studies since it would block molting and metamorphosis processes. Insecticidal activity would be caused through disruption of EcR/USP function which is involved in a multitude of physiological processes in different tissues of the insect.

3 Other Screening Systems

The success of EcRE-based reporter systems to rapidly identify compounds with high ecdysteroid activity, illustrates the usefulness of cell-based screening systems to screen for insecticidal compounds, if appropriate reporter assays can be developed. From this example, it is also clear that detailed knowledge of the molecular mechanism of action is essential for the development of robust screening systems.

3.1 Screening Systems for JH Analogs

For the other major hormone that controls insect development, namely JH, only recently the molecular mechanism of action has been revealed. The current model proposes that JH interacts with the methoprene-tolerant (Met) transcriptional activator, a basic helix-loop-helix (bHLH)-PAS protein, which presumably acts as a JH receptor (Miura et al. 2005). To maintain larval development, JH induces expression of the *Krüppel homolog 1* (*Kr-h1*) gene which inhibits the induction of the Broad-Complex zinc finger protein that acts as the switch between larval and pupal programs (Minakuchi et al. 2008; Konopova and Jindra 2008).

Based on the elucidation of this molecular mechanism, two types of assays can be designed to search for JH analogs. (1) The promoter of the *Kr-h1* gene, a primary JH-responsive gene, is cloned upstream of a reporter-polyadenylation cassette, generating a JH-responsive reporter system (Minakuchi et al. 2009). (2) The ligand-binding domain of the Met transcription factor is fused to the DNA-binding domain of Gal4, which is used in combination with the UAS reporter for Gal4 (Miura et al. 2005).

More recently, it was found that Met interacts in a JH-dependent manner with the FISC and steroid receptor co-activator, a mosquito and *Tribolium* transcriptional co-activator of the ecdysteroid receptor complex, respectively (Li et al. 2011; Zhang et al. 2011). A third screening system therefore can consist of a two-hybrid assay between fusions of relevant domains of Met and FISC with activation and DNA-binding domains of Gal4.

For the most efficient control of insect larvae, inhibition of JH action is preferred because it would induce pupation and metamorphosis, i.e. life stages that do not cause crop damage. Screening systems therefore should be preferably designed to select JH antagonists. It is noted that all available JH analogs (methoprene, kinoprene, pyriproxyfen, fenoxycarb; Dhadialla et al. 1998) are all agonists and have their most potent effects on eggs (ovicidal activity) and late larvae.

3.2 Other Nuclear Receptors

Until recently, most nuclear receptors in insects, with the exception of EcR, were considered orphan receptors for which no ligands were identified (Escriva et al. 2000). More recently, it was found that the E75 receptors bind heme within their ligand-binding domains and function as gas (CO and NO) sensors (Marvin et al. 2009). DHR96 receptors are implicated in cholesterol homeostasis and are reported to bind cholesterol (Horner et al. 2009).

A special case is USP/RXR, the heterodimer partner of EcR, which was also proposed to act as a JH receptor (Jones and Sharp 1997). However, experiments using insect cell lines did not provide evidence for direct cross talk between 20E and JH on the function of the EcR/USP complex (Soin et al. 2008). The subject of ligand binding by USP/RXR remains a debated issue (Iwema et al. 2007; Riddiford 2007). Recently, other farnesoids have been proposed to act as ligands for USP (Jones et al. 2006).

Since nuclear receptors usually are constitutive or ligand-activated transcriptional activators, the development of reporter systems is straightforward, as illustrated before for EcR/USP. Basal reporter cassettes are engineered with identified binding sites for the nuclear receptor (for native receptors) or for the Gal4 activator (for fusions with Gal4 DNA-binding domain). An example is the reporter system for the *Bombyx* HR3 and E75 receptors (Swevers et al. 2002).

3.3 G Protein-Coupled Receptors (GPCRs)

GPCRs represent the most important class of pharmacological targets (Xiao et al. 2008; Chap. 4 by Bai and Palli 2012, this book) but, paradoxically, this class of receptors has received relatively little attention as targets for insecticide action (Van Hiel et al. 2010). In many cases, activation of GPCRs results in the production of intracellular messengers such as cAMP and calcium for which fluorescent and luminescent detection methods are readily available for screening in high-throughput format (Milligan 2003; Williams 2004). Even if GPCRs do not normally produce these secondary messengers, co-expression of the promiscuous Gα15/16 proteins allows coupling of almost any GPCR to phospholipase Cβ to generate both diacylglycerol and inositol (1,4,5)-trisphosphate with subsequent activation of protein kinase C and elevation of intracellular calcium (Kostenis 2001).

Many GPCRs are involved in important physiological and developmental processes in insects and thus represent excellent targets (Van Hiel et al. 2010). Recently, a large scale RNAi-based screen was carried out in *Tribolium* to identify GPCRs involved in growth, molting and metamorphosis (Bai et al. 2011). Interestingly, a GPCR was identified that is responsive to ecdysteroids and therefore is a prime candidate to mediate the non-genomic actions of 20E (Elmogy et al. 2004; Srivastava et al. 2005).

4 Insect-Specific Metabolic Pathways for Chitin and Cuticle Synthesis

The use of insect cell lines to study mode of action and to screen for inhibitors of chitin synthesis was first suggested already 20 years ago (Spindler-Barth et al. 1989). In addition to organ cultures of imaginal discs and integument epidermis prepared from different insect species, including representatives of Lepidoptera, Coleoptera and Orthoptera (reviewed in Oberlander and Smagghe 2001), two cell lines synthesize or degrade at least parts of the cuticle. These are the epithelial cell line of the midge *Chironomus tentans* (Spindler-Barth et al. 1989) and the IAL-PID2 cell line from *P. interpunctella* (Oberlander and Silhacek 1998). Interference with chitin synthesis and chitin degradation can be measured conveniently using insect cell lines, and these assays can be supplemented with tests using homogenates of the same tissue (Palli and Retnakaran 1999). Chitin degradation by chitinases and hexosaminidases can then be measured with high sensitivity and specificity using *N*-acetylglucosamine (GlcNAc) and its oligomers coupled to a fluorogenic dye (McCearth and Gooday 1992). For the future, with the tools of biotechnology, the availability of both cDNA of the chitin producing proteins and antibody probes, there is now a real possibility of developing cell-based HTS assays for the discovery of new chitin biosynthesis inhibitors.

The feasibility of cell cultures allows investigators to determine QSARs, aiding in design and synthesis of newer and more active chitin synthesis inhibitory compounds. For example, the introduction of electron-withdrawing and hydrophobic substituents at the para-position of the phenyl (aniline) moiety of BPUs enhanced activity, whereas larger groups reduced activity. *In vitro* activities and *in vivo* larvicidal toxicities were correlated after separate consideration of the hydrophobic factors participating in absorption and transport in the insect body (Nakagawa et al. 1989a, b).

The ability to culture chitin-producing cells also helps improve the understanding of the mechanism of newer compounds. For instance in the chemical class of 2,4-diphenyl-1,3-oxazolines (Suzuki et al. 2006), the mechanism of etoxazole action could be confirmed as chitin inhibitory using epidermis cell cultures derived from *Spodoptera frugiperda* (Nauen and Smagghe 2006).

In addition to insect cuticle, the peritrophic matrix in insect midgut is closely associated with specific glycoproteins and the covalent binding of chitin (*N*-acetylglucosamine) to proteins. In this context, insecticidal lectins can be of great interest with respect to pest insect management (reviewed in Michiels et al. 2008; Vandenborre et al. 2011). Lectins are proteins of non-immune origin that interact with cells through sugar-specific binding sites (Van Damme et al. 2007). With lepidopteran midgut cell cultures (CF-203), experiments using a series of plant lectins with specificity for mannose, galactose and GlcNAc oligomers demonstrated that the lectin effects are not correlated with the carbohydrate-binding activity (Smagghe et al. 2005a). This concurs with previous experiments which revealed that lectins can elicit a variety of biological activities such as mitosis stimulation, growth inhibition and apoptosis. However, it should be emphasized that fine specificity of different lectins towards oligosaccharides and glycans can be very different even though they interact with the same monosaccharide. To complete these studies, tagged (FITC) lectin was used to visualize lectin binding to membrane receptor(s) and cellular internalization and this was investigated for different lectin family representatives (Vandenborre et al. 2008; Hamshou et al. 2010; Shahidi-Noghabi et al. 2011). Here the use of cultured midgut cells facilitates investigation of the glycosylation process in insects and the interaction of lectin with receptor proteins. Hence primary midgut cell cultures are also of use to demonstrate the importance of the microvillar area, also named as the "brush border membrane zone" (Hakim et al. 2010). We believe that new knowledge of these insect-specific systems opens the door to HTS and other approaches to discovery of novel compounds that inhibit these processes.

5 Insect Cell Lines for the Evaluation and Screening of *Bt* Insecticidal Proteins

It is a requisite that rapid evaluation of the effects of *Bt* toxins has a need for "established" midgut epithelium cell cultures. In the past decade, significant progress has been made in the preparation of primary cultures of midgut insect stem cells (reviewed by Hakim et al. 2010). Midgut epithelial cell cultures from lepidopterans and coleopterans have been established and maintained *in vitro* for periods up to 3–6 months while preserving their differentiated characteristics (Sadrud-Din et al. 1996; Smagghe et al. 2005b). These primary midgut cell cultures have been applied for the study of *Bt* endotoxin binding to the microvilli of intact epithelial cells from different lepidopteran species (Wang and McCarthy 1997; Loeb et al. 2001).

In continuation, different research groups in the world have used microscopic observations and electrophysiological studies involving the patch clamp technique and fluorescent probes, to investigate the action of *Bt* toxins on insect cells derived from different species and tissues. However, it should be remarked that there exists an inherent risk of over-interpreting results from experiments with insect cells (Gringorten 2001). Continuous cell lines assume morphological and physiological characteristics that can differ from the source of primary cultures. Their response to *Bt* toxins often does not correlate with responses of the insects from which the cells were derived. Established midgut cell lines bear little resemblance to midgut cells *in vivo* and their susceptibility to toxins correlates poorly with the susceptibility of the host insect. Generally, they are sensitive to fewer toxins than the host insect. On the other hand, primary midgut cell cultures appear to be susceptible to a broader spectrum of *Bt* toxins than the host insect, a feature that creates a bias towards overrating insecticide activity based on *in vitro* assays.

Although insect cell cultures may be poor indicators of *Bt* insecticide activity, they have been proven to be useful for the characterization of the toxin activity spectra and investigation of the membrane permeabilizing effects, particularly in determining pore size (Knowles and Ellar 1987; Potvin et al. 1998). As with the columnar cells *in vivo*, cultured insect cells respond to *Bt* toxin injury by swelling and lysis (Loeb et al. 2001). Figure 6.4 demonstrates the primary midgut cell cultures to screen for *Bt* toxicity. Upon insect midgut dissection, mature differentiated cell cultures inclusive columnar epithelial cells were obtained after collagenase treatment and cultured in modified Grace's medium. In the controls, the typical microvillar area, also named as the "brush border membrane zone", remains intact, while exposure to *Bt* toxin leads to rapid cell death with the typical loss of the microvillar zone, swelling, blebbing and cell lysis (own unpublished results). Here, the midgut cells of important pest insects can be investigated and compared to beneficial insects as pollinators and natural enemies to screen for an optimal *Bt* toxicity spectrum. Hence, in a pivotal study with CF-1 cells, a midgut cell line from neonates of the spruce budworm *C. fumiferana*, the cytolytic effect of toxin was studied in the presence of neutral solutes with different hydrodynamic radii. This work helped generate a model for membrane pore formation and colloid osmotic lysis to describe the toxin mechanism of action (Knowles and Ellar 1987).

More recently, in the study of the specific role of cadherin receptors in cytotoxicity of *Bt* toxins and their interactions with cell membrane, a cell-based system was established utilizing Hi5 insect cells that are stably expressing BT-R-1, which is the cadherin receptor for Cry1Ab toxin (Zhang et al. 2006). In this assay, the toxin oligomers in the cell membrane do not produce lytic pores and do not kill insect cells. Rather, the cell death correlates with the binding of the Cry1Ab toxin monomer to

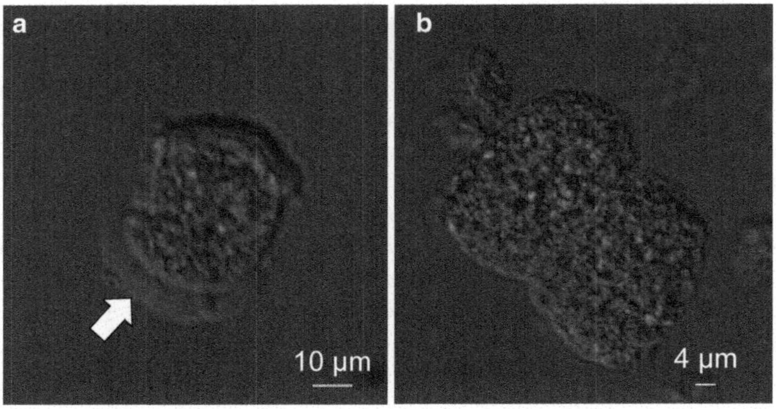

Fig. 6.4 Primary midgut cell cultures to screen for *Bacillus thuringiensis* toxicity. Upon insect midgut dissection, mature differentiated cell cultures inclusive columnar epithelial cells, were obtained after collagenase treatment and cultured in modified Grace's medium. (**a**) In the controls, the typical microvillar area, also named as the "brush border membrane zone", is indicated with a *white arrow*. (**b**) Exposure to *Bt* toxin during 4 h leads to cell death with the typical loss of the microvillar zone, swelling, blebbing and cell lysis (Dr. Silvia Caccia and Dr. Guy Smagghe, own unpublished results)

BT-R-1, which apparently activates a previously undescribed Mg^{2+}-dependent cellular signaling pathway. This unique cell-based screening system is of great use as it provides insights into how insects evolve resistance to *Bt* toxins, and allows to screen for newer safer insecticides.

Bt in bacterial and toxin formulations, and as expressed in genetically modified crop and fiber plants, is used globally to protect plants from insect damage. Understanding of the modes of *Bt* action and the insect mechanisms of *Bt* resistance will be increasingly important in future. We believe here that primary and established cell cultures will draw increasing attention in efforts to generate new knowledge and new hypotheses about *Bt*.

6 Use of Cell Lines to Detect Resistance Mechanisms to Insecticides

In this last part we want to demonstrate the use of insect cell lines to screen for and to clarify the mechanisms of resistance against insecticide/IGR compounds. Specifically here we want to focus on the DAH-based ecdysteroid agonists, like methoxyfenozide (RH-2485), that are used as important novel biorational insecticides against susceptible insects as Lepidoptera (see above). In principle, resistance mechanisms can be situated at two targets: (1) at the level of the catabolism of the ecdysteroid agonists, which will likely play a major role in the acquisition of resistance in the insects, and (2) at the level of signaling of the ecdysteroid receptor complex, consisting of EcR and the heterodimer partner receptor USP(RXR), for which however little data is available (Smagghe et al. 1998; Spindler-Barth and Spindler 1998; Dhadialla et al. 2005; Nakagawa 2005; Mosallanejad and Smagghe 2009; Nakagawa and Henrich 2009). To unravel potential mechanisms at the level of the signaling pathway, insect cell lines represent a system of choice based on the availability of a sufficient amount of homogeneous material and the ease of manipulation (Smagghe 2007; Smagghe et al. 2009). On the other hand, it can be argued that cell lines are less ideal substitutes in the case of tissues involved in defense against xenobiotics, such as fat body and midgut, since the expression of enzymes involved in detoxification is not always comparable with that of cells in culture. Thus, "resistance" to methoxyfenozide in cell lines (more likely based on altered functioning of the ecdysone regulatory pathway) may only partly account for "resistance" to methoxyfenozide in insect pest populations (more likely based on compound detoxification). However, both mechanisms of resistance may occur concomitantly in resistant insects, and in this context evaluation of the properties of resistant cell lines could provide valuable "leads" in the search of mechanisms contributing to resistance in insect populations.

In this context, ecdysteroid-responsive Se4 cell lines, which are derived from the beet armyworm *Spodoptera exigua*, an important polyphagous noctuid in agriculture, were selected for resistance by continuous exposure to 20E and methoxyfenozide in order to obtain 20E- and methoxyfenozide-resistant mutant cells (Mosallanejad et al. 2008). As shown in Fig. 6.5, the use of insect cell cultures allowed a rapid selection for very high levels of resistance (of 1,000,000-fold) where the lepidopteran cells lost their sensitivity from 0.1 nM at the start up to 100 μM methoxyfenozide at the end of the experiment over a relatively short period of about 50 passages. These authors showed that resistance in these cells was not due to a differential metabolism and uptake of methoxyfenozide and 20E compared to the sensitive cells. Because cross-resistance existed between the 20E- and methoxyfenozide-selected cells, it was hypothesized that the resistance mechanism may be at the level of the ecdysone-signaling pathway, which is the common effector pathway for both compounds (Dhadialla et al. 1998; Nakagawa 2005). Previously, this conclusion was also obtained for the prototype compound RH-5849 with *Drosophila* Kc cells (Wing 1988). Accordingly, Swevers et al. (2008) investigated the ecdysteroid signaling pathway by measuring the activity of selected transcription factors known to be involved in this cascade. The early gene *HR3* was constitutively expressed in the resistant cell lines grown in the presence of 20E and methoxyfenozide. In addition, the gene *FTZ-F1* was constitutively expressed in both resistant and sensitive Se4 cells, suggesting that its expression was not regulated by the addition of methoxyfenozide and 20E. However, the results demonstrated the existence of a normally functioning EcR/USP complex in the resistant Se4 cells. Very similar data were also obtained for resistant cell lines derived from other Lepidoptera such as Bm5 (derived from *Bombyx mori*) and CF-203 cells (derived from *Choristoneura fumiferana*) (Mosallanejad 2009). It is proposed that the resistance mechanism exists at the junction between the conserved ecdysone regulatory cascade and the differentiation program in the cell line. RNAi studies in

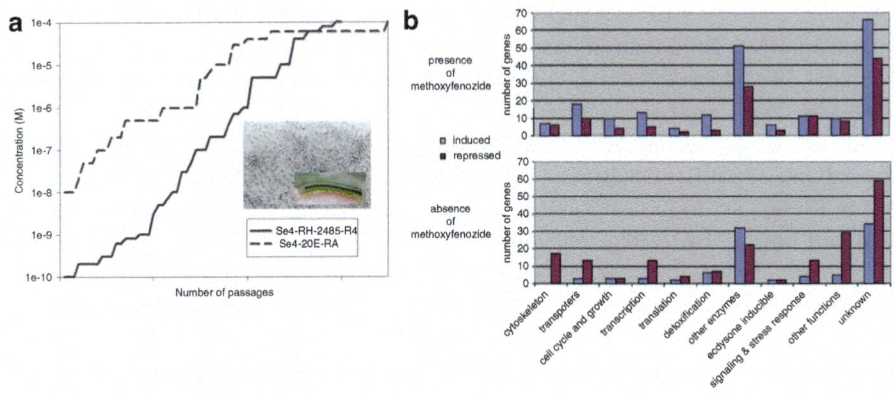

Fig. 6.5 Use of insect cells to screen and study mechanisms of insecticide resistance. (**a**) Development of resistance towards methoxyfenozide and 20-hydroxyecdysone (20E) in lepidopteran Se4 cells (*Spodoptera exigua*) for successive selection with increasing concentrations over different passages for the Se4-RH2485-R4 and the Se4-20E-R4 subclone, respectively. Indicated are the passage numbers at which growth was observed for different concentrations of compounds during the selection period. (**b**) Classification of genes differentially expressed between the methoxyfenozide-resistant dipteran S2 cells (*Drosophila melanogaster*) and the sensitive cells. *Graphs* display numbers of differentially expressed genes for different classes of genes in conditions of presence and absence of methoxyfenozide for the resistant cells. Genes that are induced and repressed are separated in different bars (Redrafted from Mosallanejad et al. 2008, 2010)

IAL-PID2 cells have confirmed the involvement of the ecdysone regulatory cascade in cell cycle arrest and morphological transformation (Siaussat et al. 2007, 2008).

In continuation of this research, and because of the fact that many of the above mentioned agriculturally important insects are not genome-sequenced, Mosallanejad et al. (2010) selected a dipteran cell line, Schneider 2 (S2) cells from the fruit fly *D. melanogaster*, for resistance towards methoxyfenozide. Although methoxyfenozide is an insecticide displaying high specificity against lepidopteran insects (Dhadialla et al. 1998), it also has considerable activity in S2 cells and is one of the most active compounds from a library of DAH-type compounds in EcR reporter assays (EC_{50} = 16.6 μM; Soin et al. 2010b). According to Nakagawa et al. (2002a, b), the IC_{50} of methoxyfenozide is less than 1 μM in the binding assay using *Drosophila* Kc cells, while the IC_{50} of PonA is about 1 nM. The advantage of the S2 insect cell line is that it represents a useful model for researchers because of the availability of the sequence of the genome of *Drosophila* and corresponding commercially available microarray slides. So the latter authors investigated the functionality of the EcR/USP complex, and in addition performed a microarray study to determine transcript profiles of genes involved in methoxyfenozide resistance. The information provided insights in altered functioning of the ecdysone-signaling pathway as contributing factor to the resistance mechanism to methoxyfenozide, resulting in a total loss of susceptibility, and in the functional link between ecdysone signaling and cell proliferation. Indeed the most striking observation was that most differentially expressed transcripts are increased in expression in the presence of methoxyfenozide, while they are decreased in expression in its absence (Fig. 6.5). In both conditions the EcR complex is not acting as a transcriptional activator since continuous presence of methoxyfenozide inactivates EcR, while in the other condition the activating ligand is absent. Since in the absence of ligand, EcR can act as a repressor (Tsai et al. 1999), it was therefore proposed that the preferential decrease in expression after removal of the selection pressure restores the repressor function of EcR in the resistant cells and that the observed deficiency of the EcR complex in continuous presence of ecdysone agonist involves both activation and repressor functions. As a consequence, the microarray study does not only provide a significant list of "leads" in the search for possible mechanisms of resistance against ecdysone agonists, but also uncovers more complex levels of EcR signaling than were previously conceived.

Next to the use of insect cell cultures, tobacco plant *Nicotiana tabacum* cell cultures have been employed to study the capacity for instance of cytochrome P450 systems to metabolize different classical and modern insecticides. Good examples are the results of Joussen et al. (2008, 2010) reporting the inhibition of the metabolism of imidacloprid in the heterologous Cyp6g1-transgenic culture by 82% in the presence of piperonyl butoxide. CYP6G1 is responsible for the resistance of *Drosophila* against imidacloprid, DDT and methoxychlor. Furthermore, it was of interest that treating *Drosophila* fruit flies with piperonyl butoxide could also weaken the observed resistance phenomena.

7 Conclusions

Because of their easy manipulation and amenability to high-throughput format, insect cell lines can represent a valuable initial screening step for compounds with biological activity. This is most clearly demonstrated with the development of cell-based screening systems for biorational IGRs as the ecdysone agonists or also named MAC insecticides. It is expected that the scope of the use of insect cell lines (as well as primary cells) will be extended, not only to other hormonal systems, but also to discover new enzyme inhibitors and insect-specific targets and pathways, and to disclose mechanisms of insecticide resistance. With the use of *in vitro* systems, highly active compounds can be identified rapidly and without delay forwarded to test their toxicity on pest insects. The discovery of new insecticides will thus be accelerated, while additional safety tests will ensure that the new compounds can be applied with minimal impact on the environment and human health.

Acknowledgements The authors acknowledge support for their research by the Fund for Scientific Research-Flanders (FWO-Vlaanderen), the Flemish agency for Innovation by Science and Technology (IWT-Vlaanderen) and the Special Research Funds of Ghent University in Belgium, and the General Secretariat for Research and Technology, Hellenic Republic Ministry of National Education and Religious Affairs in Greece.

References

Bai H, Palli SR (2012) G protein-coupled receptors as target sites for insecticide discovery. In: Ishaaya I, Palli SR, Horowitz AR (eds) Advanced technologies for managing insect pests. Springer, Dordrecht, pp 57–82

Bai H, Zhu F, Shah K, Palli SR (2011) Large-scale RNAi screen of G protein-coupled receptors involved in larval growth, molting and metamorphosis in the red flour beetle. BMC Genomics 12:338

Baret J-F, Beck Y, Billas-Massobrio I, Moras D, Griffiths AD (2010) Quantitative cell-based reporter gene assays using droplet-based microfluidics. Chem Biol 17:528–536

Beatty J, Fauth T, Callender JL, Spindler-Barth M, Henrich VC (2006) Analysis of transcriptional activity mediated by *Drosophila melanogaster* ecdysone receptor isoforms in a heterologous cell culture system. Insect Mol Biol 15:785–795

Beatty J, Smagghe G, Ogura T, Nakagawa Y, Spindler-Barth M, Henrich VC (2009) Properties of ecdysteroid receptors from diverse insect species in a heterologous cell culture system – a basis for screening novel insecticidal candidates. FEBS J 276:3087–3098

Beckage NE, Marion KM, Walton WE, Wirth MC, Tan FF (2004) Comparative larvicidal toxicities of three ecdysone agonists on the mosquitoes *Aedes aegypti, Culex quinquefasciatus*, and *Anopheles gambiae*. Arch Insect Biochem Physiol 57:111–122

Beckmann M, Haack K.-J. (2003) Insektizide für die Landwirtschaft: Chemische Schädlingsbekämpfung. Chemie in unserer Zeit 37:88–97

Birru W, Fernley RT, Graham LD, Grusovin J, Hill RJ, Hofmann A, Howell L, James PJ, Jarvis KE, Johnson WM, Jones DA, Leitner C, Liepa AJ, Lovrecz GO, Lu L, Nearn RH, O'Driscoll BJ, Phan T, Pollard M, Turner KA, Winkler DA (2010) Synthesis, binding and bioactivity of γ-methylene γ-lactam ecdysone receptor ligands: advantages of QSAR models for flexible receptors. Bioorg Med Chem 18:5647–5660

Boudjelida H, Bouaziz A, Soin T, Smagghe G, Soltani N (2005) Effects of ecdysone agonist halofenozide against *Culex pipiens*. Pestic Biochem Physiol 83:115–123

Cherbas L, Cherbas P (1997) "Parahomologous" gene targeting in *Drosophila* cells: an efficient homology-dependent pathway of illegitimate recombination near a target site. Genetics 145:349–358

Cherbas P, Cherbas L, Lee SS, Nakanishi K (1988) 26-[125I]iodoponasterone A is a potent ecdysone and a sensitive radioligand for ecdysone receptors. Proc Natl Acad Sci USA 85:2096–2100

Christopherson KS, Mark MR, Bajaj V, Godowski PJ (1992) Ecdysteroid-dependent regulation of genes in mammalian cells by a *Drosophila* ecdysone receptor and chimeric transactivators. Proc Natl Acad Sci USA 89:6314–6318

Dhadialla TS, Carlson GR, Le DP (1998) New insecticides with ecdysteroidal and juvenile hormone activity. Annu Rev Entomol 43:545–569

Dhadialla TS, Retnakaran A, Smagghe G (2005) Insect growth and development disrupting insecticides. In: Gilbert LI, Iatrou K, Gill S (eds) Comprehensive insect molecular science, vol 6. Elsevier/Pergamon, New York, pp 55–116

Dinan L (1995) A strategy for the identification of ecdysteroid receptor agonists and antagonists from plants. Eur J Entomol 92:271–283

Dinan L, Bourne P, Whiting P, Dhadialla TS, Hutchinson TH (2001) Screening of environmental contaminants for ecdysteroid agonist and antagonist activity using the *Drosophila melanogaster* B(II) cell *in vitro* assay. Environ Toxicol Chem 20:2038–2046

Elmogy M, Iwami M, Sakurai S (2004) Presence of membrane ecdysone receptor in the anterior silk gland of the silkworm *Bombyx mori*. Eur J Biochem 271:3171–3179

Escriva H, Delaunay F, Laudet V (2000) Ligand binding and nuclear receptor evolution. Bioessays 22:717–727

Fahrbach SE, Smagghe G, Velarde RA (2012) Insect nuclear receptors. Annu Rev Entomol 57:83–106

Gaw S-Y (1958) Culturing all types of silkworm tissues using the monolayer culture. Chin Sci Bull 7:219–220

Goodman CL, El Sayed GN, McIntosh AH, Grasela JJ, Stiles B (2001) Establishment and characterization of insect cell lines from 10 lepidopteran species. In Vitro Cell Dev Biol Anim 37:67–373

Grace TDC (1962) Establishment of four strains of cells from insect tissues grown *in vitro*. Nature (London) 195:788–789

Graham LD, Johnson WM, Pawlak-Skrzecz A, Eaton RE, Bliese M, Howell L, Hannan GN, Hill RJ (2007) Ligand binding by recombinant domains from insect ecdysone receptors. Insect Biochem Mol Biol 37:611–626

Gringorten JL (2001) Ion balance in the lepidopteran midgut and insecticidal action of *Bacillus thuringiensis*. In: Ishaaya I (ed) Biochemical sites of insecticide action and resistance. Springer, Dordrecht, pp 167–207

Hakim RS, Baldwin K, Smagghe G (2010) Regulation of midgut growth, development, and metamorphosis. Annu Rev Entomol 55:593–608

Hamshou M, Smagghe G, Shahidi-Noghabi S, De Geyter E, Lannoo N, Van Damme EJM (2010) Insecticidal properties of *Sclerotinia sclerotiorum* agglutinin and its interaction with insect tissues and cells. Insect Biochem Mol Biol 40:883–890

Hannan GN, Hill RJ, Dedos SG, Swevers L, Iatrou K, Tan A, Parthasarathy R, Bai H, Zhang Z, Palli SR (2009) Applications of RNA interference in ecdysone research. In: Smagghe G (ed) Ecdysone, structures and functions. Springer, Dordrecht, pp 205–227

Harada T, Nakagawa Y, Ogura T, Yamada Y, Ohe T, Miyagawa H (2011) Virtual screening for ligands of the insect molting hormone receptor. J Chem Inf Model 51:296–305

Harmatha J, Dinan L, Lafont R (2002) Biological activities of a specific ecdysteroid dimmer and of selected monomeric structural analogues in the B(II) assay. Insect Biochem Mol Biol 32:181–185

Henrich VC (2005) The ecdysteroid receptor. In: Gilbert LI, Iatrou K, Gill S (eds) Comprehensive insect molecular science, vol 3. Elsevier/Pergamon, New York, pp 243–282

Henrich VC, Burns E, Yelverton DP, Christensen E, Weinberger C (2003) Juvenile hormone potentiates ecdysone receptor-dependent transcription in a mammalian cell culture system. Insect Biochem Mol Biol 33:1239–1247

Horner MA, Pardee K, Liu S, King-Jones K, Lajoie G, Edwards A, Krause HM, Thummel CS (2009) The *Drosophila* DHR96 nuclear receptor binds cholesterol and regulates cholesterol homeostasis. Genes Dev 23:2711–2716

Iwema T, Billas IML, Beck Y, Bonneton F, Nierengarten H, Chaumot A, Richards G, Laudet V, Moras D (2007) Structural and functional characterization of a novel type of ligand-independent RXR-USP receptor. EMBO J 26:3770–3782

Jackson TA, Richer JK, Bain DL, Takimoto GS, Tung L, Horwitz KB (1997) The partial agonist activity of antagonist-occupied steroid receptors is controlled by a novel hinge domain-binding coactivator l7/spa and the corepressors N-CoR or SMRT. Mol Endocrinol 11:693–705

Jones G, Sharp PA (1997) Ultraspiracle: an invertebrate nuclear receptor for juvenile hormones. Proc Natl Acad Sci USA 94:13499–13503

Jones G, Jones D, Teal P, Sapa A, Wozniak M (2006) The retinoid-X receptor ortholog, ultraspiracle, binds with nanomolar affinity to an endogenous morphogenetic ligand. FEBS J 273:4983–4996

Joussen N, Heckel DG, Haas M, Schuphan I, Schmidt B (2008) Metabolism of imidacloprid and DDT by P450 GYP6G1 expressed in cell cultures of *Nicotiana tabacum* suggests detoxification of these insecticides in Cyp6g1-overexpressing strains of *Drosophila melanogaster*, leading to resistance. Pest Manag Sci 64:65–73

Joussen N, Schuphan I, Schmidt B (2010) Metabolism of methoxychlor by the P450-monooxygenase CYP6G1 involved in insecticide resistance of *Drosophila melanogaster* after expression in cell cultures of *Nicotiana tabacum*. Chem Biodivers 7:722–735

King-Jones K, Thummel CS (2005) Nuclear receptors – a perspective from *Drosophila*. Nat Rev Genet 6:311–323

Knowles BH, Ellar DJ (1987) Colloid-osmotic lysis is a general feature of the mechanism of action of *Bacillus thuringiensis* δ-endotoxin with different insect specificity. Biochim Biophys Acta 924:509–518

Koelle MR, Talbot WS, Segraves WA, Bender MT, Cherbas P, Hogness DS (1991) The *Drosophila EcR* gene encodes an ecdysone receptor, a new member of the steroid receptor superfamily. Cell 67:59–77

Konopova B, Jindra M (2008) Broad-Complex acts downstream of Met in juvenile hormone signaling to coordinate primitive holometabolan metamorphosis. Development 135:559–568
Kostenis E (2001) Is Gα16 the optimal tool for fishing ligands of orphan G-protein coupled receptors? Trends Pharmacol Sci 22:560–564
Kumar MB, Fujimoto T, Potter DW, Deng Q, Palli SR (2002) A single point mutation in ecdysone receptor leads to increased ligand specificity: implications for gene switch applications. Proc Natl Acad Sci USA 99:14710–14715
Kumar MB, Potter DW, Hormann RE, Edwards A, Tice CM, Smith HC, Dipietro MA, Polley M, Lawless M, Wolohan PRN, Kethidi DR, Palli SR (2004) Highly flexible ligand binding pocket of ecdysone receptor. A single amino acid change leads to discrimination between two groups of nonsteroidal ecdysone agonists. J Biol Chem 279:27211–27218
Li M, Mead EA, Zhu J (2011) Heterodimer of two bHLH-PAS proteins mediates juvenile hormone-induced gene expression. Proc Natl Acad Sci USA 108:638–643
Loeb MJ, Martin PAW, Hakim RS, Goto S, Takeda M (2001) Regeneration of cultured midgut cells after exposure to sublethal doses of toxin from two strains of *Bacillus thuringiensis*. J Insect Physiol 47:599–606
Long SH, McIntosh AH, Grasela JJ, Goodman CL (2002) The establishment of a Colorado potato beetle (Coleoptera: Chrysomelidae) pupal cell line. Appl Entomol Zool 37:447–450
Lynn DE (2007) Available lepidopteran insect cell lines. In: Murhammer DW (ed) Methods in molecular biology series. Baculovirus and insect cell expression protocols. Springer, New York, pp 117–144
Maeda S, Kamita SG, Kondo A (1993) Host range expansion of *Autographa californica* nuclear polyhedrosis virus (NPV) following recombination of a 0.6 kilobase-pair DNA fragment originating from Bombyx mori NPV. J Virol 67:6234–6238
Marvin KA, Reinking JL, Lee AJ, Pardee KM, Krause HM, Burstyn JN (2009) Nuclear receptors *Homo sapiens* Rev-erbβ and *Drosophila melanogaster* E75 are thiolate-ligated heme proteins, which undergo redox-mediated ligand switching and bind CO and NO. Biochemistry 48:7056–7071
McCearth KJ, Gooday GW (1992) A rapid and sensitive microassay for determination of chitinolytic activity. J Microbiol Methods 14:229–237
Michiels K, Van Damme EJM, Smagghe G (2008) Plant-insect interactions: what can we learn from plant lectins? Arch Insect Biochem Physiol 73:193–212
Mikitani K (1996) A new nonsteroidal chemical class of ligand for the ecdysteroid receptor 4,5-di-*tert*-butyl-4-hydroxy-*N*-isobutyl-benzamide shows apparent insect molting hormone activities at molecular and cellular levels. Biochem Biophys Res Commun 227:427–432
Milligan G (2003) High-content assays for ligand regulation of G-protein-coupled receptors. Drug Discov Today 8:579–585
Minakuchi C, Nakagawa Y, Kamimura M, Miyagawa H (2003) Binding affinity of nonsteroidal ecdysone agonists against the ecdysone receptor complex determines the strength of their molting hormone activity. Eur J Biochem 270:4095–4104
Minakuchi C, Nakagawa Y, Soya Y, Miyagawa H (2004) Preparation of functional ecdysteroid receptor proteins (EcR and USP) using a wheat germ cell-free protein synthesis system. J Pestic Sci 29:189–194
Minakuchi C, Ogura T, Miyagawa H, Nakagawa Y (2007) Effects of the structures of ecdysone receptor (EcR) and ultraspiracle (USP) on the ligand-binding activity of the EcR/USP heterodimer. J Pestic Sci 32:379–384
Minakuchi C, Zhou X, Riddiford LM (2008) Krüppel homolog 1 (Kr-h1) mediates juvenile hormone action during metamorphosis of *Drosophila melanogaster*. Mech Dev 125:91–105
Minakuchi C, Namiki T, Shinoda T (2009) Krüppel homolog 1, an early juvenile hormone-response gene downstream of Methoprene-tolerant, mediates its anti-metamorphic action in the red flour beetle *Tribolium castaneum*. Dev Biol 325:341–350
Miura K, Oda M, Makita S, Chinzei Y (2005) Characterization of the *Drosophila* Methoprene-tolerant gene product. Juvenile hormone binding and ligand-dependent gene regulation. FEBS J 272:1169–1178

Mosallanejad H (2009) Resistance mechanisms for methoxyfenozide: *in vitro* and *in vivo* approaches. Ph.D. thesis, Ghent University, Ghent, Belgium

Mosallanejad H, Smagghe G (2009) Biochemical mechanisms of methoxyfenozide resistance in the cotton leafworm Spodoptera littoralis. Pest Manag Sci 65:736–737

Mosallanejad H, Soin T, Smagghe G (2008) Selection for resistance to methoxyfenozide and 20-hydroxyecdysone in cells of the beet armyworm, Spodoptera exigua. Arch Insect Biochem Physiol 67:36–49

Mosallanejad H, Badisco L, Swevers L, Soin T, Knapen D, Vanden Broeck J, Smagghe G (2010) Ecdysone signaling and transcript signature in *Drosophila* cells resistant against methoxyfenozide. J Insect Physiol 56:1973–1985

Nakagawa Y (2005) Nonsteroidal ecdysone agonists. Vitam Horm 73:131–173

Nakagawa Y, Harada T (2012) Advanced screening to identify novel pesticides. In: Ishaaya I, Palli SR, Horowitz AR (eds) Advanced technologies for managing insect pests. Springer, Dordrecht, pp 135–163

Nakagawa YY, Henrich VC (2009) Arthropod nuclear receptors and their role in molting. FEBS J 276:6128–6157

Nakagawa Y, Akagi T, Iwamura H, Fujita T (1989a) Quantitative structure–activity studies of benzoylphenylurea larvicides. VI. Comparison of substituent effects among activities against different insect species. Pestic Biochem Physiol 33:144–157

Nakagawa Y, Matsutani M, Kurihara N, Nishimura K, Fujita T (1989b) Quantitative structure-activity studies of benzoylphenylurea larvicides. VIII. Inhibition of N-acetylglucosamine incorporation into the cultured integument of Chilo suppressalis Walker. Pestic Biochem Physiol 43:141–151

Nakagawa Y, Hattori K, Shimizu B-I, Akamatsu M, Miyagawa H, Ueno T (1998) Quantitative structure-activity studies of insect growth regulators XIV. Three-dimensional quantitative structure-activity relationship of ecdysone agonists including dibenzoylhydrazine analogs. Pestic Sci 53:267–277

Nakagawa Y, Hattori K, Minakuchi C, Kugimiya S, Ueno T (2000) Relationships between structure and molting hormonal activity of tebufenozide, methoxyfenozide, and their analogs in cultured integument system of *Chilo suppressalis*. Steroids 65:117–123

Nakagawa Y, Minakuchi C, Takahashi K, Ueno T (2002a) Inhibition of [(3)H]ponasterone A binding by ecdysone agonists in the intact Kc cell line. Insect Biochem Mol Biol 32:175–180

Nakagawa Y, Minakuchi C, Ueno T (2002b) Inhibition of [^{3}H]ponasterone A binding by ecdysone agonists in the intact Sf-9 cell line. Steroids 65:537–542

Nakagawa Y, Takahashi K, Kishikawa H, Ogura T, Minakuchi C, Miyagawa H (2005) Classical and three-dimensional QSAR for the inhibition of [^{3}H]ponasterone A binding by diacylhydrazine-type ecdysone agonists to insect Sf-9 cells. Bioorg Med Chem 13:1333–1340

Nauen R, Smagghe G (2006) Mode of action of etoxazole. Pest Manag Sci 62:375–382

No P, Yao T-P, Evans RM (1996) Ecdysone-inducible gene expression in mammalian cells and transgenic mice. Proc Natl Acad Sci USA 93:3346–3351

Oberlander H, Silhacek DL (1998) New perspectives on the mode of action of benzoylphenylurea insecticides. In: Ishaaya I, Degheele D (eds) Insecticides with novel modes of action. Springer, Berlin, pp 92–105

Oberlander H, Smagghe G (2001) Imaginal discs and tissue cultures as targets for insecticide action. In: Ishaaya I (ed) Biochemical sites of insecticide action and resistance. Springer, Berlin, pp 133–150

Ogura T, Minakuchi C, Nakagawa Y, Smagghe G, Miyagawa H (2005a) Molecular cloning, expression analysis and functional confirmation of ecdysone receptor and ultraspiracle from the Colorado potato beetle *Leptinotarsa decemlineata*. FEBS J 272:4114–4128

Ogura T, Nakagawa Y, Minakuchi C, Miyagawa H (2005b) QSAR for binding affinity of substituted dibenzoylhydrazines to intact Sf9 cells. J Pestic Sci 30:1–6

Palli SR, Retnakaran A (1999) Molecular and biochemical aspects of chitin synthesis inhibition. In: Jollés P, Muzzarelli RAA (eds) Chitin and chitinases. Birkhäuser Publishing, Basel, pp 85–98

Palli SR, Kapitskaya MZ, Kumar MB, Cress DE (2003) Improved ecdysone-receptor based inducible gene regulation system. Eur J Biochem 270:1308–1315

Palli SR, Hormann RE, Schlattner U, Lezzi M (2005a) Ecdysteroid receptors and their applications in agriculture and medicine. Vitam Horm 60:59–100

Palli SR, Tice CM, Margam VM, Clark AM (2005b) Biochemical mode of action and differential activity of new ecdysone agonists against mosquitoes and moths. Arch Insect Biochem Physiol 58:234–242

Potvin L, Laprade R, Schwartz JL (1998) Cry1Ac, a *Bacillus thuringiensis* toxin, triggers extracellular Ca^{2+} influx ad Ca^{2+} release from intracellular stores in Cf1 cells (*Choristoneura fumiferana*, Lepidoptera). J Exp Biol 201:1851–1858

Retnakaran A, Hiruma K, Palli SR, Riddiford LM (1995) Molecular analysis of the mode of action of RH-5992, a lepidopteran specific, non-steroidal ecdysteroid agonist. Insect Biochem Mol Biol 25:109–117

Retnakaran A, Macdonald A, Tomkins WL, Davis CN, Brownwright AJ, Palli SR (1997) Ultrastructural effects of a non-steroidal ecdysone agonist, RH-5992, on the sixth instar larva of the spruce budworm, *Choristoneura fumiferana*. J Insect Physiol 43:55–68

Retnakaran A, Krell P, Feng Q, Arif B (2003) Ecdysone agonists: mechanism and importance in controlling insect pests of agriculture and forestry. Arch Insect Biochem Physiol 54:187–199

Riddiford LM (2007) Juvenile hormone action: a 2007 perspective. J Insect Physiol 54:895–901

Riddiford LM, Cherbas P, Truman JW (2000) Ecdysone receptors and their biological actions. Vitam Horm 60:1–73

Sadrud-Din SY, Loeb MJ, Hakim RS (1996) In vitro differentiation of isolated stem cells from the midgut of *Manduca sexta* larvae. J Exp Biol 199:319–325

Sawada Y, Yanai T, Nakagawa H, Tsukamoto Y, Yokoi S, Yanagi M, Sugisaki H, Toya T, Kato Y, Watanabe T, Masui A (2003) Synthesis and insecticidal activity of benzoheterocyclic analogues of N-benzoyl-*N*-(*tert*-butyl) benzohydrazide. Part 2: introduction of substituents on the benzene rings of the benzoheterocycle moiety. Pest Manag Sci 59:36–48

Shahidi-Noghabi S, Van Damme EJM, De Vos WH, Smagghe G (2011) Internalization of *Sambucus nigra* agglutinins I and II in insect midgut CF-203 cells. Arch Insect Biochem Physiol 76:211–222

Siaussat D, Bozzolan F, Porcheron P, Debernard S (2007) Identification of steroid hormone signalling pathway in insect cell differentiation. Cell Mol Life Sci 64:365–376

Siaussat D, Bozzolan F, Porcheron P, Debernard S (2008) The 20-hydroxyecdysone-induced signalling pathway in G2/M arrest of *Plodia interpunctella* imaginal wing cells. Insect Biochem Mol Biol 38:529–539

Smagghe G (2007) Insect cell lines as tools in insecticide mode of action research. In: Ishaaya I, Nauen R, Horowitz AR (eds) Insecticides design using advanced technologies. Springer, Dordrecht, pp 263–304

Smagghe G, Degheele D (1995) Biological activity and receptor-binding of ecdysteroids and the ecdysteroid agonists RH-5849 and RH-5992 in imaginal wing discs of Spodoptera exigua (Lepidoptera: Noctuidae). Eur J Entomol 92:333–340

Smagghe G, Eelen H, Verschelde E, Richter K, Degheele D (1996) Differential effects of non-steroidal ecdysteroid agonists in Coleoptera and Lepidoptera: analysis of evagination and receptor binding in imaginal discs. Insect Biochem Mol Biol 26:687–695

Smagghe G, Dhadialla TS, Derycke S, Tirry L, Degheele D (1998) Action of the ecdysteroid agonist tebufenozide in susceptible and artificially selected beet armyworm. Pestic Sci 54:27–34

Smagghe G, Carton B, Decombel L, Tirry L (2000) Toxicity of four dibenzoylhydrazines correlates with evagination-induction in the cotton leafworm. Pestic Biochem Physiol 68:49–58

Smagghe G, Dhadialla TS, Lezzi M (2002) Comparative toxicity and ecdysone receptor affinity of nonsteroidal ecdysone agonists and 20-hydroxyecdysone in *Chironomus tentans*. Insect Biochem Mol Biol 32:187–192

Smagghe G, Ryckaert J, Soin T, Caputo G, Van Damme EJM (2005a) Effect of plant lectins on growth of insect midgut cells. In Vitro Cell Dev Biol Anim 41:34

Smagghe G, Vanhassel W, Moeremans C, De Wilde D, Goto S, Loeb MJ, Blackburn MB, Hakim RS (2005b) Stimulation of midgut stem cell proliferation and differentiation by insect hormones and peptides. Ann N Y Acad Sci 1040:472–475

Smagghe G, Goodman CL, Stanley D (2009) Insect cell culture and applications in research and pest control management. In Vitro Cell Dev Biol Anim 45:93–105

Smith HC, Cavanaugh CK, Friz JL, Thompson CS, Saggers JA, Michelotti EL, Garcia J, Tice CM (2003) Synthesis and SAR of cis-1-benzoyl-1,2,3,4-tetrahydroquinoline ligands for control of gene expression in ecdysone responsive systems. Bioorg Med Chem Lett 13:1943–1946

Sohi SS, Palli SR, Cook BJ, Retnakaran A (1995) Forest insect cell lines responsive to 20-hydroxyecdysone and two nonsteroidal ecdysone agonists, RH-5849 and RH-5992. J Insect Physiol 41:457–464

Soin T, Swevers L, Mosallanejad H, Efrose R, Labropoulou V, Iatrou K, Smagghe G (2008) Juvenile hormone analogs do not affect directly the activity of the ecdysteroid receptor complex in insect culture cell lines. J Insect Physiol 54:429–438

Soin T, Masatoshi I, Swevers L, Rougé P, Janssen CR, Smagghe G (2009) Towards Coleoptera-specific high-throughput screening systems for compounds with ecdysone activity: development of EcR reporter assays using weevil (*Anthonomus grandis*)-derived cell lines and *in silico* analysis of ligand binding to *A. grandis* ligand-binding pocket. Insect Biochem Mol Biol 39:523–534

Soin T, De Geyter E, Mosallanejad H, Iga M, Martín D, Ozaki S, Kitsuda S, Harada T, Miyagawa H, Stefanou D, Kotzia G, Efrose R, Labropoulou V, Geelen D, Iatrou K, Nakagawa Y, Janssen CR, Smagghe G, Swevers L (2010a) Assessment of species specificity of molting accelerating compounds in Lepidoptera: comparison of activity between *Bombyx mori* and *Spodoptera littoralis* by *in vitro* reporter and *in vivo* toxicity assays. Pest Manag Sci 66:526–535

Soin T, Swevers L, Kotzia G, Iatrou K, Janssen CR, Rougé P, Harada T, Nakagawa Y, Smagghe G (2010b) Comparison of the activity of non-steroidal ecdysone agonists between dipteran and lepidopteran insects, using cell-based EcR reporter assays. Pest Manag Sci 66:1215–1229

Spindler-Barth M, Spindler K-D (1998) Ecdysteroid resistant subclones of the epithelial cell line from *Chironomus tentans*. I. Selection and characterization of the resistant clones. In Vitro Cell Dev Biol Anim 34:116–122

Spindler-Barth M, Spindler K-D, Londershausen M, Thomas H (1989) Inhibition of chitin synthesis in an insect cell line. Pestic Sci 25:115–121

Srivastava DP, Yu EJ, Kennedy K, Chatwin H, Reale V, Hamon M, Smith T, Evans PD (2005) Rapid, nongenomic responses to ecdysteroids and catecholamines mediated by a novel *Drosophila* G-protein-coupled receptor. J Neurosci 25:6145–6155

Stiles B, McDonald IC, Gerst JW, Adams TS, Newman SM (1992) Initiation and characterization of 5 embryonic-cell lines from the cotton boll weevil *Anthonomus grandis* in a commercial serum-free medium. In Vitro Cell Dev Biol Anim 28:355–363

Suhr ST, Gil EB, Senut M-C, Gage FH (1998) High level transactivation by a modified *Bombyx* ecdysone receptor in mammalian cells without exogenous retinoid X receptor. Proc Natl Acad Sci USA 95:7999–8004

Suzuki J, Tanji I, Ota Y, Toda K, Nakagawa Y (2006) QSAR of 2,4-diphenyl-1,3-oxazolines for ovicidal activity against the two-spotted spider mite *Tetranychus urticae*. J Pestic Sci 31:409–416

Swevers L, Cherbas L, Cherbas P, Iatrou K (1996) *Bombyx* EcR (BmEcR) and *Bombyx* USP (BmCF1) combine to form a functional ecdysone receptor. Insect Biochem Mol Biol 26:217–221

Swevers L, Ito K, Iatrou K (2002) The BmE75 nuclear receptors function as dominant repressors of the nuclear receptor BmHR3A. J Biol Chem 277:41637–41644

Swevers L, Kravariti L, Ciolfi S, Xenou-Kokoletsi M, Ragoussis N, Smagghe G, Nakagawa Y, Mazomenos B, Iatrou K (2004) A cell-based high-throughput screening system for detecting ecdysteroid agonists and antagonists in plant extracts and libraries of synthetic compounds. FASEB J 18:134–136

Swevers L, Soin T, Mosallanejad H, Iatrou K, Smagghe G (2008) Ecdysteroid signaling in ecdysteroid-resistant cell lines from the polyphagous noctuid pest *Spodoptera exigua*. Insect Biochem Mol Biol 38:825–833

Terentiou P, Blanman M, Bradbrook D, Kaser G, Koohnan J (1993) Biological activity and receptor-binding of ecdysteroids in imaginal discs of *Calliphora vicina:* a comparison. Insect Biochem Mol Biol 23:131–136

Thomas HE, Stunnenberg HG, Stewart AF (1993) Heterodimerization of the *Drosophila* ecdysone receptor with retinoid X receptor and *ultraspiracle*. Nature 362:471–475

Thummel CS (1997) Dueling orphans–interacting nuclear receptors coordinate Drosophila metamorphosis. Bioessays 19:669–672

Tice CM, Hormann RE, Thompson CS, Friz JL, Cavanaugh CK, Michelotti EL, Garcia J, Nicolas E, Albericio F (2003a) Synthesis and SAR of alpha-acylaminoketone ligands for control of gene expression. Bioorg Med Chem Lett 13:475–478

Tice CM, Hormann RE, Thompson CS, Friz JL, Cavanaugh CK, Saggers JA (2003b) Optimization of α-acylaminoketone ecdysone agonists for control of gene expression. Bioorg Med Chem Lett 13:1883–1886

Tsai CC, Kao HY, Yao TP, McKeown M, Evans RM (1999) SMRTER, a *Drosophila* nuclear receptor coregulator, reveals that EcR-mediated repression is critical for development. Mol Cell 4:175–186

Van Damme EJM, Rougé P, Peumans WJ (2007) Carbohydrate-protein interactions: plant lectins. In: Kamerling JP, Boons GJ, Lee YC, Suzuki A, Taniguchi N, Voragen AJG (eds) Comprehensive glycoscience. From chemistry to systems biology, vol 3. Elsevier, New York, pp 563–599

Van Hiel MB, Van Loy T, Poels J, Vandersmissen HP, Verlinden H, Badisco L, Vanden Broeck J (2010) Neuropeptide receptors as possible targets for development of insect pest control agents. Adv Exp Med Biol 692:211–226

Vandenborre G, Lannoo N, Smagghe G, Daniel E, Breite A, Soin T, Jacobsen L, Van Damme EJM (2008) Cell-free expression and functionality analysis of the tobacco lectin. In Vitro Cell Dev Biol Anim 44:228–235

Vandenborre G, Smagghe G, Van Damme EJM (2011) Plant lectins as defence proteins against phytophagous insects. Phytochemistry 72:1538–15550

Verhaegen Y, Parmentier K, Swevers L, Rougé P, Soin T, De Coen W, Cooreman K, Smagghe G (2010) The brown shrimp (*Crangon crangon* L.) ecdysteroid receptor complex: cloning, structural modeling of the ligand-binding domain and functional expression in an EcR-deficient *Drosophila* cell line. Gen Comp Endocrinol 168:415–423

Wagner BL, Norris JD, Knotts TA, Weigel NL, McDonnell DP (1998) The nuclear corepressors NCoR and SMRT are key regulators of both ligand- and 8-bromo-cyclic amp-dependent transcriptional activity of the human progesterone receptor. Mol Cell Biol 18:1369–1378

Wang P, McCarthy WJ (1997) Cytolytic activity of *Bacillus thuringiensis* Cry1C and Cry1AC toxins to *Spodoptera* sp. midgut epithelial cells in vitro. In Vitro Cell Dev Biol Anim 33:315–323

Wheelock CE, Nakagawa Y, Harada T, Oikawa N, Akamatsu M, Smagghe G, Stefanou D, Iatrou K, Swevers L (2006) High-throughput screening of ecdysone agonists using a reporter gene assay followed by 3-D QSAR analysis of the molting hormonal activity. Bioorg Med Chem 14:1143–1159

Williams CM (1969) Third generation pesticides. Sci Am 217:13–17

Williams C (2004) cAMP detection methods in HTS: selecting the best from the rest. Nat Rev Drug Discov 3:125–135

Wing KD (1988) RH 5849, a nonsteroidal ecdysone agonist: effects on a *Drosophila* cell line. Science 241:467–469

Xiao SH, Reagan JD, Lee PH, Fu A, Schwandner R, Zhao X, Knop J, Beckmann H, Young SW (2008) High throughput screening for orphan and liganded GPCRs. Comb Chem High Throughput Screen 11:195–205

Yao T-P, Forman BM, Jiang Z, Cherbas L, Chen J-D, McKeown M, Cherbas P, Evans RM (1993) Functional ecdysone receptor is the product of *EcR* and *Ultraspiracle* genes. Nature 366:476–479

Zhang XB, Candas M, Griko NB, Taussig R, Bulla LA (2006) A mechanism of cell death involving an adenylyl cyclase/PKA signaling pathway is induced by the Cry1Ab toxin of *Bacillus thuringiensis*. Proc Natl Acad Sci USA 103:9897–9902

Zhang ZL, Xu JJ, Sheng ZT, Sui Y, Palli SR (2011) Steroid receptor co-activator is required for juvenile hormone signal transduction through a bHLH-PAS transcription factor, methoprene tolerant. J Biol Chem 286:8437–8447

Chapter 7
Advanced Screening to Identify Novel Pesticides

Yoshiaki Nakagawa and Toshiyuki Harada

Abbreviations

DAH	diacylhydrazine
DBH	dibenzoylhydrazine
PoA	ponasterone A
QSAR	quantitative structure-activity relationship
HTS	high throughput screening
CoMFA	comparative molecular field analysis
E	ecdysone

1 Introduction

Prior to the twentieth century, scientists have tried to identify active components from plants and other organisms to utilize them as pharmaceuticals. In the first half of the twentieth century, drug design was executed by modification of natural products and their preceding biochemical metabolites. In the 2nd half of the twentieth century, a rational approach based on quantitative structure-activity relationship (QSAR) was introduced for drug design (Fujita et al. 1964; Hansch et al. 1962, 1963; Hansch and Fujita 1964). After the birth of the Hansch-Fujita approach, currently called

Y. Nakagawa (✉)
Division of Applied Life Sciences, Graduate School of Agriculture, Kyoto University, Kyoto 606-8502, Japan
e-mail: naka@kais.kyoto-u.ac.jp

T. Harada
Health and Crop Sciences Research Laboratory, Sumitomo Chemical Company, Tokyo, Japan
e-mail: haradat5@sc.sumitomo-chem.co.jp

I. Ishaaya et al. (eds.), *Advanced Technologies for Managing Insect Pests*,
DOI 10.1007/978-94-007-4497-4_7, © Springer Science+Business Media Dordrecht 2012

"classical QSAR", various QSAR procedures such as three dimensional (3D) QSAR (Akamatsu 2002; Cramer III et al. 1988) and multi-dimensional QSAR (Vedani and Dobler 2002) have been developed and used for drug design. Recently, synthetic chemists may not always derive QSAR equations in their decision-making process for drug design, but they must be aware of the importance of the physicochemical properties such as hydrophobicity and steric hindrance. In addition to QSAR, other computer-aided drug design approaches such as *in silico* screening, and high throughput screening (HTS) have become common in the drug design industry (Jain and Nicholls 2008).

The classical and popular strategy in drug design is "me-too synthesis" Me-too drugs are similar compounds to those already marketed, which are synthesized and optimized based on SAR study. By modification of an active molecule, an increase in potency, better specificity profile, improved safety, or formulation, are sought. In parallel, synthetic chemists have obtained novel bioactive compounds (i.e., seeds) by random screening, because *de novo* design of novel bioactive compounds without information of their structure-activity relationship is rather difficult. Recently, *in silico* screening methods are easy to use for medicinal chemists, but not for chemists in agrochemical companies. This is probably due to the fact that agrochemical industry directly find out novel *in vivo* active compounds. The problem for agrochemical chemists is that often the molecular target is not known. If it is known, then the same approaches as used by medicinal chemists can be used by agrochemical chemists. Consequently, with the elucidation of 3D structures of target receptors and enzymes related to mode of action of pesticides, *in silico* screening has begun to find new agrochemicals.

In this chapter, we summarize the classical QSAR of the non-steroidal ecdysone agonists diacylhydrazines (DAHs) including dibenzoylhydrazines (DBHs), and briefly review the application of advanced computer-aided drug design to design insecticides. Even though QSARs may not be good for lead generation, they are useful for optimizing potency or other desirable properties. In addition, QSAR results for ecdysteroids were successfully included in *in silico* screening. About half of this chapter is related to QSARs, focusing on the design of non-steroidal ecdysone agonists and ecdysteroids (Sect. 2). In the sections (in the second half), recent computer-aided drug discovery methods such as HTS and *in silico* screening as well as homology modeling and bioinformatics that are sometimes required for *in silico* screening, are briefly reviewed. Unfortunately, since there are only a few publications on the application of HTS and *in silico* screening to problems in agrochemical design, some pharmaceutical examples are given in Sects. 3 and 4. In particular, *in silico* screening was conducted for finding novel compounds that specifically bind to ecdysone receptors.

2 QSARs

Rational drug design has been achieved by the application of computational techniques combining of "lead generation" and "lead optimization". As stated above, classical QSAR (Hansch and Fujita 1964) (Fujita 2011) and multi-dimensional

QSARs (Vedani and Dobler 2002) are employed for drug discovery. QSAR is a linear free energy relationship (LFER) approach, because the descriptors are derived from rates or equilibrium constants (Fujita 1990). In the Hansch-Fujita approach, the hydrophobicity, octanol/water partition coefficient (log P) or substituent parameter TL, was introduced as a LFER parameter with electronic and steric parameters as expressed by Eq. 7.1.

$$BA = \Sigma\{f(\text{steric}) + f(\text{electrostatic}) + f(\text{hydrophobicity})\} + c \qquad (7.1)$$

In particular, they found that the biological activity varies parabolically with respect to log P, rather than linearly. Therefore, the squared term of log P is used in the QSAR equation. Meanwhile, this parabolic model is sometimes substituted with the bilinear model (Kubinyi 1977). Early QSAR studies were mainly focused on analyzing the effects of aromatic ring substituents on biological activity (*in vivo* and *in vitro*) to describe the steric, electronic, and lipophilic characteristics of substituents. For structures lacking physicochemical descriptor variables, the Free-Wilson method has been used (Free and Wilson 1964). Although the Hansch-Fujita approach has been widely used for a half century, there is a drawback that more compounds than variables are required (five times as many compounds as descriptors) to derive a significant QSAR. Furthermore, the structural modifications best understood and interpreted through QSAR are generally limited to the aromatic ring substitution and core structure partial modification. To compensate for such weaknesses of classical QSAR, the 3D QSAR method, a comparative molecular field analysis (CoMFA), was introduced (Cramer III et al. 1988). The 3D results might be helpful not only in designing novel compounds, but also in understanding the molecular mechanism of action of a particular ligand (Wheelock et al. 2006). However, we have to carefully consider 3D QSAR results, because outputs of 3D models change depending upon compound superpositions.

2.1 *Classical QSAR*

DAH-type compounds, which also produce larvicidal activity, bind to ecdysone receptors, and competitively inhibit the binding of ecdysteroids, such as 20-hydroxyecdysone (20E) and ponasterone A (PoA) to their receptors (Nakagawa 2005). Four DAH-type compounds, tebufenozide, methoxyfenozide, chromafenozide and halofenozide (Fig. 7.1), have been used as insecticides in agriculture.

The basic compound is unsubstituted DBH (RH5849; *N-t*-butyl-*N*-benzoyl- *N′*-benzoylhydrazine: IUPAC name: *N-t*-butyl-*N′*-benzoyl benzohydrazide), which was discovered by the Rohm and Haas company in USA (Hsu 1991; Wing 1988). By optimizing the substitution pattern, tebufenozide was discovered (Hsu et al. 1997), which is selectively toxic to Lepidoptera. A short time later, methoxyfenozide (Carlson et al. 2001), and halofenozide were also developed by Rohm and Haas company. Methoxyfenozide is slightly more toxic to Lepidoptera than tebufenozide, and halofenozide is toxic to Coleoptera as well as Lepidoptera. In the meantime, a Japanese group developed chromafenozide; also selectively toxic to Lepidoptera (Sawada et al. 2003; Tanaka et al. 2001). Tebufenozide, methoxyfenozide, and

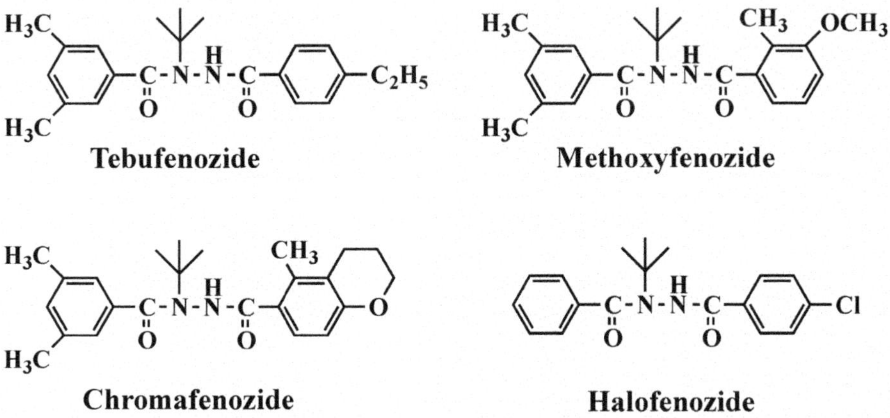

Fig. 7.1 Chemical structure of commercial ecdysone agonists

Fig. 7.2 Basic structure of DBHs

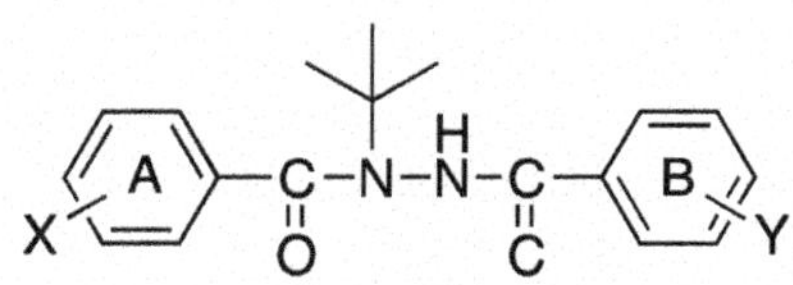

halofenozide are thought to be developed based on the QSAR approach, while chromafenozide was designed by a docking simulation, and is reviewed in Sect. 2.

Our group had executed QSAR studies for DBHs to examine the essential physicochemical properties for larvicidal activity (Oikawa et al. 1994a, b) and molting hormonal activity (Nakagawa et al. 1995b; Oikawa et al. 1993). Moreover, we were able to predict the high potency of tebufenozide and methoxyfenozide, before the announcement of their structures. First, we synthesized DBH analogs with various substituents (X) at various positions of the A-ring (Y = H) in Fig. 7.2, and quantitatively analyzed substituent effects on larvicidal activity against the rice stem borer *Chilo suppressalis* (Oikawa et al. 1994a). For the mono-substituted compounds, we derived a significant correlation as shown in Eq. 7.2.

2.1.1 Substitution at A-Ring Moiety (Mono-substitution)

$$\begin{aligned} pLD_{50} &= 0.977 \log P + 1.280\sigma_I^{ortho}(X) - 0.480\Delta V_w^{meta}(X) - 0.890\Delta V_w^{para}(X) \\ &\quad + 3.616 \qquad n = 27, s = 0.300, r = 0.899, F_{4,22} = 23.29 \end{aligned} \quad (7.2)$$

In Eq. 7.2 and following equations, log P is the hydrophobicity parameter that was either experimentally measured or calculated by MacLogP (Hansch et al. 1995; Leo 1993). σ_I is the electronic parameter (Charton 1981), and V_w is the steric parameter (Bondi 1964). The superscript attached to each parameter means that the parameter value for the substituent (X) at the corresponding position was used in analysis. In the original papers, Δlog P was used as a hydrophobicity parameter instead of log P (Oikawa et al. 1994a); therefore, the constant value was different from that in Eqs. 7.2 and 7.5. According to this correlation equation (Eq. 7.2), the hydrophobic compounds with the electron-withdrawing group at *ortho*-position were thought to be potent. Thus, we introduced hydrophobic substituents multiply at A-ring moiety, and reanalyzed QSAR for all compounds to formulate Eq. 7.3.

2.1.2 Substitution at A-Ring Moiety (All Compounds)

$$\begin{aligned} pLD_{50} &= 0.879 \log P + 1.504\,\Sigma\sigma_I^{ortho}(X) - 0.325\,\Sigma\Delta V_w^{meta}(X) \\ &\quad -0.815\,\Delta V_w^{para}(X) - 2.501\,I_{2,6}(X) - 0.935\,I_{2,3,5}(X) + 3.792 \\ &\quad n = 46, s = 0.337, r = 0.904, F_{4,22} = 29.03 \end{aligned} \tag{7.3}$$

In Eq. 7.3 and other equations, Σ means the sum of parameter values for di-*ortho*- and di-*meta*-substitutions. As shown in Eq. 7.3, indicator parameters, $I_{2,6}$ and $I_{2,3,5}$, for multiple substitution patterns became significant: $I_{2,6}$ takes 1 for 2,6-disubstitution and otherwise 0, and $I_{2,3,5}$ is 2 for compounds with 2,3,5-tri and 2,3,4,5-tetra substitution pattern, and 1 for 2,3- and 2,5-disubstitution, and otherwise zero. Equation 7.3 indicates that the activity decreases 300 times if both *ortho*-positions are substituted, and the activity of compounds with the 2,3,5-trisubstitution pattern decreases 100 times.

Based on the above QSAR results, we fixed the A-ring substituent to 2-Cl (Obsd 6.83; Calcd 6.78 in Eq. 7.3) and varied the substituents of the B-ring, although the 3,5-Cl_2-substituted compound (Obsd 7.07, Calcd 6.92 in Eq. 7.3) was slightly more potent than the 2-Cl compound. QSARs for the substituent effects for the B-ring moiety were given as shown below. Equation 7.4 is formulated for mono-substituted compounds and Eq. 7.5 for all compounds including the mono-substituted and 14 disubstituted compounds.

2.1.3 Substitution at B-Ring Moiety (Mono-substituted Compounds)

$$\begin{aligned} pLD_{50} &= 0.717 \log P - 0.880\,\Delta L^{ortho}(Y) - 0.975\,\Delta V_w^{meta}(Y) - 0.589\,\Delta L^{para}(Y) \\ &\quad +4.925 \qquad n = 30, s = 0.254, r = 0.912, F_{4,25} = 30.91 \end{aligned} \tag{7.4}$$

2.1.4 Substitution at B-Ring Moiety (All Compounds)

$$pLD_{50} = 0.722 \log P - 0.740\, \Sigma\Delta L^{ortho}(Y) - 0.868\, \Sigma\Delta V_w^{meta}(Y) - 0.485\, \Delta L^{para}(Y) + 4.783 \quad n = 44,\ s = 0.284,\ r = 0.896,\ F_{4,39} = 39.50 \tag{7.5}$$

In Eqs. 7.4 and 7.5, L is the STERIMOL parameter, which expresses the length of substituents (Verloop 1983). No indicator variable was necessary to derive Eq. 7.5, even though the disubstituted compounds are included. The steric effects of 3- and 4-substiuent for 2,3- and 2,4-disubstituted compounds were ignored to formulate Eq. 7.5, because the addition of those steric effects made the correlation worse. Probably, the unfavorable steric effects recognized for mono-substituted compounds became insignificant in the 2,3- and 2-4-disubstituted compounds. It is likely due to the fact that the receptor wall was pushed away by the ortho-substituent and the steric interaction is erased. Since Eqs. 7.3 and 7.5 have similar coefficients for log P term, they were combined as shown in Eq. 7.6.

$$\begin{aligned} pLD_{50} = {} & 0.791 \log P + 1.406\sigma_I^{ortho}(X) - 0.282\, \Sigma\Delta V_w^{meta}(X) - 0.749\Delta V_w^{para}(X) \\ & -2.404 I_{2,6}(X) - 0.895 I_{2,3,5}(X) - 0.791\, \Sigma\Delta L^{ortho}(Y) \\ & -0.935\, \Sigma\Delta V_w^{meta}(Y) - 0.548\Delta L^{para}(Y) + 4.002 \\ & n = 89, s = 0.313, r = 0.898, F_{9.79} = 36.48 \end{aligned} \tag{7.6}$$

As shown in Eqs. 7.5 and 7.6, introduction of the hydrophobic substituents (Cl, Br, I, CF_3, i-Pr, t-Bu) at the *para*-position of B-ring is less unfavorable to the larvicidal activity among *ortho*-, *meta*- and *para*-positions. With respect to the A-ring moiety, the unfavorable steric effect is less for *meta*-substitution. Thus, the relatively smaller and hydrophobic substituents such as Cl, Br, and CH_3 were introduced at the *meta*-position of the A-ring to derive 3,5-Cl_2, 3,5-Br_2, and 3,5-$(CH_3)_2$ analogs with various substituents at 4-position of B-ring moiety (Table 7.1).

As shown in Table 7.1, tebufenozide (**4**) was seven times more potent than the corresponding unsubstituted compound (**1**). Although the substitution of Et (**4**) with CH_3 (**3**) and i-Pr (**6**) did not greatly change the larvicidal activity, the activity significantly decreased by substitution with Cl (**2**) and n-Pr (**5**). All 3,5-disubstituted analogs (**2**–**8**) containing a 4-substituent on the B-ring moiety were 10–40 times higher in activity than values calculated by Eq. 7.6. These poor predictions will be improved by including these compounds in a new QSAR.

We also measured the molting hormone activity in the cultured integument, in which the detoxification of compounds is thought to be insignificant compared to that in the whole insect system (Nakagawa et al. 1995b). For the molting hormone activity of a mono-substituted compound, we could derive the significant correlation Eq. 7.7 using V_w instead of L. If we use the same parameter set as that used to derive Eq. 7.6, the correlation becomes worse (s = 0.377, r = 0.875).

Table 7.1 Larvicidal activity of tebufenozide analogs against *Chilo suppressalis*

	Compound			pLD_{50} (mmol/insect)		
No	A (X)	B (Y)	log P[a]	Obsd	Calcd[b]	Δ[c]
1	3,5-$(CH_3)_2$	H	3.39	6.43	6.05	0.38
2	3,5-$(CH_3)_2$	4-Cl	4.31	6.96	5.80	1.16
3	3,5-$(CH_3)_2$	4-CH_3	3.95	7.54	6.05	1.49
4[d]	3,5-$(CH_3)_2$	4-CH_3CH_2	4.39	7.32	5.72	1.60
5	3,5-$(CH_3)_2$	4-*n*-Pr	4.86	6.94	5.65	1.29
6	3,5-$(CH_3)_2$	4-*i*-Pr	4.91	7.32	6.13	1.19
7	3,5-Cl_2	4-CH_3CH_2	5.25	7.67	6.50	1.17
8	3,5-Br_2	4-CH_3CH_2	5.51	7.52	6.53	0.99

[a]See Ref. Oikawa et al. (1994b)
[b]Calculated by Eq. 7.6
[c]The difference between observed and calculated values
[d]Tebufenozide

$$\begin{aligned} pEC_{50} &= 1.015\log P + 1.403\sigma_I^{ortho}(X) - 0.523\Delta V_w^{meta}(X) - 1.078\Delta V_w^{para}(X) \\ &\quad -0.899\Delta V_w^{ortho}(Y) - 1.170\Delta V_w^{meta}(Y) - 0.766\Delta V_w^{para}(Y) + 4.056 \\ &\quad n = 37,\ s = 0.339,\ r = 0.900, F_{7,29} = 17.68 \end{aligned} \tag{7.7}$$

In our further study, we reanalyzed these 37 compounds by adding the electronic parameter (σ^0) and the steric parameter (V_w) for the *ortho*-substituent of the A-ring moiety to formulate the better correlation Eq. 7.8,

$$\begin{aligned} pEC_{50} &= 0.977\log P + 2.891\sigma_I^{ortho}(X) - 0.783\sigma^0(X) - 0.591\Delta V_w^{ortho}(X) \\ &\quad -0.464\Delta V_w^{meta}(X) - 1.056\Delta V_w^{para}(X) - 1.002\Delta V_w^{ortho}(Y) \\ &\quad -1.247\Delta V_w^{meta}(Y) - 0.850\Delta V_w^{para}(Y) + 4.377 \\ &\quad n = 37,\ s = 0.288,\ r = 0.934, F_{9,27} = 20.43 \end{aligned} \tag{7.8}$$

The prediction of the 3,5-dimethyl analog activity is better in Eq. 7.8, although the prediction of the larvicidal activity of 3,5-dimethyl analogs by Eq. 7.6 was not effective as described above. Since the unfavorable steric effect for the *meta*-substituents is a minimum regarding the A-ring ($-0.464\ \Delta V_w^{meta}$) the introduction of the hydrophobic electron donating and a less bulky CH_3 group ($\pi = 0.56$, $\sigma^0 = -0.12$) at the *meta* position was thought to be preferred at the A-ring moiety. Electron withdrawing substituents, Cl ($\pi = 0.71$) and Br ($\pi = 0.86$), seem to be electronically unfavorable, because the coefficient of σ^0 is negative (−0.783) in Eq. 7.8. The increments of the hydrophobicity by Cl or Br atoms probably compensated for the unfavorable electronic effects compared to the substituent (CH_3) of tebufenozide.

We also evaluated the ligand-receptor binding affinity of ecdysone agonists in insect cells (Nakagawa et al. 2000, 2002a). Ogura and co-workers quantitatively analyzed the binding activity of 20 DBHs (X=2–Cl) with various substituents (Y) at the *para*-position of B-ring as shown in Eq. 7.9 (Ogura et al. 2005). The electronic effect at the B-ring moiety on the larvicidal and molting hormonal activity was not significant, while an electron-donating group at the B-ring is favored for receptor binding. A similar electronic effect is also significant in the analysis of the compound's hormonal activity with various substituents at the A-ring moiety. Probably, the presence of the electron donating groups at both benzene rings is favorable for their intrinsic activity.

$$pIC_{50} = 0.607 \log P - 0.822\sigma - 0.367\Delta B_5 + 4.377$$
$$n = 17,\ s = 0.243,\ r = 0.914, F_{3,13} = 46.026 \qquad (7.9)$$

From these QSAR results, we could conclude that molecular hydrophobicity is the most important parameter not only for the larvicidal activity (*in vivo*) but also for hormonal and binding activity (*in vitro*). According to the QSAR for *in vitro* activity, less bulky electron donating groups such as CH_3 and CH_3CH_2 are intrinsically favored. The SAR of commercial insecticides such as tebufenozide and methoxyfenozide are reasonably explained by these QSARs. Even though we could not design chromafenozide in our QSAR approach, the approximate activity of chromafenozide was predicted by our QSARs.

We also performed classical QSAR analyses for additional larvicidal activity against beet armyworms, *Spodoptera exigua* (Nakagawa et al. 2002b; Smagghe et al. 1999), and Colorado potato beetles *Leptinotarsa decemlineata* (Nakagawa et al. 1999, 2001). The QSAR equations for larvicidal activity against *S. exigua* were very similar to those for *C. suppressalis* (Nakagawa et al. 2002b; Smagghe et al. 1999). However, the QSAR equations for larvicidal activity against *L. decemlineata* were very different from those for larvicidal activity against *C. suppressalis* and *S. exigua* (Nakagawa et al. 1999, 2001). These three larvicidal activities were also quantitatively compared using 3D QSAR (Hormann et al. 2008).

2.2 *Three Dimensional QSAR*

2.2.1 Diacylhydrazines

CoMFA is often used among various multi-dimensional QSARs (3D QSARs), because a variety of structures can be combined for analysis (Akamatsu 2002). Although the activity of the designed compounds can be calculated using QSAR equations, the extrapolation is sometimes poor compared with classical QSAR. The problem is that QSAR results substantially change depending upon the superposition pattern. We used CoMFA to predict the possible superposition between ecdysteroids and non steroidal ecdysone agonists, DAHs, before the disclosure of the crystal

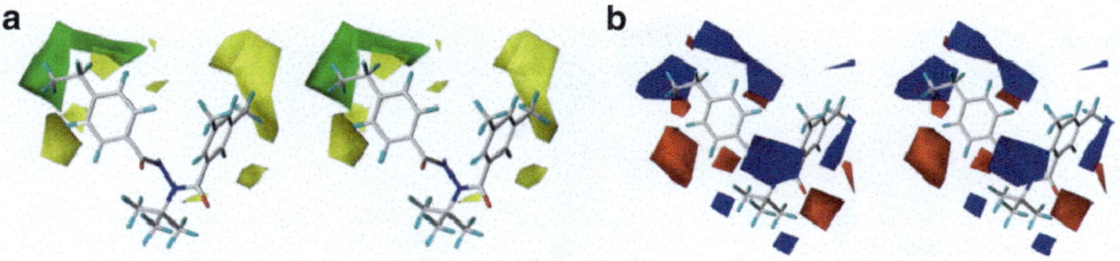

Fig. 7.3 CoMFA steric (**a**) and electrostatic (**b**) field maps (Reproduced from the Ref. Wheelock et al. 2006 with permission of Elsevier)

structures of ecdysone receptors (Billas et al. 2003). Even though the actual superposition between ecdysteroids and DAHs was slightly different from what we proposed from our QSAR study, the idea that the side chain moiety of ecdysteroids corresponds to one of the benzoyl moieties was valid (Nakagawa et al. 1995a, 1998). We proposed that the quaternary carbon atom of DAH's t-butyl group corresponds to the C20 of PoA (Nakagawa et al. 1995a, 1998; Shimizu et al. 1997), but it was demonstrated that the *t*-butyl group of DAH matched the terminal *i*-Pr moiety of PoA in an analysis of the crystal structures (Billas et al. 2003).

We quantitatively analyzed the hormonal activity of 158 diacylhydrazines using CoMFA, and favorable/unfavorable steric effects and negative/positive electrostatic interactions for their activity are shown in Fig. 7.3 (Wheelock et al. 2006). CoMFA statistics are indicated in Eq. 7.10. The addition of hydrophobic parameters log P and/or the squared log P term did not improve the correlation.

$$\begin{aligned} pEC_{50} &= \text{CoMFA} + 4.041 \quad n = 158,\ s = 0.554,\ r^2 = 0.737, \\ & F_{4,153} = 107.199,\ q^2 = 0.477,\ S_{CV} = 0.781,\ m = 4 \end{aligned} \tag{7.10}$$

Green regions indicate that the presence of bulky substituents are favorable for activity, and yellow regions indicate that the presence of bulky substituents are not favored for activity. The space surrounding the A-ring is restricted, and the region surrounding the CH_3CH_2 group of the B-ring moiety is quite large. Even though no CoMFA steric fields appeared around the t-butyl moiety, yellow fields seem to exist because the activity is significantly decreased by the substitution of the *t*-Bu group with Ph and Ph (4-F). This decreased activity is probably due to the fact that the number of compounds with different group at *t*-Bu moiety is small, only 7% of the total number of compounds (n = 158) used for CoMFA. In order to analyze the structure-activity relationship for this moiety, 11 active compounds with different alkyl groups (*i*-Pr, *n*-Bu, *i*-Bu, *sec*-Bu, *t*-Bu, 2-Me-Butyl, *t*-Amyl, Ph, Ph(4-F), *cyc*-Hex, $-CH_2$-*cyc*-Hex) were submitted to the classical QSAR analysis, given the QSAR shown in Eq. 7.11 and justified at the 92.6% F-test level (unpublished).

$$\begin{aligned} pEC_{50} &= 1.223 \log P - 0.790\, L + 6.406 \\ & n = 11,\ s = 0.539,\ r = 0.692, F_{2,8} = 3.672\,(> 92.6\%) \end{aligned} \tag{7.11}$$

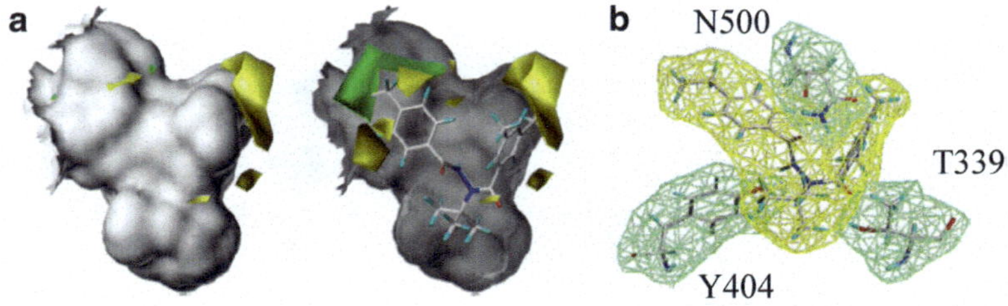

Fig. 7.4 Steric and electrostatic interactions for ligand-receptor binding. (**a**) CoMFA steric fields ovelayed on the binding pocket of BmEcR. (**b**) Amino acid residues participating for the hydrogen bonding between tebufenozide and BmEcR (Modified from the figures of Ref. Wheelock et al. 2006 in Bioorg. Med. Chem. with the permission of Elsevier, Ltd.)

The replacement of L with B_5 gave a poor and insignificant correlation ($pEC_{50}=-0.24 \log P+0.32\ B_5$: $s=0.706$, $r=0.324$, $F_{2,8}=0.469>0.358$).

In order to verify the CoMFA results, the CoMFA steric and electrostatic maps (Fig. 7.3) were overlain on the ligand binding pocket of BmEcR constructed from the X-ray crystal structure of HvEcR using a homology modeling technique, full automatic modeling system (FAMS; Sect. 3) (Ogata and Umeyama 2000). As drawn in Fig. 7.4, the sterically favored region is located inside the pocket, and the unfavorable region is located outside the pocket, which is consistent with the CoMFA results. In addition, the binding pocket accommodating the *t*-butyl group is fairly spacious, indicating that a bulkier group is allowed at this position. The presence of the large pocket is understandable from Eq. 7.11 for this moiety, because no significant correlation was found with the maximum substituent width (B_5). According to Eq. 7.11, if a long alkyl group is introduced, it causes unfavorable steric interactions. The 3D QSAR model constructed for 158 compounds and the classical QSAR models reasonably modeled the ligand receptor interaction. CoMFA electrostatic fields are also understandable, being close to the hydrogen-bond forming amino acid residues (T339, Y404, N500).

2.2.2 Ecdysteroids

A series of PoA analogs with various steroidal skeletons were synthesized to examine the structure-activity relationship for their steroid moiety. Their ligand-receptor binding activity was measured in intact Kc cells and the activity values of all ecdysteroids are listed in Table 7.2.

As shown in Table 7.2, ecdysteroids containing varied steroid skeletons having less than three functional groups such as –OH and =O (**22**, **27**, **29**) were inactive. Compounds **32** and **33** have a steroid skeleton moiety of the plant steroid hormone, castasterone (Yokota et al. 1982), with a side chain moiety of PoA (**17**), and compound **33** is the enantiomer of compound **32** (Watanabe et al. 2004). The stereochemistry

Table 7.2 Binding activity of ecdysteroids to the molting hormone receptors of Kc cells and HB number evaluated in the ligand-receptor complex model

No	Compound	pIC50	HB	No	Compound	pIC50	HB
9	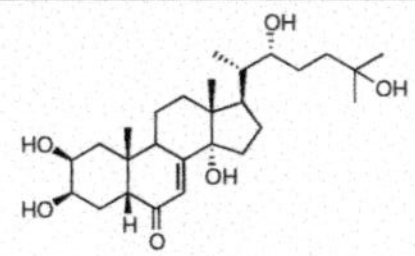	5.59	6	**18**	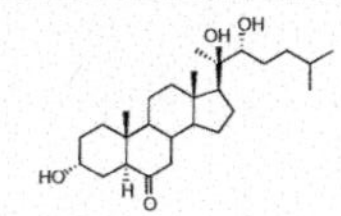	6.10	4
10	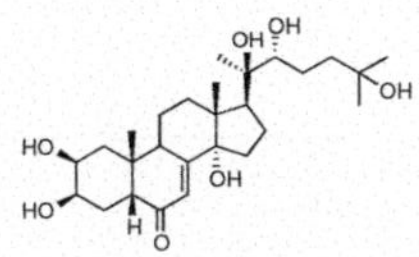	7.34	7	**19**	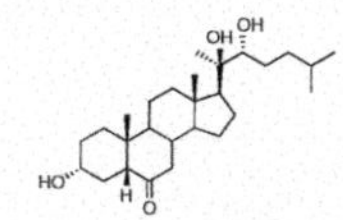	4.05	4
11	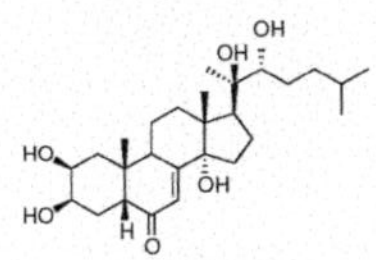	8.89	10	**20**	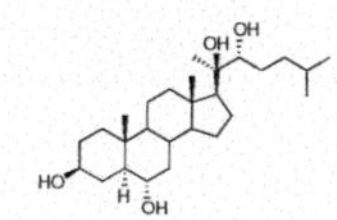	<3.61	3
12	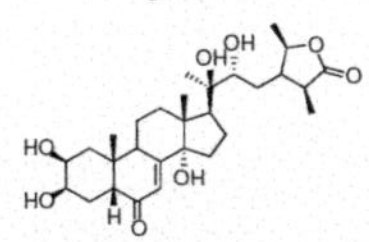	7.21	7	**21**	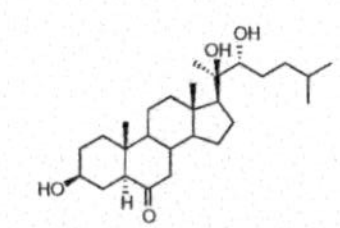	5.02	4
19	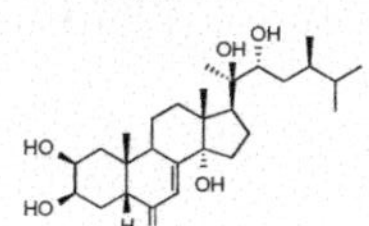	6.95	7	**22**	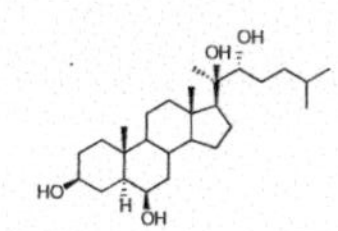	<3.61	3
13	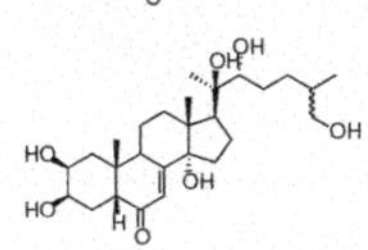	7.04	8	**23**	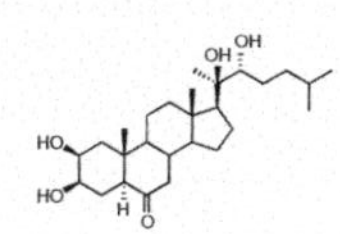	4.84	6

(continued)

Table 7.2 (continued)

No	Compound	pIC50	HB	No	Compound	pIC50	HB
14	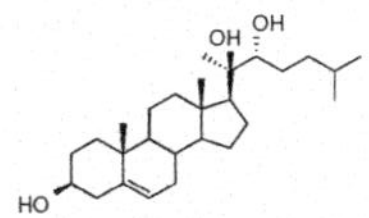	4.38	4	**24**	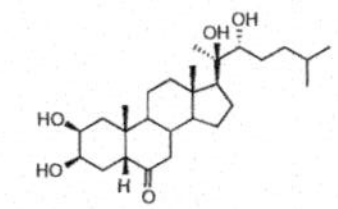	7.23	5
15	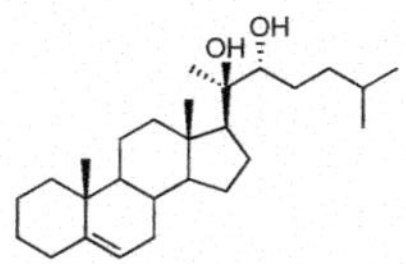	<3.61	1	**25**	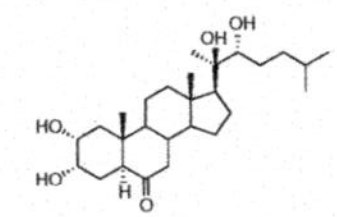	6.49	4
16	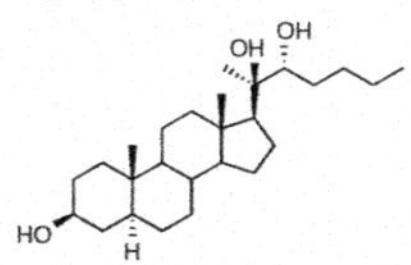	4.38	4	**26**	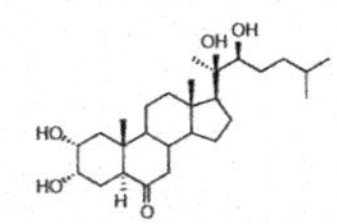	4.41	4
17	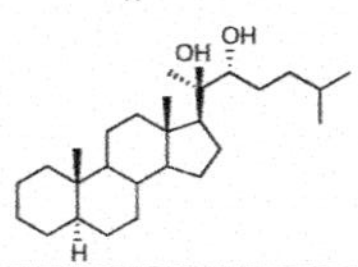	<3.61	1				

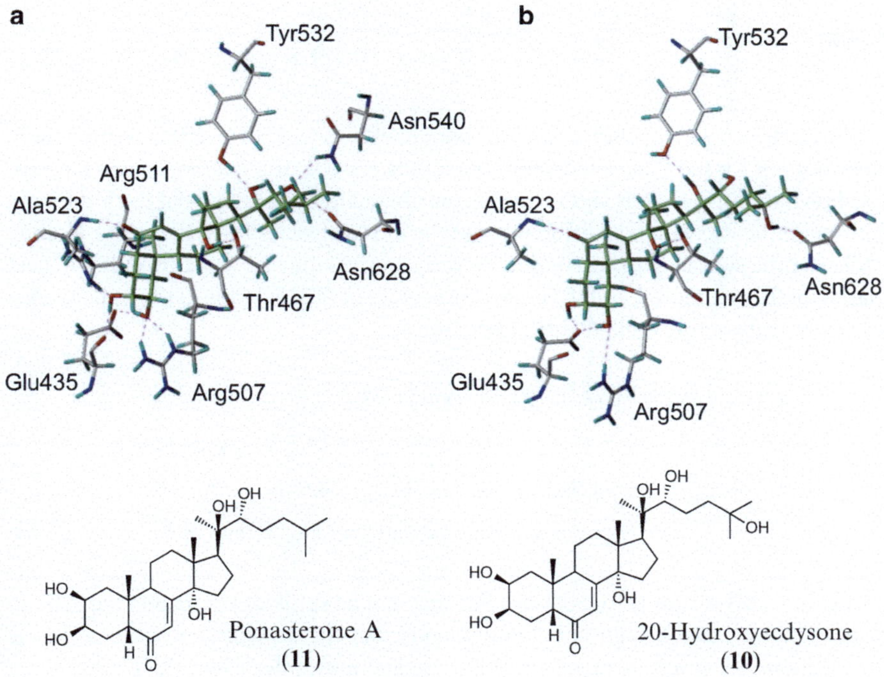

Fig. 7.5 Hydrogen-bonds (HBs) in the ligand-receptor docking models of (**a**) PoA and (**b**) 20E (Modified from the figure of Ref. Harada et al. 2009 in Bioorg. Med. Chem. with the permission of Elsevier, Ltd.)

of 20- and 22-OH groups of compound **26** is identical to that of PoA (**11**). When the 22-OH group of **26–27** was inverted, the binding activity decreased 100 times, and the molting hormonal activity was lost (Watanabe et al. 2004). Although the number of PoA (**11**) functional groups is less than that of 20E, the binding activity of PoA (**11**) is 35-folds higher compared with 20E. In addition, PoA (**11**) is 2,000 times more potent than ecdsyone (E; **9**) although the number of functional groups of PoA (**11**) is same as that of E (**9**). This result may not be consistent with the structure-activity relationship demonstrated by Arai et al. (2008). Therefore, Harada and co-workers tried to count the actual number of hydrogen bonds (HBs) observed in the ligand-receptor docking model (Harada et al. 2009). Table 7.2 lists the number of HBs observed for each ligand-receptor complex of DmEcR constructed from HvEcR using FAMS (Ogata and Umeyama 2000). As shown in Fig. 7.5, ten HBs were observed for PoA (**11**, A in Fig. 7.5), seven for 20E (**10**; B in Fig. 7.5), and six for E (**9**).

Interestingly, two amino acid residues (Arg511 and Asn540) that make HBs in the PoA-EcR complex model did not participate in the 20E HB formation. The relationship between the binding activity and the number of HBs was analyzed to formulate the significant Eq. 7.12.

$$\mathrm{pIC}_{50} = 0.609\,\mathrm{HB} + 2.589$$
$$n = 15, s = 0.920, r = 0.791, F_{1,13} = 21.753 \quad (7.12)$$

Since electrostatic, steric, and hydrophobic effects as well as HBs are thought to be important for the ligand-receptor interaction, we constructed the CoMFA model. As shown in Eq. 7.13, HB significantly contributed to the activity, but the log P term was insignificant, which is different from the QSAR results for DBHs. The binding pocket of the steroid skeleton moiety of ecdysteroids may be less hydrophobic than that of the B-ring moiety of DBHs. A-ring moiety of DBHs binds to the same pocket as that of ecdysteroid's alkyl side chain moiety.

$$\mathrm{pIC}_{50} = 0.549\,\mathrm{HB} + [\mathrm{CoMFA\ term}] + 2.82$$
$$n = 15,\ q^2 = 0.543,\ S_{press} = 1.017,\ \mathrm{component} = 1;$$
$$s = 0.880,\ r = 0.811,\ F_{1,13} = 24.978$$
$$\mathrm{Steric} = 8\%,\ \mathrm{electrostatic} = 14\%,\ \mathrm{HB} = 78\% \quad (7.13)$$

These CoMFA results for ecdysteroids may not be sufficient to predict the activity of ecdysteroids and design novel compounds, but these results are utilized in the following *in silico* screening of novel EcR ligand molecules (Sect. 3.1).

3 High Throughput Screening (HTS) and *In Silico* Screening

Both natural products and biologically active synthetic compounds have been used for a long time as lead structures in hit finding and generation, because the discovery of novel chemistry (i.e., "lead generation") was not easy. Recently, HTS and *in silico* (virtual) screening are often used to find lead compounds for pharmaceutical development, and these techniques came into the drug discovery field over two decades ago (Walters et al. 1998). It has been stated that drug discovery is becoming more difficult, as finding a good drug is like finding a needle in a haystack (Bomgardner 2011); i.e., the increase of the database size makes the haystack bigger, which makes finding the needle even more difficult. In addition, the increased chemical library size creates higher costs for synthesis and bioassays. According to the survey by Keseru and Makara (2009), the overall success rate of HTS-based lead discovery has remained 45–55%, despite its enormous financial investment.

Other advanced approaches such as combinatorial chemistry with high-throughput screening (HTS) and *in silico* screening have been introduced as methods for "lead generation." Recently, primary screening has shifted from low-throughput *in vivo* pharmacology to modern high-throughput *in vitro* biochemical screens. Combinatorial chemistry and HTS originally started in peptide chemistry. Using their original screening system from their combinatorial library, Miyashita and co-workers recently found a novel peptide which activates the plant immune system (Miyashita et al. 2011).

HTS is often used in the pharmaceutical industry. HTS technology is very simple and easy to understand, but it has some disadvantages such as the limitation of the combinatorial synthesis of low molecules and the utilization of a known screening system. In order to compensate the shortage of target oriented synthesis (TOS)-guided HTS, the "Diversed-Oriented Synthesis (DOS)" method (Burke and Schreiber 2004; Schreiber 2000) and the "Biology Oriented Synthesis (BIOS)" were developed (Wilk et al. 2010). In contrast to retrosynthetic TOS, DOS is a forward synthetic analysis. An important goal in DOS is to develop efficient synthesis pathways that yield products with varied 3D information. As stated above, lead compounds are often taken from the chemical structure of natural products and biologically active compounds, and are candidates in TOS. To perform DOS, the solid phase method is first introduced for peptide synthesis, which was later adapted to non-peptidic small molecules. Solid phase organic synthesis has become widely used in recent years. Thomas and colleagues discovered a novel anti-methicillin-resistant *Staphylococcus aureus* (MRSA) agent using DOS (Thomas et al. 2008). In BIOS, known synthetic methods are used in the synthesis of the core structure and attachment of substituents, and new diastereo- and enantio-selective synthetic methods are in demand. Here, building blocks are commercially available or have to be synthesized.

Another novel method comparable to HTS is Fragment-based Screening (FBS) for small molecules (MW < 300) (Hajduk and Greer 2007). Since molecular weight usually increases in the optimization process, smaller molecules are better as lead compounds to lower the synthetic cost. However, very sensitive assay systems are required for FBS, because the affinity of small molecules to proteins is generally low. Nuclear magnetic resonance (NMR), mass spectrometry (MS), isothermal titration calorimetry (ITC), surface plasmon resonance (SPR) have all been applied to detect fragments bound to target macromolecules. It has been suggested that a weak binder in itself can be a drug candidate due to its dynamic behavior and mild "influence" on a receptor (Ohlson 2008). Recently, Duong-Thi and co-workers developed a weak affinity chromatography method that is based on weak zonal affinity separation of small molecules as an alternative tool for FBS (Duong-Thi et al. 2011).

3.1 *In Silico Screening for Drug Discovery*

An early *in silico* drug discovery was published in Journal of Medicinal Chemistry (Horvath 1997), where 2,500 molecular structures were used for binding to trypanothione reductase (TR) to detect putative TR ligands. In this algorithm, all 2-D structures were fully converted to the corresponding 3D structures to explore the conformational space of the ligand, and to calculate enthalpy and entropy for the ligand-protein interaction. This *in silico* approach was used for predicting new pharmacological leads (Smellie et al. 1991) and 3D database searching (Martin 1992). In the twenty-first century, *in silico* screening is making dramatic advances day-by-day, because the computational cost and time has dramatically decreased with the advanced processing ability of hardware. In 2009, the *in silico* procedure is only 6% of the lead generation (Keseru and Makara 2009), but it is rapidly expanding.

In silico screening procedure is classified as ligand-based VS (LBVS) and structure-based virtual screening (SBVS) (or high-throughput docking; HTD). In LBVS, the construction of the structure of the ligand molecule can be performed using 2D techniques or 3D techniques using calculated conformations for the queries – pharmacophore tools and ROCS (OpenEye Scientific Inc., USA) can both work well without knowing the bound conformation of any active molecules. 2D-LBVS is much faster than any 3D technique, and LBVS using tools like ROCS, which is much faster than docking. ROCS can process 25 molecules per second. The fastest docking tools require 3–5 s for each molecule. Research from Merck & Co. showed that 3D LBVS hits are almost as diverse as hits from SBVS and many more hits are found by LBVS (Mcgaughey et al. 2007).

On the other hand, screening does not depend on the ligand structure in SBVS, although it requires a longer calculation time and higher accuracy. The probability of novel structure discovery is greater and applicable for searching an unknown ligand molecule for its orphan receptor. The tertiary structures of proteins such as receptors and enzymes are essential for SBVS, and these are obtained from crystal structures or homology modeling construction procedures (Sect. 3).

Recently, virtual screening was performed against the urokinase receptor (Wang et al. 2011), glycogen synthase kinase 3-beta (Osolodkin et al. 2011), GPCR (Sage et al. 2011), cytochrome P450 (Mo et al. 2012). Since inhibitors of glycogen synthase kinase 3-beta possibly show a therapeutic role for diabetes, Alzheimer's disease, bipolar disorder, and some others, screening for this enzyme is desirable. Other screenings such as those for pharmaceuticals targeting SARS (Mukherjee et al. 2011) and malaria (Shah et al. 2011) are also of interest.

In agrochemical area, trihydroxynaphthalene reductase (3HNR), an essential enzyme in the biosynthesis of fungal melanin was studied using VS (Andersson et al. 1996), because 3HNR represents an emerging target for new fungicide development (Tsuji et al. 1997) and DHN-melanin synthesis does not occur in host organisms. Brunskole and co-workers used homology modeling of 3HNR because the crystal structure of 3HNR from *Curvularis lunata* is unknown (Brunskole et al. 2008). The model was created from 3HNR from *Magnaporthe grisea* (74% sequence identity) and 17β-HSDcl (58%). Brunskole Švegelj and colleagues used OpenEye software to screen 280,000 compounds from the National Cancer Institute (NCI) for 3HNR inhibitors (Brunskole Svegelj et al. 2011). In the first screening, the number of compounds to be screened was reduced to roughly 72,000 compounds using Filter (OpenEye Scientific Software Inc), where the molecular weight, log P, and H-bond donor/acceptor count were considered. All compounds with atoms other than H, C, N, O, F, S, Cl and Br and reactive functional groups were eliminated in this first screening. In the next step, 3D similarity searches using ROCS (1,3,8-triydroxynaphthalene: substrate for 3HNR is used as query) were executed. The results were ranked according to the "combo" score, which considers similarities in molecular shape and color (atom types) to obtain 5,000 compounds. In the third step, these 5,000 compounds were submitted to the docking simulation with the active site using FlexX 3.1 (BiosolveIT GmbH) and ranked. From this survey, 19 compounds were submitted to the *in vitro* assay, and two compounds showed the significant inhibition (>50%) at 50 μM.

3.2 *In Silico Screening for the Discovery of Novel Ecdysone Agonists*

As shown above, commercial DBH-type compounds are selectively toxic against Lepidoptera (Carlson et al. 2001; Hsu et al. 1997; Nakagawa 2005; Tanaka et al. 2001), and the reasoning for this selective toxicity is known by comparing the crystal structures of EcR bound to DBH and ponasterone A (Billas et al. 2003). According to this analysis, ecdysteroids and DBH-type compounds partially share the binding site of EcR's ligand binding domain (Billas et al. 2003). Here, the primary sequence of the ligand binding pockets accommodating one of the DBH benzoyl moieties seems to be rather different between Lepidoptera and other insect orders (Nakagawa and Henrich 2009). Since 20E is a common natural molting hormone in most insects and the SAR for steroidal molting hormone agonists are similar among insect species, PoA > 20E, MaA, Cya > E (Nakagawa and Henrich 2009) and PoA > 20E > Ino > E (Minakuchi et al. 2003; Tohidi-Esfahani et al. 2011). We tested VS against the PoA-bound EcR (Billas et al. 2003; Carmichael et al. 2005; Iwema et al. 2007) to obtain novel non-steroidal compounds with a broad activity spectrum (Harada et al. 2011).

In the first step of our screening, three million compounds in the Namiki Database (Namiki Shoji, Co., Ltd., Tokyo, Japan) were first reduced to 2.1 million compounds based on their molecular weight (300–500) using Filter (Harada et al. 2011). In the next step, 170 conformers were generated for each compound by OMEGA (OpenEye Scientific Software Inc) to obtain 350 million conformers in total. In the third step, the similarity of each conformer to the EcR-bound PoA structure (PDB code: 1R1K) was calculated based on Shape Tanimoto (ST: 0–1) and Scaled Color (SC: corresponding to Chemical Similarity: 0–1) by ROCS. The 20,000-screened compounds were ranked based on the combination of ST and SC values. ST coefficient of 20E (**10**) to PoA (**11**) is 0.95, and the SC score is focused on the overlap of HB donor/acceptor, anionic/cationic charge, hydrophobic region, and ring structure.

To reduce the 20,000 compounds, we used an in-house Python script considering the overlaps of six HB acceptors and one hydrophobic moiety (Harada et al. 2011). After superimposing each compound to a template molecule, PoA, we checked the match of selected heavy atoms of the database molecules (O, N, S, Cl, Br, F, and P) to the six oxygen atoms of PoA by this script. Finally, 237 compounds were visually inspected to select 24 compounds with a focus on their chemical diversity. The final

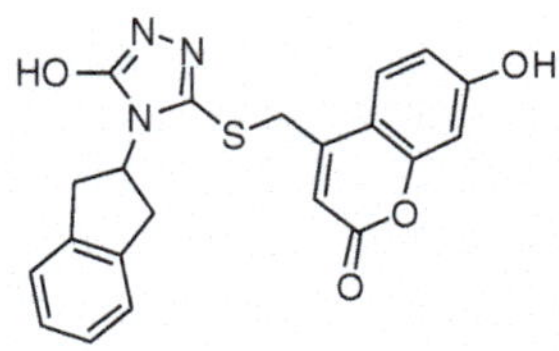

28

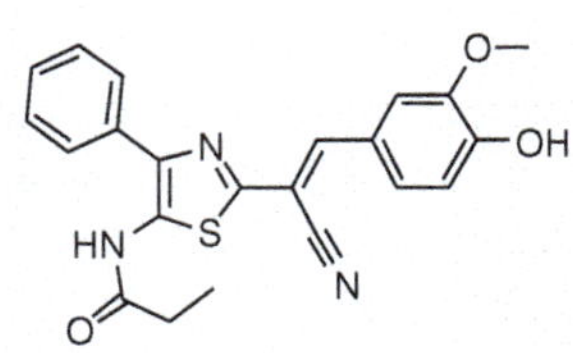

29

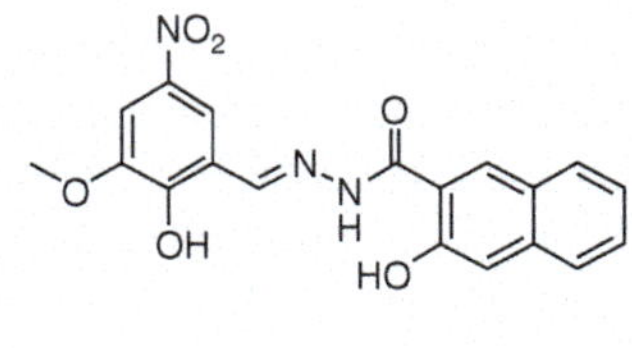

30

Fig. 7.6 Compounds found by *in silico* screening

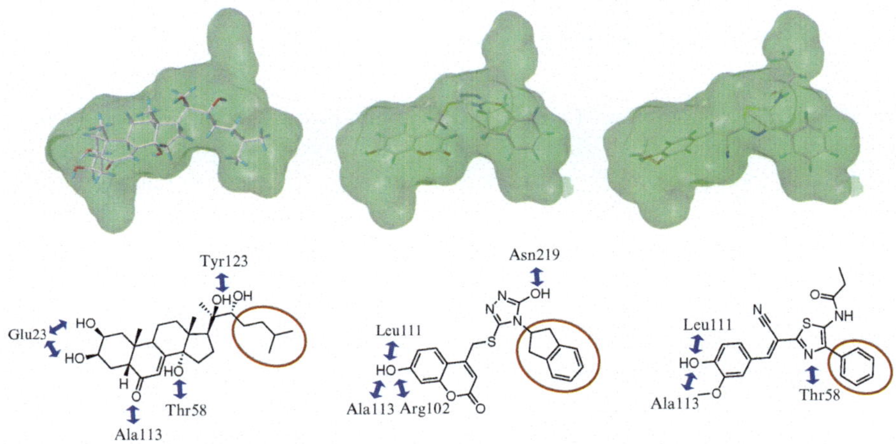

Ponasterone A (11) **28** **29**

Fig. 7.7 Docking of PoA and the compounds found by in silico screening to the modeled EcR (Reproduced from the publication in J. Chemical Information and Modeling, 2011 with the permission from ACS)

24 compounds screened were submitted for a competitive binding assay using lepidopteran and coleopteran cells. Finally, three compounds (Fig. 7.6) showed a specific binding activity to both cells.

In our further study, we constructed a 3D LBD structure of EcR from *Spodoptera frugiperda* (Chen et al. 2002) from considering the crystal structure of HvEcR-LBD. We conducted this analysis using a homology modeling software, FAMS, developed by Ogata and Umeyama (2000). To validate that the two novel potent ligand molecules can bind to the pocket of the modeled SfEcR-LBD, an *in silico* docking simulation was performed using two docking programs: GOLD (CCDC) and FRED (OpenEye Scientific Software Inc). The docking models constructed using GOLD for the two potent compounds and PoA are shown in Fig. 7.7. The substructures marked by the red circle are hydrophobic moieties.

As described above, three DBHs are rationally designed with the QSAR approach, but chromafenozide was designed based on a ligand-receptor docking model

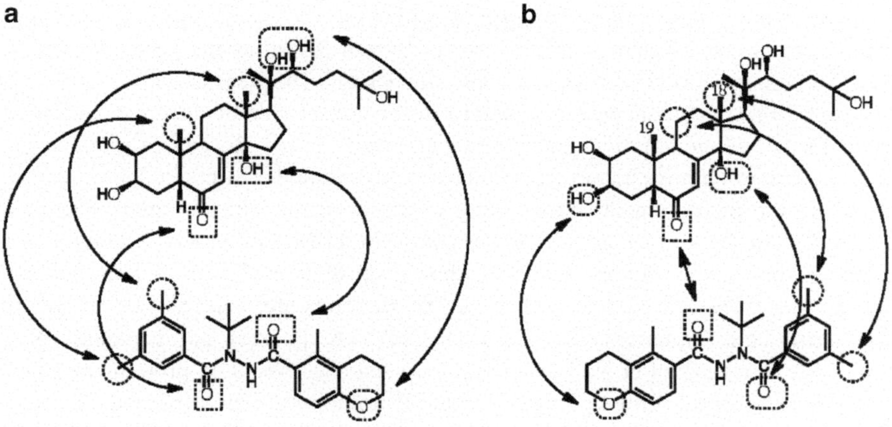

Fig. 7.8 Superposition between chromafenozide and 20-hydroxyecdysone (Modified from the figure of Ref. Kasuya et al. 2003)

(Kasuya et al. 2003). Since the crystal structure of EcR was not available at that time, crystal structures of the human steroid receptors were used for the modeling of EcR. Even though the superposition model proposed by Kasuya et al. (2003) was different from the superposition between DAHs and ecdysteroids when solved by X-ray analysis, they were able to discover the potent compound chromafenozide. In their approach, two possible superpositions (a and b in Fig. 7.8) were used in the docking simulation. In both superpositions, one of the methyl groups of the A-ring moiety is matched to C18 of 20E, and the other methyl group on either C19 or C11 of 20E. Two carbonyl groups of chromafenozide are matched to the carbonyl (C6) and 14-OH of 20E. On the other hand, the B ring moiety of DBH was placed on either the A-ring or the space surrounding C-20 of 20E, in which the oxygen atom of the chroman ring is placed near 3-OH and 20- or 22-OH. Differing from our hypothesis (Nakagawa et al. 1995a, 1998) and the real superposition solved by X-ray analysis (Billas et al. 2003), the side chain moiety of ecdysteroids did not correspond to any portion of chromafenozide. At any rate, chromafenozide was successfully designed based on the superposition between the chemical structure of DBH published in 1988 (Wing 1988) and 20E, although this superposition varied from the alignment revealed by crystallography.

4 Homology Modeling

As shown above, protein structures are important for advanced drug design, but the number of protein structures is limited. The number of protein 3D structures stored in the protein data bank (PDB) is increasing at an accelerated pace, but the number of 3D structures (43,836 in Jan 2009; >70,000 in Dec 2010) is small compared to primary sequence numbers (DDBJ: 20,742,576 in June 2011). Therefore, we need to predict the proper protein structures, which can be created by two methods "Template Based Modeling (TBM)" using known 3D structures, and "Template Free Modeling (TFM)," also called "*de novo* modeling". The present *de novo* modeling concept is that the whole protein structure can be constructed by the combination of partially homologous polypeptide fragments, using either Rosetta (Simons et al. 1997, 1999) or iterative Tasser fragment assembly methods (Bujnicki 2006). At this moment, the prediction of 3D structure by *de novo* modeling is inferior to that derived from the similar known 3D structure. In TBM, homologous proteins are found from known primary sequences based on their evolutionary homologous relationship, where it is thought that the 3D structures are conserved even though the amino acid sequences vary. Generally, FASTA (Pearson and Lipman 1988) and BLAST (Altschul et al. 1990) homology searches based on protein sequence comparisons have been employed. Advanced homology search methods such as PSI-BLAST (Altschul et al. 1997) and IMPALA (Schaffer et al. 1999), RPS-BLAST (Marchler-Bauer et al. 2002) have been developed to search distant sequences with profiling analysis. Furthermore, profile-profile comparison is performed to search for distantly related proteins using software (Pearson and Lipman 1988) such as FORTE (Tomii and Akiyama 2004) and SP3 (Zhou and Zhou 2005). Most recently, HHsearch (HMM-HMM comparison) (Zhou and Zhou 2005) can compare Hidden Markov Models (HMM) (Eddy 1996).

The second step of homology modeling is the coordination of 3D structures. Although a number of protein modeling systems are used in the world, we used FAMS. In the FAMS system, we can accurately predict the 3D structures not only for the main chain, but also for side chains. The FAMS method was judged superb in "Critical Assessment of Techniques for Protein Structure Prediction (CASP)." The prediction of the modeled proteins' structure is validated by stereochemistry quality with consideration of factors such as acceptable bond length, bond angle and dihedral, and the presence/absence of atom repulsion. The dihedral angle of the main chain is evaluated by a Ramachandran plot (Ramachandran et al. 1963). The constructed model was evaluated by PROCHECK (Laskowski et al. 1993), Verify3D (Luthy et al. 1992), ProSA (Laskowski et al. 1993; Wiederstein and Sippl 2007), and ProQ (Elofsson et al. 2003; Wallner et al. 2003). It was also validated by CIRCLE (Iwadate et al. 2010; Iwema et al. 2007; Kanou et al. 2010a, b), the extension of Verify3D (Terashi et al. 2005).

Even though it is becoming easy to crystallize proteins including membrane proteins such as G-protein coupled receptor (GPCR), it takes time to crystallize novel proteins. In addition, the crystal structures may not be functional structural forms. Since protein numbers with unknown 3D structure are growing steadily, protein modeling is needed not only for drug discovery, but for the disclosure of the molecular mechanisms at present and in the future.

In drug discovery studies, GPCR, nuclear receptors (NRs), ion channels phosphatases, kinases, and proteases are important target proteins. Among various proteins, the GPCR family is one of the largest and most diverse protein groups involved in many physiological processes. Therefore, GPCRs are an attractive target for pharmacological intervention to modify these processes in normal and pathological cell states. Although 3D structures solved by X-ray analysis and NMR are thought to be feasible for SBVS, only a few crystal GPCR structures are known to date.

The first crystallized GPCR was bovine rhodopsin (Okada et al. 2000) with other GPCR structures predicted based on this structure. In 2007, the structure of another GPCR, human-β2 adrenergic receptor (β2-AR) (Cherezov et al. 2007; Rasmussen et al. 2007) was solved. The β2-AR thereby became the first structurally characterized human trans-membrane protein. Vilar and co-workers demonstrated that docking-based virtual screening techniques are not only applicable to crystal structures but also to the modeled structures (Vilar et al. 2011). This finding is very important because now the applicability area extends to the vast majority of receptors, for which crystal structures are not available. The results definitely encourage computer-aided technique application to the discovery of novel GPCR ligands. Moreover, the crystal structure of histamine receptor H1 bound to the antihistamine drug doxepin was recently disclosed (Shimamura et al. 2011), and is another important target for drug discovery.

5 Informatics

Powerful and efficient data mining methods are required to treat the enormous amounts of data generated by genomics, proteomics, and metabolomics, as well as combinatorial chemistry and HTS. In this context, the use of information technology and management (i.e., "informatics") has become a critical part of the drug discovery process. Among informatics methods, "chemoinformatics" is very important to solve chemical problems (Agrafiotis et al. 2007; Engel 2006a, b). Chemoinformatics is the mixing of information resources to transform data into information and information into knowledge, making better decisions faster in drug discovery.

It has been thought that the completion of the human genome's final draft might allow the discovery of many new drugs based on the new gene structures revealed (ca. 22,000 genes). This genome resolution gives us a comprehensive understanding of life with looking at interactions among various factors, and we can search genes and proteins causing diseases. On the other hand, it is very difficult to find the seeds for biologically active compounds by experimentally exploring an enormous number of low molecular weight compounds. Therefore, the many combinations between biological material and small molecules are important for the drug design. As we have mentioned above, VS methods have been developed to reduce the time and money needed for screening large numbers of compounds. With the progress of chemical genomics, an enormous amount of data relating to protein-ligand interactions will accumulate, and novel VS methods are anticipated. In this context, "chemical genomics-based virtual screening (CGBVS)" was introduced for practical use as a novel *in silico* screening method (Okuno 2008). It was demonstrated that the prediction of the β2-AR kinase activity by CGBVS is superior to that by LBVS and SBVS (Yabuuchi et al. 2011).

As mentioned above, we can treat a larger and larger numbers of compounds due to the constant increases in computer power available. Although the rate-determining step used to be creating accurate databases and excluding incorrect data, databases are now easily constructed by linking international journal papers published online. The most important aspect is to have an eye to finding a good database, because we can become misled by retrieving poorly curated or incorrect data. An important database in this area was constructed by KEGG (Kyoto Encyclopedia of Genes and Genomics), a project that was started in 1995, in which database search and alignment service are offered worldwide (http://www.genome.jp/kegg/) (Aoki-Kinoshita 2006; Kanehisa et al. 2006). In KEGG, metabolic and signal transduction networks are connected to gene and protein databases to comprehensively understand cellular dynamics and physiology, which is fruitful for drug discovery and chemotherapy. At present, KEGG is connected to medical (drugs) and disease information in its genome net service (KEGG DRUG and KEGG DISEASE).

Another important technology is OMICS (Wheelock and Miyagawa 2006), where genes, gene expression, proteins as well as their metabolites (metabolome) are comprehensively studied. OMICS information seems to be clinically helpful in tailor-made remedies and the individualization of drug therapies. Since the probable number of target molecules is thought to be about 2,000–3,000, OMICS information is fruitful for searching unidentified drug-target genes. Genes expressed in diseased tissue and therapeutic genes depressed in diseased tissue, are targets for drug discovery.

In OMICS technology, diseases can be classified, and the prognosis predicted using a gene expression system. A decade ago, it was reported that acute myeloid leukemia and acute lymphoblastic leukemia are distinguishable without previous knowledge of these classes (Golub et al. 1999). Rosenwald and co-workers used DNA microarrays to formulate a molecular predictor of survival after chemotherapy for diffuse large-B-cell lymphoma (Rosenwald et al. 2002). Using gene expression profiling, Chang and co-workers predicted the therapeutic response to docetaxel in patients with breast cancer where the differential patterns of expression of 92 genes correlated with docetaxel response (Chang et al. 2003). Ayers and co-workers reported that transcriptional profiling has the potential to identify a gene expression pattern in breast cancer (Ayers et al. 2004). Another clinical application of OMICS, proteome analysis may allow early cancer detection. In this approach, the abnormal expression of low molecular weight proteins was examined by MALDI-TOF MS for early cancer detection.

6 Conclusion

In this Chapter, we reviewed the literature related to the rational design of molting hormone agonists using QSAR and virtual (*in silico*) screening. QSARs of known steroidal ecdysone agonists and synthetic ecdysteroids were first summarized, then we showed the discovery of novel non-steroidal compounds that specifically bind to the molting hormone receptor using LBVS. We also briefly introduced other important technologies such as homology modeling and informatics that are necessary for medicinal drug design.

The development of method for drug discovery is making great progress. They are grouped into two major approaches, random screening and rational approaches, both enabled by computer hardware and software systems advancement. In random screening, combinatorial synthesis and HTS are performed using robots. The rational approach began with the introduction of classical QSAR (Hansch-Fujita approach). Before QSAR, drug discovery based on the search of natural products (1st generation) was common. In 2nd generation drug design, the creation of agonists/antagonists toward target proteins such as enzymes and receptors was possible by computational advances including the development of a drug design software. Recently, orphan receptor ligand molecules and novel gene product (3rd generation drug design) searching is in progress. In future work (4th generation drug design), the relationship between genome and proteome will be very helpful in understanding cellular physiology, and current medical care will shift to tailor-made care by analyzing individual genome variation. Even though the number of pharmacological targets including pesticide targets such as plants, insects, and fungi is increasing, it can be systematically decreased using bioinformatics and genome technology.

Acknowledgements We thank Prof. Tsunemi Yamashita for his time to edit the English. We also give sincere thanks to Dr. Paul Hawkins for providing invaluable suggestions for this review.

References

Agrafiotis DK, Bandyopadhyay D, Wegner JK, Vlijmen H (2007) Recent advances in chemoinformatics. J Chem Inf Model 47:1279–1293

Akamatsu M (2002) Current state and perspectives of 3D-QSAR. Curr Top Med Chem 2: 1381–1394

Altschul SF, Gish W, Miller W, Myers EW, Lipman DJ (1990) Basic local alignment search tool. J Mol Biol 215:403–410

Altschul SF, Madden TL, Schaffer AA, Zhang J, Zhang Z, Miller W, Lipman DJ (1997) Gapped BLAST and PSI-BLAST: a new generation of protein database search programs. Nucleic Acids Res 25:3389–3402

Andersson A, Jordan D, Schneider G, Lindqvist Y (1996) Crystal structure of the ternary complex of 1,3,8-trihydroxynaphthalene reductase from Magnaporthe grisea with NADPH and an active-site inhibitor. Structure 4:1161–1170

Aoki-Kinoshita KF (2006) Overview of KEGG applications to omics-related research. J Pestic Sci 31:296–299

Arai H, Watanabe B, Nakagawa Y, Miyagawa H (2008) Synthesis of ponasterone A derivatives with various steroid skeleton moieties and evaluation of their binding to the ecdysone receptor of Kc cells. Steroids 73:1452–1464

Ayers M, Symmans WF, Stec J, Damokosh AI, Clark E, Hess K, Lecocke M, Metivier J, Booser D, Ibrahim N et al (2004) Gene expression profiles predict complete pathologic response to neoadjuvant paclitaxel and fluorouracil, doxorubicin, and cyclophosphamide chemotherapy in breast cancer. J Clin Oncol 22:2284–2293

Billas IM, Iwema T, Garnier JM, Mitschler A, Rochel N, Moras D (2003) Structural adaptability in the ligand-binding pocket of the ecdysone hormone receptor. Nature 426:91–96

Bomgardner MM (2011) Germinating pesticide. Chemical and Engineering News:13–17

Bondi A (1964) van der Waals volumes and radii. J Phys Chem 68:441–451

Brunskole M, Stefane B, Zorko K, Anderluh M, Stojan J, Lanisnik Rizner T, Gobec S (2008) Towards the first inhibitors of trihydroxynaphthalene reductase from Curvularia lunata: synthesis of artificial substrate, homology modelling and initial screening. Bioorg Med Chem 16: 5881–5889

Brunskole Svegelj M, Turk S, Brus B, Lanisnik Rizner T, Stojan J, Gobec S (2011) Novel inhibitors of trihydroxynaphthalene reductase with antifungal activity identified by ligand-based and structure-based virtual screening. J Chem Inf Model 51:1716–1724

Bujnicki JM (2006) Protein-structure prediction by recombination of fragments. Chembiochem 7:19–27

Burke MD, Schreiber SL (2004) A planning strategy for diversity-oriented synthesis. Angew Chem Int Ed Engl 43:46–58

Carlson GR, Dhadialla TS, Hunter R, Jansson RK, Jany CS, Lidert Z, Slawecki RA (2001) The chemical and biological properties of methoxyfenozide, a new insecticidal ecdysteroid agonist. Pest Manag Sci 57:115–119

Carmichael JA, Lawrence MC, Graham LD, Pilling PA, Epa VC, Noyce L, Lovrecz G, Winkler DA, Pawlak-Skrzecz A, Eaton RE et al (2005) The X-ray structure of a hemipteran ecdysone receptor ligand-binding domain: comparison with a lepidopteran ecdysone receptor ligand-binding domain and implications for insecticide design. J Biol Chem 280:22258–22269

Chang JC, Wooten EC, Tsimelzon A, Hilsenbeck SG, Gutierrez MC, Elledge R, Mohsin S, Osborne CK, Chamness GC, Allred DC et al (2003) Gene expression profiling for the prediction of therapeutic response to docetaxel in patients with breast cancer. Lancet 362:362–369

Charton M (1981) Electrical effect substituent constants for correlation analysis. Prog Phys Org Chem 13:119–251

Chen JH, Turner PC, Rees HH (2002) Molecular cloning and induction of nuclear receptors from insect cell lines. Insect Biochem Mol Biol 32:657–667

Cherezov V, Rosenbaum DM, Hanson MA, Rasmussen SG, Thian FS, Kobilka TS, Choi HJ, Kuhn P, Weis WI, Kobilka BK et al (2007) High-resolution crystal structure of an engineered human beta2-adrenergic G protein-coupled receptor. Science 318:1258–1265

Cramer III RD, Patterson DE, Bunce JD (1988) Comparative molecular field analysis (CoMFA). 1. Effect of shape on binding of steroids to carrier proteins. J Am Chem Soc 110:5959–5967

Duong-Thi MD, Meiby E, Bergstrom M, Fex T, Isaksson R, Ohlson S (2011) Weak affinity chromatography as a new approach for fragment screening in drug discovery. Anal Biochem 414:138–146

Eddy SR (1996) Hidden Markov models. Curr Opin Struct Biol 6:361–365

Elofsson A, Wallner B, Fang HS (2003) Automatic consensus-based fold recognition using Pcons, ProQ, and pmodeller. Proteins-Struct Funct Genet 53:534–541

Engel T (2006a) Basic overview of chemoinformatics. J Chem Inf Model 46:2267–2277

Engel T (2006b) Johann Gasteiger–Germany's pioneer in chemoinformatics. J Chem Inf Model 46:2191–2192

Free SM Jr, Wilson JW (1964) A mathematical contribution to structure-activity studies. J Med Chem 7:395–399

Fujita T (1990) The extrathermodynamic approach to drug design. In: Ramsden CA (ed) Comprehensive medicinal chemistry. Pergamon, Oxford, pp 497–560

Fujita T (2011) In memoriam Professor Corwin Hansch: birth pangs of QSAR before 1961. J Comput Aided Mol Des 25:509–517

Fujita T, Hansch C, Iwasa J (1964) New substituent constant Π derived from partition coefficients. J Am Chem Soc 86:5175–5180

Golub TR, Slonim DK, Tamayo P, Huard C, Gaasenbeek M, Mesirov JP, Coller H, Loh ML, Downing JR, Caligiuri MA et al (1999) Molecular classification of cancer: class discovery and class prediction by gene expression monitoring. Science 286:531–537

Hajduk PJ, Greer J (2007) A decade of fragment-based drug design: strategic advances and lessons learned. Nat Rev Drug Discov 6:211–219

Hansch C, Fujita T (1964) ρ–σ–Π analysis. Method for correlation of biological activity and chemical structure. J Am Chem Soc 86:1616–1626

Hansch C, Maloney PP, Fujita T (1962) Correlation of biological activity of phenoxyacetic acids with Hammett substituent constants and partition coefficients. Nature 194:178–180

Hansch C, Streich M, Geiger F, Muir RM, Maloney PP, Fujita T (1963) Correlation of biological activity of plant growth regulators and chloromycetin derivatives with Hammett constants and partition coefficients. J Am Chem Soc 85:2817–2824

Hansch C, Leo AJ, Hoekman D (1995) Exploring QSAR-hydrophobic, electronic, and steric constants. American Chemical Society, Washington, DC

Harada T, Nakagawa Y, Akamatsu M, Miyagawa H (2009) Evaluation of hydrogen bonds of ecdysteroids in the ligand-receptor interactions using a protein modeling system. Bioorg Med Chem 17:5868–5873

Harada T, Nakagawa Y, Ogura T, Yamada Y, Ohe T, Miyagawa H (2011) Virtual screening for ligands of the insect molting hormone receptor. J Chem Inf Model 51:296–305

Hormann RE, Smagge G, Nakagawa Y (2008) Multidimensional quantitative structure-activity relationships of diacylhydrazine toxicity in Spodoptera exigua, Chilo suppressalis, and Leptinotarsa decemlineata. QSAR Comb Sci 27:1098–1112

Horvath D (1997) A virtual screening approach applied to the search for trypanothione reductase inhibitors. J Med Chem 40:2412–2423

Hsu AC-T (1991) 1,2-Diacyl-1-alkylhydrazines, a new class of insect growth regulators. In: Baker DR, Fenyes JG, Moberg WK (eds) Synthesis and chemistry of agrochemicals II. American Chemical Society, Washington, DC, pp 478–490

Hsu AC-T, Fujimoto TT, Dhadialla TS (1997) Structure-activity study and conformational analysis of RH-5992, the first commercialized nonsteroidal ecdysone agonist. In: Hedin PA, Hollingworth RM, Masler EP, Miyamoto J, Thompson DG (eds) Phytochemicals for pest control. American Chemical Society, Washington, DC, pp 206–219

Iwadate M, Kanou K, Terashi G, Umeyama H, Takeda-Shitaka M (2010) Method for predicting homology modeling accuracy from amino acid sequence alignment: the power function. Chem Pharm Bull(Tokyo) 58:1–10

Iwema T, Billas IM, Beck Y, Bonneton F, Nierengarten H, Chaumot A, Richards G, Laudet V, Moras D (2007) Structural and functional characterization of a novel type of ligand-independent RXR-USP receptor. EMBO J 26:3770–3782

Jain AN, Nicholls A (2008) Recommendations for evaluation of computational methods. J Comput Aided Mol Des 22:133–139

Kanehisa M, Goto S, Hattori M, Aoki-Kinoshita KF, Itoh M, Kawashima S, Katayama T, Araki M, Hirakawa M (2006) From genomics to chemical genomics: new developments in KEGG. Nucleic Acids Res 34:D354–D357

Kanou K, Hirata T, Iwadate M, Terashi G, Umeyama H, Takeda-Shitaka M (2010a) HUMAN FAMSD-BASE: high quality protein structure model database for the human genome using the FAMSD homology modeling method. Chem Pharm Bull(Tokyo) 58:66–75

Kanou K, Hirata T, Terashi G, Umeyama H, Takeda-Shitaka M (2010b) New protein structure model evaluation methods that include a side-chain consensus score for the protein modeling. Chem Pharm Bull(Tokyo) 58:180–190

Kasuya A, Sawada Y, Tsukamoto Y, Tanaka K, Toya T, Yanagi M (2003) Binding mode of ecdysone agonists to the receptor: comparative modeling and docking studies. J Mol Model 9:58–65

Keseru GM, Makara GM (2009) The influence of lead discovery strategies on the properties of drug candidates. Nat Rev Drug Discov 8:203–212

Kubinyi H (1977) Quantitative structure-activity-relationships. 7. Bilinear model, a new model for nonlinear dependence of biological-activity on hydrophobic character. J Med Chem 20: 625–629

Laskowski RA, Macarthur MW, Moss DS, Thornton JM (1993) Procheck – a program to check the stereochemical quality of protein structures. J Appl Crystallogr 26:283–291

Leo AJ (1993) Calculating log P_{oct} from structures. Chem Rev 93:1281–1306

Luthy R, Bowie JU, Eisenberg D (1992) Assessment of protein models with three-dimensional profiles. Nature 356:83–85

Marchler-Bauer A, Panchenko AR, Shoemaker BA, Thiessen PA, Geer LY, Bryant SH (2002) CDD: a database of conserved domain alignments with links to domain three-dimensional structure. Nucleic Acids Res 30:281–283

Martin YC (1992) 3D database searching in drug design. J Med Chem 35:2145–2154

Mcgaughey GB, Sheridan RP, Bayly CI, Culberson JC, Kreatsoulas C, Lindsley S, Maiorov V, Truchon JF, Cornell WD (2007) Comparison of topological, shape, and docking methods in virtual screening. J Chem Inf Model 47:1504–1519

Minakuchi C, Nakagawa Y, Miyagawa H (2003) Validity analysis of a receptor binding assay for ecdysone agonists using cultured intact insect cells. J Pestic Sci 28:55–57

Miyashita M, Oda M, Ono Y, Komoda E, Miyagawa H (2011) Discovery of a small peptide from combinatorial libraries that can activate the plant immune system by a jasmonic acid signaling pathway. Chembiochem 12:1323–1329

Mo SL, Liu WF, Chen Y, Luo HB, Sun LB, Chen XW, Zhou ZW, Sneed KB, Li CG, Du YM et al (2012) Ligand- and protein-based modeling studies of the inhibitors of human cytochrome P450 2D6 and a virtual screening for potential inhibitors from the Chinese herbal medicine, Scutellaria baicalensis (Huangqin, Baikal Skullcap). Comb Chem High Throughput Screen 15:36–80

Mukherjee P, Shah F, Desai P, Avery M (2011) Inhibitors of SARS-3CLpro: virtual screening, biological evaluation, and molecular dynamics simulation studies. J Chem Inf Model 51: 1376–1392

Nakagawa Y (2005) Nonsteroidal ecdysone agonists. Vitam Horm 73:131–173

Nakagawa Y, Henrich VC (2009) Arthropod nuclear receptors and their role in molting. FEBS J 276:6128–6157

Nakagawa Y, Shimizu B, Oikawa N, Akamatsu M, Nishimura K, Kurihara N, Ueno T, Fujita T (1995a) Three-dimensional quantitative structure-activity analysis of steroidal and dibenzoylhydrazine-type ecdysone agonists. In: Hansch C, Fujita T (eds) Classical and three-dimensional QSAR in agrochemistry. American Chemical Society, Washington, DC, pp 288–301

Nakagawa Y, Soya Y, Nakai K, Oikawa N, Nishimura K, Ueno T, Fujita T, Kurihara N (1995b) Quantitative structure-activity studies of insect growth regulators. XI. Stimulation and inhibition of *N*-acetylglucosamine incorporation in a cultured integument system by substituted *N*-*tert*-butyl-*N*, N′-dibenzoylhydrazines. Pestic Sci 43:339–345

Nakagawa Y, Hattori K, Shimizu B, Akamatsu M, Miyagawa H, Ueno T (1998) Quantitative structure-activity studies of insect growth regulators XIV. Three dimensional quantitative structure-activity relationship of ecdysone agonists including dibenzoylhydrazine analogs. Pestic Sci 53:267–277

Nakagawa Y, Smagghe G, Kugimiya S, Hattori K, Ueno T, Tirry L, Fujita T (1999) Quantitative structure-activity studies of insect growth regulators: XVI. Substituent effects of dibenzoylhydrazines on the insecticidal activity to Colorado potato beetle Leptinotarsa decemlineata. Pestic Sci 55:909–918

Nakagawa Y, Minakuchi C, Ueno T (2000) Inhibition of [^{3}H]ponasterone A binding by ecdysone agonists in the intact Sf-9 cell line. Steroids 65:537–542

Nakagawa Y, Smagghe G, Van Paemel M, Tirry L, Fujita T (2001) Quantitative structure-activity studies of insect growth regulators: XVIII. Effects of substituents on the aromatic moiety of dibenzoylhydrazines on larvicidal activity against the Colorado potato beetle Leptinotarsa decemlineata. Pest Manag Sci 57:858–865

Nakagawa Y, Minakuchi C, Takahashi K, Ueno T (2002a) Inhibition of [^{3}H]ponasterone A binding by ecdysone agonists in the intact Kc cell line. Insect Biochem Mol Biol 32:175–180

Nakagawa Y, Smagghe G, Tirry L, Fujita T (2002b) Quantitative structure-activity studies of insect growth regulators: XIX. Effects of substituents on the aromatic moiety of dibenzoylhydrazines on larvicidal activity against the beet armyworm Spodoptera exigua. Pest Manag Sci 58:131–138

Ogata K, Umeyama H (2000) An automatic homology modeling method consisting of database searches and simulated annealing. J Mol Graph Model 18:258–272

Ogura T, Nakagawa Y, Minakuchi C, Miyagawa H (2005) QSAR for binding affinity of substituted dibenzoylhydrazines to intact Sf-9 cells. J Pestic Sci 30:1–6
Ohlson S (2008) Designing transient binding drugs: a new concept for drug discovery. Drug Discov Today 13:433–439
Oikawa N, Nakagawa Y, Soya Y, Nishimura K, Kurihara N, Ueno T, Fujita T (1993) Enhancement of *N*-acetylglucosamine incorporation into the cultured integument of *Chilo suppressalis* by molting hormone and dibenzoylhydrazine insecticides. Pestic Biochem Physiol 47:165–170
Oikawa N, Nakagawa Y, Nishimura K, Ueno T, Fujita T (1994a) Quantitative structure-activity analysis of larvicidal 1-(substituted benzoyl)-2-benzoyl-1-*tert*-butylhydrazines against *Chilo suppressalis*. Pestic Sci 41:139–148
Oikawa N, Nakagawa Y, Nishimura K, Ueno T, Fujita T (1994b) Quantitative structure-activity studies of insect growth regulators X. Substituent effects on larvicidal activity of 1-*tert*-butyl-1-(2-chlorobenzoyl)-2-(substituted benzoyl)hydrazines against *Chilo suppressalis* and design synthesis of potent derivatives. Pestic Biochem Physiol 48:135–144
Okada T, Le Trong I, Fox BA, Behnke CA, Stenkamp RE, Palczewski K (2000) X-Ray diffraction analysis of three-dimensional crystals of bovine rhodopsin obtained from mixed micelles. J Struct Biol 130:73–80
Okuno Y (2008) In silico drug discovery based on the integration of bioinformatics and chemoinformatics. Yakugaku Zasshi 128:1645–1651
Osolodkin DI, Palyulin VA, Zefirov NS (2011) Structure-based virtual screening of glycogen synthase kinase 3beta inhibitors: analysis of scoring functions applied to large true actives and decoy sets. Chem Biol Drug Des 78:378–390
Pearson WR, Lipman DJ (1988) Improved tools for biological sequence comparison. Proc Natl Acad Sci USA 85:2444–2448
Ramachandran GN, Ramakrishnan C, Sasisekharan V (1963) Stereochemistry of polypeptide chain configurations. J Mol Biol 7:95–99
Rasmussen SG, Choi HJ, Rosenbaum DM, Kobilka TS, Thian FS, Edwards PC, Burghammer M, Ratnala VR, Sanishvili R, Fischetti RF et al (2007) Crystal structure of the human beta2 adrenergic G-protein-coupled receptor. Nature 450:383–387
Rosenwald A, Wright G, Chan WC, Connors JM, Campo E, Fisher RI, Gascoyne RD, Muller-Hermelink K, Smeland EB, Staudt LM (2002) The use of molecular profiling to predict survival after chemotherapy for diffuse large-B-cell lymphoma. N Engl J Med 346:1937–1947
Sage C, Wang R, Jones G (2011) G-protein coupled receptors virtual screening using genetic algorithm focused chemical space. J Chem Inf Model 51:1754–1761
Sawada Y, Yanai T, Nakagawa H, Tsukamoto Y, Tamagawa Y, Yokoi S, Yanagi M, Toya T, Sugizaki H, Kato Y et al (2003) Synthesis and insecticidal activity of benzoheterocyclic analogues of *N*′-benzoyl-*N*-(*tert*-butyl)benzohydrazide: Part 3. Modification of N-tert-butylhydrazine moiety. Pest Manag Sci 59:49–57
Schaffer AA, Wolf YI, Ponting CP, Koonin EV, Aravind L, Altschul SF (1999) IMPALA: matching a protein sequence against a collection of PSI-BLAST-constructed position-specific score matrices. Bioinformatics 15:1000–1011
Schreiber SL (2000) Target-oriented and diversity-oriented organic synthesis in drug discovery. Science 287:1964–1969
Shah F, Mukherjee P, Gut J, Legac J, Rosenthal PJ, Tekwani BL, Avery MA (2011) Identification of novel malarial cysteine protease inhibitors using structure-based virtual screening of a focused cysteine protease inhibitor library. J Chem Inf Model 51:852–864
Shimamura T, Shiroishi M, Weyand S, Tsujimoto H, Winter G, Katritch V, Abagyan R, Cherezov V, Liu W, Han GW et al (2011) Structure of the human histamine H1 receptor complex with doxepin. Nature 475:65–70
Shimizu B, Nakagawa Y, Hattori K, Nishimura K, Kurihara N, Ueno T (1997) Molting hormonal and larvicidal activities of aliphatic acyl analogs of dibenzoylhydrazine insecticides. Steroids 62:638–642

Simons KT, Kooperberg C, Huang E, Baker D (1997) Assembly of protein tertiary structures from fragments with similar local sequences using simulated annealing and Bayesian scoring functions. J Mol Biol 268:209–225

Simons KT, Ruczinski I, Kooperberg C, Fox BA, Bystroff C, Baker D (1999) Improved recognition of native-like protein structures using a combination of sequence-dependent and sequence-independent features of proteins. Proteins 34:82–95

Smagghe G, Nakagawa Y, Carton B, Mourad AK, Fujita T, Tirry L (1999) Comparative ecdysteroid action of ring-substituted dibenzoylhydrazines in *Spodoptera exigua*. Arch Insect Biochem Physiol 41:42–53

Smellie AS, Crippen GM, Richards WG (1991) Fast drug-receptor mapping by site-directed distances – a novel method of predicting new pharmacological leads. J Chem Inf Comput Sci 31:386–392

Tanaka K, Tsukamoto Y, Sawada Y, Kasuya A, Hotta H, Ichinose R, Watanabe T, Toya T, Yokoi S, Kawagishi A et al (2001) Chromafenozide: a novel lepidopteran insect control agent. Annu Rep Sankyo Res Lab 53:1–49

Terashi G, Takeda-Shitaka M, Takaya D, Komatsu K, Umeyama H (2005) Searching for protein-protein interaction sites and docking by the methods of molecular dynamics, grid scoring, and the pairwise interaction potential of amino acid residues. Proteins-Struct Funct Bioinform 60:289–295

Thomas GL, Spandl RJ, Glansdorp FG, Welch M, Bender A, Cockfield J, Lindsay JA, Bryant C, Brown DF, Loiseleur O et al (2008) Anti-MRSA agent discovery using diversity-oriented synthesis. Angew Chem Int Ed Engl 47:2808–2812

Tohidi-Esfahani D, Lawrence MC, Graham LD, Hannan GN, Simpson AM, Hill RJ, Tohidi-Esfahani D, Lawrence MC, Graham LD, Hannan GN, Simpson AM, Hill RJ (2011) Isoforms of the heteropteran Nezara viridula ecdysone receptor: protein characterisation, RH5992 insecticide binding and homology modelling. Pest Manag Sci 67:1457–1467

Tomii K, Akiyama Y (2004) FORTE: a profile-profile comparison tool for protein fold recognition. Bioinformatics 20:594–595

Tsuji G, Takeda T, Furusawa I, Horino O, Kubo Y (1997) Carpropamid, an anti-rice blast fungicide, inhibits scytalone dehydratase activity and appressorial penetration in Colletotrichum lagenarium. Pestic Biochem Physiol 57:211–219

Vedani A, Dobler M (2002) Multidimensional QSAR: moving from three- to five-dimensional concepts. Quant Struct-Act Relatsh 21:382–390

Verloop A (1983) The STERIMOL approach: further development of the method and new applications. In: Miyamoto J, Kearney PC (eds) Pesticide chemistry, human welfare and environment. Pergamon Press, Oxford, pp 339–344

Vilar S, Ferino G, Phatak SS, Berk B, Cavasotto CN, Costanzi S (2011) Docking-based virtual screening for ligands of G protein-coupled receptors: not only crystal structures but also in silico models. J Mol Graph Model 29:614–623

Wallner B, Fang H, Elofsson A (2003) Automatic consensus-based fold recognition using Pcons, ProQ, and Pmodeller. Proteins 53(Suppl 6):534–541

Walters WP, Stahl MT, Murcko MA (1998) Virtual screening – an overview. Drug Discov Today 3:160–178

Wang F, Li J, Sinn AL, Knabe WE, Khanna M, Jo I, Silver JM, Oh K, Li L, Sandusky GE et al (2011) Virtual screening targeting the urokinase receptor, biochemical and cell-based studies, synthesis, pharmacokinetic characterization, and effect on breast tumor metastasis. J Med Chem 54:7193–7205

Watanabe B, Nakagawa Y, Ogura T, Miyagawa H (2004) Stereoselective synthesis of (*22R*)- and (*22S*)-castasterone/ponasterone A hybrid compounds and evaluation of their molting hormone activity. Steroids 69:483–493

Wheelock CE, Miyagawa H (2006) The omicization of agrochemical research. J Pestic Sci 31:240–244

Wheelock CE, Nakagawa Y, Harada T, Oikawa N, Akamatsu M, Smagghe G, Stefanou D, Iatrou K, Swevers L (2006) High-throughput screening of ecdysone agonists using a reporter gene assay followed by 3-D QSAR analysis of the molting hormonal activity. Bioorg Med Chem 14:1143–1159

Wiederstein M, Sippl MJ (2007) ProSA-web: interactive web service for the recognition of errors in three-dimensional structures of proteins. Nucleic Acids Res 35:W407–W410

Wilk W, Zimmermann TJ, Kaiser M, Waldmann H (2010) Principles, implementation, and application of biology-oriented synthesis (BIOS). Biol Chem 391:491–497

Wing KD (1988) RH 5849, a nonsteroidal ecdysone agonist: effects on a *Drosophila* cell line. Science 241:467–469

Yabuuchi H, Niijima S, Takematsu H, Ida T, Hirokawa T, Hara T, Ogawa T, Minowa Y, Tsujimoto G, Okuno Y (2011) Analysis of multiple compound-protein interactions reveals novel bioactive molecules. Mol Syst Biol 7:472

Yokota T, Arima M, Takahashi N (1982) Castasterone, a new phytosterol with plant-hormone potency, from chestnut insect gall. Tetrahedron Lett 23:1275–1278

Zhou H, Zhou Y (2005) Fold recognition by combining sequence profiles derived from evolution and from depth-dependent structural alignment of fragments. Proteins 58:321–328

Chapter 8
Arthropod Genomics and Pest Management Targeting GPCRs

Cornelis J.P. Grimmelikhuijzen and Frank Hauser

1 Introduction

Insects are the largest animal group on Earth, comprising more than 1.4 million species, which is a very large number when compared to mammals (5,000 species). Together with the other arthropods, such as crustaceans, chelicerates (spiders and mites), and myriapods (centipedes), insects constitute more than 85% of all living animal species (Fig. 8.1). In accordance with this large number, insects play a crucial role in ecology, being an important part of the food web, and being crucial for the pollination of more than 70% of all flowering plants. These pollinator activities are not equally carried out by all insect groups, but are especially prominent among hymenopterans, such as the various honeybees and bumblebees (Fig. 8.2). Also for agriculture, the pollination activities of insects are extremely important and it has been calculated that the value of honey bee pollination in the United States alone is already 15 billion dollars worth of crop yearly (Morse and Calderone 2000). However, insects can also be serious agricultural pests, destroying up to 30% of our potential annual harvest (http://www.gnb.ca/0078/ForestPestControl-e.asp). In addition, some insects are important plant disease vectors, transmitting pathogenic plant viruses, causing, again, considerable crop losses (Oerke and Dehne 2004).

Insects are also medically important, because some of them are vectors for serious diseases such as malaria, elephantiasis (lymphatic filariasis), sleeping sickness, yellow fever, dengue fever, West Nile disease, Chagas disease, and many more. There are yearly 300–500 million cases of malaria and more than one million deaths (mostly children under the age of 5). But also the other insect-borne diseases are equally serious.

C.J.P. Grimmelikhuijzen (✉) • F. Hauser
Center for Functional and Comparative Insect Genomics, Cell and Neurobiology, Department of Biology, University of Copenhagen, Universitetsparken 15, DK-2100 Copenhagen, Denmark
e-mail: cgrimmelikhuijzen@bio.ku.dk; fhauser@bio.ku.dk

I. Ishaaya et al. (eds.), *Advanced Technologies for Managing Insect Pests*,
DOI 10.1007/978-94-007-4497-4_8,

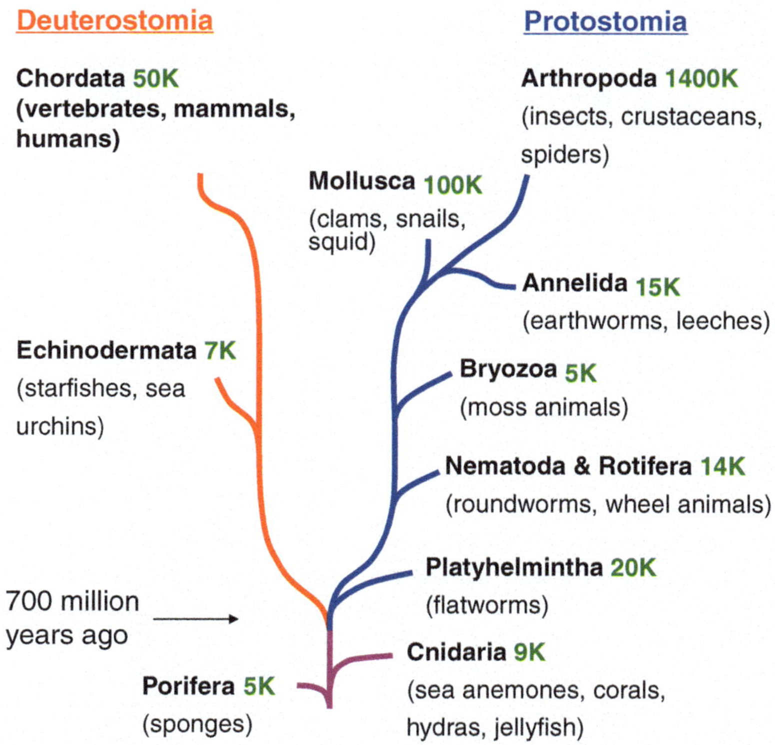

Fig. 8.1 Schematic representation of the evolution of multicellular animals. Two major evolutionary branches are shown, the Protostomia, to which arthropods and most other invertebrates belong (indicated by a *blue line*), and the Deuterostomia to which vertebrates and some minor groups of invertebrates belong (indicated by a *red line*). Note that most animal species (>85%) are arthropods. The proto- and deuterostomian branches split about 700 million years ago (Douzery et al. 2004) (Modified from Grimmelikhuijzen et al. 2009 with permission)

Elephantiasis, for example, disables more than 150 million people worldwide and 1.1 billion people, 16% of the world's population, are at risk of being infected.

Finally, insects can be excellent model animals for the study of basic biological questions, such as the molecular mechanism of development (Nusslein-Volhard and Wieschaus 1980). One such model is *Drosophila melanogaster*, which is indispensable for the advancement of modern biology, due to its short generation times (12 days) and the availability of mutants for virtually each of its 14,000 genes (Adams et al. 2000).

2 Arthropod Genome Projects

Because insects are so important, a large number of insect genomes have been sequenced during the last 12 years, or are in the pipeline of being sequenced (Fig. 8.2). In fact, *D. melanogaster* was the second animal to be sequenced in 2000 (Adams et al. 2000),

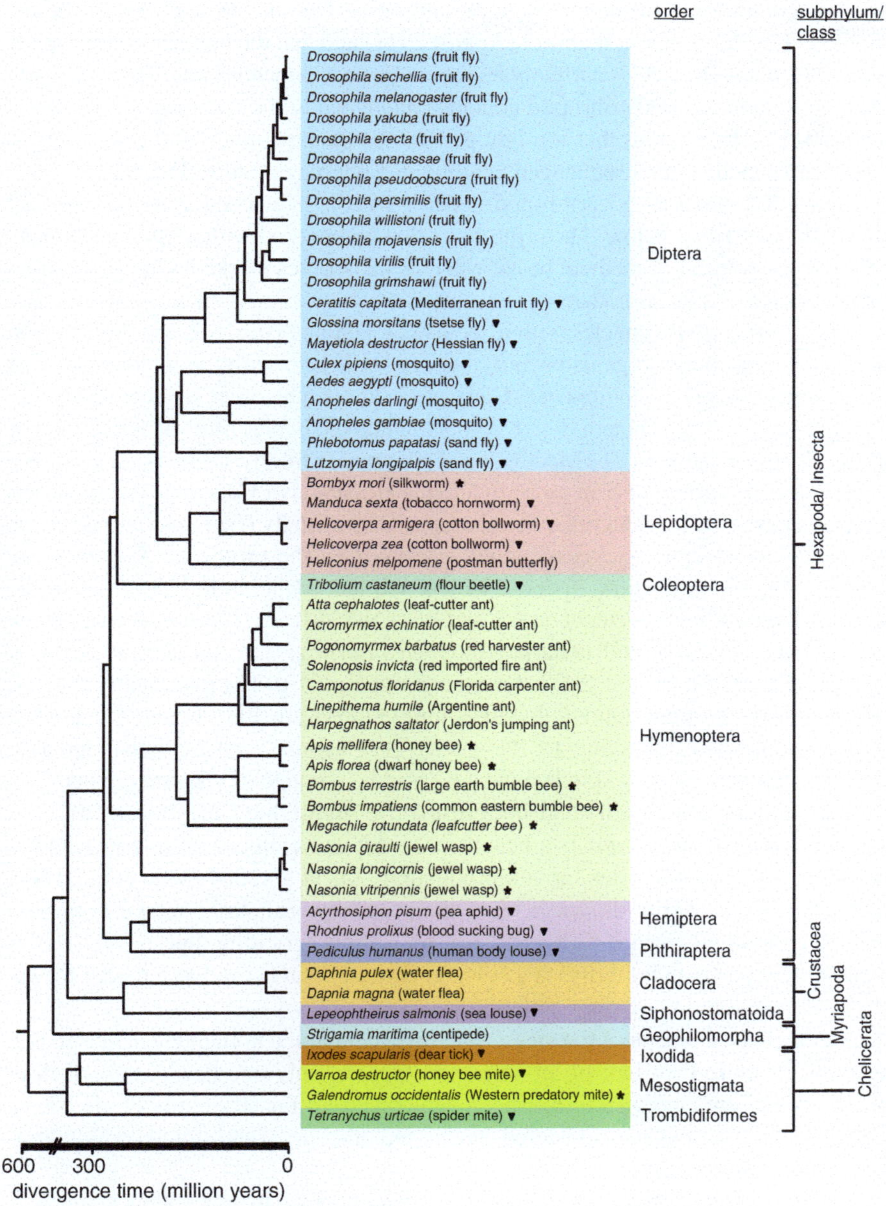

Fig. 8.2 An overview of the arthropod species with a sequenced genome. Please note that the number of sequenced arthropods is currently increasing in an exponential way so that this figure only gives the status of August 2011

after the nematode *Caenorabditis elegans* in 1998 (The *C. elegans* Sequencing Consortium 1998), but before humans in 2001 (Lander et al. 2001). In addition to insects, Fig. 8.2 also shows the other arthropods with a sequenced genome. We have included them, because several of the other arthropods, such as mites and ticks, also can be agriculturally or medically important pests. One has to realize that the list shown in Fig. 8.2 is only a snapshot (August 2011) and that the number of arthropods with a "Genome Project" is currently increasing in an exponential way. There are even plans to sequence 5,000 arthropod genomes, launched in the i5K project (Robinson et al. 2011). This implies that it might perhaps be the last time that it is possible to list all arthropods with a sequenced genome in a one-page figure (Fig. 8.2).

Figure 8.2 contains 53 arthropod species with a sequenced genome and will shortly be described below. The arthropods that are marked with a filled arrowhead (▼) are agricultural or medical pests, whereas the species marked with an asterisks (★) are beneficial arthropods.

The 12 *Drosophila* species with a sequenced genome (Fig. 8.2, top, highlighted in blue) are all model organisms and, together, will help with the interpretation of the genome from *D. melanogaster* (Clark et al. 2007). The other sequenced insects, belonging to the order Diptera (Fig. 8.2, top, highlighted in blue), are all agricultural or medically important pests. The Medfly (the Mediterranean fruitfly, *Ceratitis capitata*) is a pest for a wide variety of fruit crops, while the Hessian fly *(Mayetiola destructor)* is a serious pest of wheat and other cerial crops. The tsetse fly *(Glossina morsitans)* is the vector for sleeping sickness, which is caused by the protozoan *Trypanosoma brucei.* The four mosquito species with a sequenced genome transmit various serious diseases, such as elephantiasis and West Nile disease *(Culex pipiens),* yellow fever *(Aedes aegypti),* and malaria *(Anopheles darlingi* and *Anopheles gambiae)* (Holt et al. 2002; Nene et al. 2007; Arensburger et al. 2010). The two sandfly species, *(Phlebotomus papatasi* and *Lutzomyia longipalpis)* are the vectors for several protozoan species belonging to the genus *Leishmania,* causing Leishmaniasis, which is characterized by skin lesions, fever, and a damage of visceral organs.

The Lepidoptera (moths and butterflies) are highlighted in pink in Fig. 8.2. The silkworm *Bombyx mori* is a beneficial insect and cultured on a large scale in Asian and African countries for silk production (Xia et al. 2004; Mita et al. 2004). The tobacco hornworm *Manduca sexta,* the corn earworm, *Helicoverpa armigera,* and the cotton bollworm *Helicoverpa zea,* are serious agricultural pests. *Heliconius melpomene* (postman butterfly) is a colorful butterfly that is used as a model for the study of the evolution of wing pattern formation.

Although the Coleoptera (beetles), which are highlighted in light green in Fig. 8.2, account for more than half of all arthropod species, only one beetle, the red flour beetle *Tribolium castaneum,* has been sequenced so far (Richards et al. 2008). This insect is a serious pest for stored cereals and other dried and stored commodities for human consumption.

The Hymenoptera (bees, wasps, and ants) are highlighted in light yellow in Fig. 8.2. The Hymenoptera comprise many insects with different types of sociality (colony formation and reproductive division with queens, workers and/or soldiers). Seven different ant species (Atta *cephalotes, Acromyrmex echinatior, Pogonomyrmex barbatus, Solenopsis invicta, Camponotus floridianus, Linepithema humile,* and *Harpegnatos saltator)* have been sequenced, which all are models for the study of eusociality (Bonasio et al. 2010; Suen et al. 2011; Wurm et al. 2011; Smith et al. 2011a, b; Nygaard et al. 2011). Some ants even carry out farming and "invented" agriculture long before humans did, such as the leafcutter ants *(A. cephalotes* and *A. echinatior)* (Suen et al. 2011; Nygaard et al. 2011). Honeybees and bumblebees have different forms of sociality and are highly active pollinators and, thus, highly beneficial for agriculture. Within these social bees, four species have been sequenced *(Apis mellifera, Apis florae, Bombus terrestris, Bombus impatiens),* while a fifth bee, the leafcutter bee *(Megachile rotundata),* is a solitary species (Fig. 8.2) (Weinstock et al. 2006). Also, many wasps are social insects but, so far, only three solitary species, belonging to the genus *Nasonia (N. giraulti, N. longicornis, N. vitripennis)* have been sequenced (Fig. 8.2) (Werren et al. 2010). *Nasonia* wasps are parasitic and such wasps lie eggs inside other insects (eggs injected into host eggs, larvae, pupae, adults), some of which are agricultural pests (for example lepidopteran larvae and aphids). Parasitic wasps can, thus, be used to control the populations of agricultural pests and they are, in fact, already widely applied in organic agriculture.

All the insects discussed above are holometabolans (insects with a metamorphosis between the larval and adult developmental stages). The three remaining insects of Fig. 8.2 (highlighted in violet and blue) are all hemimetabolous (where there is no gross difference between larval and adult stages). The pea aphid, *Acyrthosiphon pisum,* (order Hemiptera) was sequenced, because it is an important agricultural pest for a large variety of green food plants and, also, because it is transmitting a large number of pathogenic plant viruses (Richards et al. 2010). The blood sucking bug *Rhodnius prolixus* is a vector for the protozoan *Trypanosoma cruzi* that causes Chagas disease, which is occurring in rural areas of South America. The human body louse, *Pediculus humanus,* (order Phtiraptera), highlighted in blue, has been sequenced, because it is medically important, transmitting epidemic typhus, relapsing fever, and trench fever (Kirkness et al. 2010).

The branchiopods (order Cladocera, subphylum Crustacea), highlighted in orange in Fig. 8.2, are small salt- and freshwater animals like the water fleas, and are regarded to be the ancestor group of insects (Glenner et al. 2006). The genomes from two water flea species, *Daphnia pulex* and *Daphnia magna,* have been sequenced, because these animals are models for freshwater ecology and other environmental issues (Colbourne et al. 2011). The sea louse *Lepeophtheirus salmonis* (order Siphonostomatoida), highlighted in purple is another crustacean with a sequenced genome. It is an exoparasite feeding on the epidermis and blood of marine fish, especially wild and farmed salmon and, therefore, a severe pest for salmon aquaculture.

The centipede *Strigamia maritima* (order Geophilomorpha, subphylum Myriapoda, highlighted in grey-blue in Fig. 8.2) diverged from the other arthropods 400 million years ago. It has been sequenced, because of its evolutionary position, giving invaluable information about the evolution of arthropods and arthropod segment numbers. *Strigamia* has no less than 45–53 trunk segments, bearing 90–106 ft (Arthur and Chipman 2005).

The tick *Ixodes scapularis* (subphylum Chelicerata, highlighted in brown in Fig. 8.2) is medically important, because it is a vector for *Borellia* and other pathogens, which cause diseases such as Lyme disease, and tick-borne encephalitis. The mite *Varroa destructor,* highlighted in bright yellow, is another chelicerate and the most important parasite for bees. An infection by *Varroa* mites can cause the collapse of a large number of bee colonies, which has a strong economic impact on bee keeping and crop pollination. The Western predatory mite *Galendromus occidentalis* is the natural enemy of a variety of other mites, among them spider mites (see below) and can be used to control the populations of agriculturally important pest mites. The spider mite *Tetranychus urticae* (highlighted in green) attacks a wide variety of ornamental and food plants and is an important agricultural pest.

3 How Can Arthropod Genome Project Be of Use for Agriculture and Crop Protection?

This paragraph only concerns the agriculturally important arthropods, which are either beneficial (10 species marked with ★ in Fig. 8.2) or pests (20 species marked with ▼in Fig. 8.2).

The genomes from the beneficial arthropods represent invaluable resources to better understand these animals. This improved understanding may lead to the identification of genes that make the animals more robust to parasitism (honey bees), less aggressive (honey bees), or simply larger and more effective (silkworms). If the arthropods are used for the biological control of pests, it would be important to identify genes involved in host seeking behavior and host specificity. The obvious next step would be to create transgenic animals that have improved qualities, for example a larger silkworm that produces more silk, silk with different physical properties, or bulk amounts of recombinant proteins woven as silk (Omenetto and Kaplan 2010; Ma et al. 2011; Tomita 2011). It will be clear that scientists will need many years to fully exploit a newly sequenced arthropod genome and to convert that knowledge into useful high-tech applications.

Also for the arthropod pests listed in Fig. 8.2 their sequenced genomes will in first instance be used to better understand their biology. An improved understanding of their biology will lead to obvious drug targets and to the possibility to reduce their populations. However, already before the complete understanding of the biology of certain pest arthropods, drug targets can be identified and exploited. This means that we can start with this type of work as soon as a genome sequence has become available.

Historically, there are only a few insecticide targets available in insects and other arthropods. The best known target is acetylcholinesterase, which is blocked by various organophosphate pesticides. However, the acetylcholinesterases, which break down the neutrotransmitter acetylcholine (into acetic acid and choline), occur in all higher eukaryotes with a nervous system (all proto- and deuterostomians from Fig. 8.1), implying that a drug blocking this family of enzymes cannot be very selective and, thus, not very safe for the environment. This is dramatically illustrated by the fact that Tabun and Sarin, which are highly deadly nerve gases used for chemical warfare, are both organophosphates, chemically related to the traditional pesticides. Not all classes of pesticides are targeted at acetylcholinesterases and many of them, such as the pyrethroids, act at neuronal sodium channels.

These channels are, again, occurring in all higher eukaryotes depictured in Fig. 8.1 and it is, therefore, hard to obtain a desirable selectivity.

A new generation of insecticides should be selective and, thereby, safe for the environment. Finding a specific insecticide target, occurring in only a small group of insects is now possible after the genomes from 53 arthropods are becoming freely accessible (Fig. 8.2). We would like to proposed G protein-coupled receptors (GPCRs) as suitable new insecticide targets and the rationale behind this is explained in the next chapter.

Instead of spraying insecticides, insect pests could be hit more precisely. For example, insect drug targets (proteins) could be genetically down-regulated by using transgenic host plants that produce double-stranded RNA (dsRNA) directed against the insect target mRNA (Gordon and Waterhouse 2007). This RNAi approach is very promising because it is environmentally friendly: The agent is not sprayed, but only produced in the host plants. It is also potentially more selective, because in this case the selectivity is based on nucleotide hybridization and the nucleotide triplets coding for the target protein's amino acid residues are much less conserved (and therefore more variable among insects) than the amino acid residues themselves. For developing RNAi-based pest control, it is extremely important to have access to the genomes of these pest insects, so that the nucleotide sequences of the target genes can be quickly identified. However, although a very promising approach, it is still very uncertain whether it works. For a large group of agricultural pest insects, the Lepidoptera, for example, RNAi does not appear to be effective for most of them (Terenius et al. 2011).

4 G Protein-Coupled Receptors (GPCRs) as Targets for a New Generation of Insecticides

Biogenic amines, neuropeptides, protein hormones and their G protein-coupled receptors (GPCRs) occupy a high hierarchical position in the physiology of arthropods and other animals and steer central processes such as reproduction, development, and feeding (Nassel and Winther 2010). GPCRs are transmembrane proteins that cross the cell membrane seven times. Their extracellular portions bind to signal molecules that are transported by the blood stream or are diffusing in the local extracellular space. GPCRS are highly "drugable" proteins, because (1) drugs have only to reach the blood stream and it is not necessary for them to cross several membrane layers, including the cell membrane, and (2) they are located at the start of an intracellular second messenger cascade that strongly amplifies the signal. This drugabilty of GPCRs is illustrated by the fact that 40–50% of all pharmaceutical drugs act on human GPCRs.

In the last few years, we and other research groups have identified about 70% of all biogenic amine, neuropeptide, and protein hormone GPCRs from *Drosophila* and several other insects (Hauser et al. 2006, 2008; Stafflinger et al. 2008; Hansen et al. 2010, 2011; Collin et al. 2011; Yapici et al. 2008; Kim et al. 2010; Yamanaka

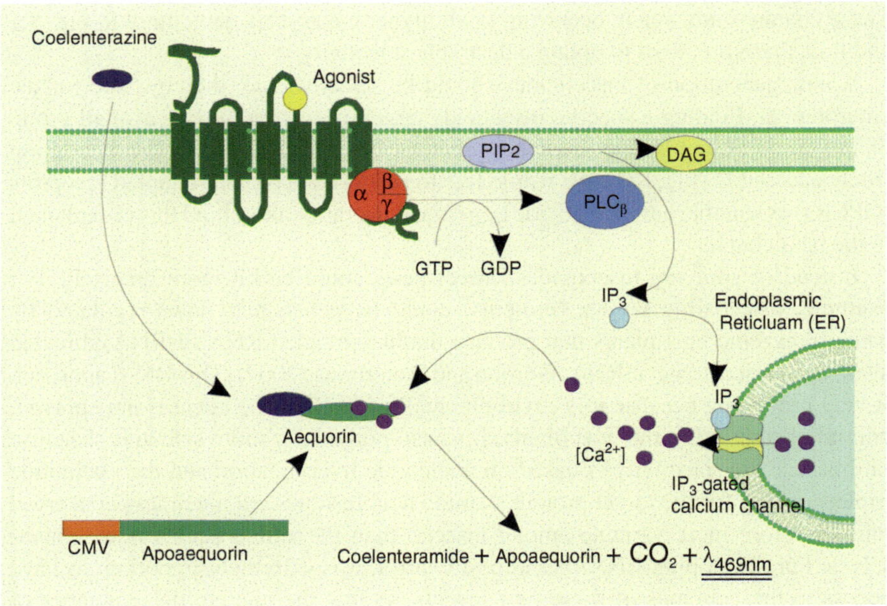

Fig. 8.3 Schematic drawing of the assay system used in our research group. In the upper part the CHO cell membrane is shown, expressing the insect GPCR (*dark green*), which binds to both the extracellular agonist (*yellow*) and the intracellular G protein, G-16 (*red*). Upon receptor activation, the alpha subunit of G-16 dissociates from the beta/gamma subunits and activates phospholipase C (PLC-beta) which initiates an IP_3/Ca^2+ cascade. The increased Ca^{2+} concentration stimulates aequorin to emit light of 469 nm (bioluminescence). This system was first published by Stables and coworkers for mammalian GPCRs (Stables et al. 1997) and successfully modified and applied by us for GPCRs from insects (Lenz et al. 2001; Secher et al. 2001) (Reproduced from Hauser et al. 2006 with permission)

et al. 2010; Horodyski et al. 2011). The techniques we used to identify these GPCRs include the annotations of their genes in the newly sequenced arthropod genomes (Hauser et al. 2006, 2008) and the cloning, expression, and characterization of the receptors in cells in cell culture, or in *Xenopus* oocytes (Hauser et al. 2006, 2008). Figure 8.3 shows the cell culture bioassay system that our research group has developed for insect GPCRs since 2000 (Lenz et al. 2001; Secher et al. 2001; Staubli et al. 2002; Cazzamali and Grimmelikhuijzen 2002), after it was originally published for mammalian GPCRs in 1997 (Stables et al. 1997). In our bioassay (Fig. 8.3), receptor activation by its ligand leads to an IP_3/Ca^{2+} second messenger cascade that results in light emission (bioluminescence) of the cells expressing the receptor. This bioluminescence can easily be measured and quantified (Lenz et al. 2001; Secher et al. 2001; Stables et al. 1997).

During the last 15 years we have been involved in the annotation and molecular characterization of biogenic amine, neuropeptide, and protein hormone GPCRs (Hauser et al. 1997, 1998, 2006, 2008) and their ligands (Nygaard et al. 2011; Li et al. 2008; Hauser et al. 2010; Dircksen et al. 2011) in arthropods. The picture that

Table 8.1 The core set of neuropeptide genes in arthropods. The core set of neuropeptide genes found in *Drosophila melanogaster* (fruit fly), *Aedes aegypti* (yellow fever mosquito), *Bombyx mori* (silkmoth), *Tribolium castaneum* (flour beetle), *Acromyrmex echinatior* (leaf-cutter ant), *Apis mellifera* (honey bee), *Nasonia vitripennis* (parasitic wasp), *Acyrthosiphon pisum* (pea aphid) and *Daphnia pulex* (water flea). Numbers indicate gene numbers in each species

neuropeptide	*Drosophila*	*Aedes*	*Bombyx*	*Tribolium*	*Acromyrmex*	*Apis*	*Nasonia*	*Acyrthosiphon*	*Daphnia*
AKH	1	1	2	2	1	1	1	1	1
AST-C	1	1	1	1	1	1	1	1	1
AST-CC	1	1	1	1	1	1	1	1	2
Bursicon-α	1	1	1	1	1	1	1	1	1
Bursicon-β	1	1	1	1	1	1	1	1	1
CCAP	1	1	1	1	1	1	1	1	1
CCHamide-1	1	1	1	1	1	1	1	1	1
DH (Calc.-like)	1	1	1	1	1	1	1	1	1
DH (CRF-like)	1	1	1	1	1	1	1	1	1
EH	1	5	1	1	1	1	1	3	2
ETH	1	1	1	1	1	1	1	1	1
ILP-B	5	6	38	2	1	1	1	7	1
ITP	1	1	1	1	1	1	1	1	1
Myosuppressin	1	1	1	1	1	1	1	1	1
SIFamide	1	1	2	1	1	1	1	1	1
sNPF	1	2	1	1	1	1	1	1	1
Tachykinin	1	1	1	1	1	1	1	1	1
	22	28	57	20	18	18	18	26	20

Abbreviations: *AKH* adipokinetic hormone, *AST* allatostatin, *CCAP* crustacean cardio-active peptides, *DH (Calc.-like)* calcitonin-like diuretic hormone, *DH (CRF-like)* corticotropin releasing factor-like diuretic hormone, *EH* eclosion hormone, *ETH* ecdysis triggering hormone, *ILP* insulin-like peptide, *ITP* ion transport peptide, *sNPF* short neuropeptide F

emerges from this long-term work is the following: Some neuropeptide signaling systems (neuropeptides plus GPCRs) occur in all arthropods that we investigated. We call these 17 neuropeptide genes and their corresponding GPCR genes, the "core set" (Table 8.1) (Hauser et al. 2010; Nygaard et al. 2011). About 29 neuropeptide genes (and their corresponding GPCR genes) belong to the "variable set" (Table 8.2) (Hauser et al. 2010; Nygaard et al. 2011). Members of this set occur in some arthropod groups, but are absent in others. It is possible to trace the evolution of these members of the variable set and to see which evolutionary branches have lost a variable set member and which branches have conserved it. For example, proctolin (Table 8.2) is present in *Tribolium*, pea aphid, and some of the Diptera (Fig. 8.2, top), but absent in many other arthropods. Thus, an antagonist/agonist of the proctolin receptor would be a rather selective insecticide. The presence of the "variable set" of neuropeptides, therefore, gives us the possibility to develop insect-specific insecticides that hit a certain insect group, but spare most of the other insect species. The existence of the "core set" of neuropeptides warns us that, if antagonist/agonist insecticides are

Table 8.2 The variable set of neuropeptide genes in arthropods. The variable set of neuropeptide genes present (yes) or absent (no) in *Drosophila melanogaster* (fruit fly), *Aedes aegypti* (yellow fever mosquito), *Bombyx mori* (silkmoth), *Tribolium castaneum* (flour beetle), *Acromyrmex echinatior* (leaf-cutter ant), *Apis mellifera* (honey bee), *Nasonia vitripennis* (parasitic wasp), *Acyrthosiphon pisum* (pea aphid) and *Daphnia pulex* (water flea)

neuropeptide	*Drosophila*	*Aedes*	*Bombyx*	*Tribolium*	*Acromyrmex*	*Apis*	*Nasonia*	*Acyrthosiphon*	*Daphnia*
ACP	no	yes	yes	yes	no	no	yes	yes	no
ADF-b	no	no	no	yes	no	no	no	no	no
Allatotropin	no	yes	yes	yes	no	no	no	yes	yes
AST-A	yes	yes	yes	no	yes	yes	yes	yes	yes
AST-B	yes	yes	yes	yes	no	no	no	yes	yes
Capa	yes	yes	yes	yes	no	yes	no	yes	yes
CCHamide-2	yes	yes	yes	yes	yes	yes	yes	yes	no
Corazonin	yes	yes	yes	no	yes	yes	yes	no	yes
FMRFa	yes	yes	yes	yes	no	yes	no	yes	yes
GPA2	yes	yes	yes	yes	no	no	no	yes	yes
GPB5	yes	yes	yes	yes	no	no	no	yes	yes
ILP-A	yes	yes	no	yes	no	no	no	no	no
ILP-C	yes	yes	no	yes	yes	yes	yes	yes	yes
Inotocin	no	no	no	yes	yes	no	yes	no	yes
Kinin	yes	yes	yes	no	no	no	no	yes	no
Neuroparsin	no	yes	yes	yes	yes	yes	yes	no	yes
NPF	yes	yes	yes	no	yes	yes	yes	yes	yes
NPLP-1	yes	yes	yes	yes	yes	yes	no	yes	no
NPLP-2	yes	no	no	no	no	yes	no	no	no
NPLP-3	yes	no	no	no	no	yes	no	no	no
NPLP-4	yes	no	no	no	no	no	no	no	no
Orcokinin	no	yes	yes	no	yes	yes	yes	yes	yes
PDF	yes	yes	yes	no	yes	yes	yes	no	yes
Proctolin	yes	no	no	yes	no	no	no	yes	yes
PTTH	yes	yes	yes	yes	yes	no	yes	no	no
Pyrokinin	yes	yes	yes	yes	yes	yes	yes	yes	no
RYamide	yes	yes	yes	yes	no	yes	yes	yes	yes
Sex peptide	yes	no	no	no	no	no	no	no	no
Sulfakinin	yes	yes	yes	yes	yes	yes	no	no	yes

Abbreviations: *ACP* adipokinetic hormone/corazonin-related neuropeptide, *ADF* antidiuretic factor, *AST* allatostatin, *GPA2* glycoprotein hormone A2, *GPB5* glycoprotein hormone B5, *ILP* insulin-like peptide, *NPF* neuropeptide F, *NPLP* neuropeptide-like precursor, *PDF* pigment dispersing factor, *PTTH* prothoracicotropic hormone

being developed against their receptors, all insects and probably all arthropods might be affected. An insecticide directed a member of the "core set" neuropeptide GPCRs, however, might still be relatively safe for the environment, because more than 50% of the insect GPCRs have no counterparts in the Deuterostomia and lower Protostomia (Fig. 8.1).

After selecting certain insect neuropeptide receptors as potential insecticide targets, it would be essential to test, whether an antagonist or agonist would indeed kill the target insect. For the antagonist effect, RNAi could be applied (injection of dsRNA directed against the receptor mRNA) (Bai et al. 2011). For the agonist effect, the peptides themselves could be injected.

Finding a small molecule non-peptide antagonist or agonist would require high-throughput screening of large chemical libraries on microtiter plates, containing cells expressing the receptors (Fig. 8.3) followed by lead compound optimization. This is a routine procedure for pharmaceutical industry carrying out drug discovery and can also be applied for the search of new classes of insecticides.

5 Conclusions

The availability of more than 50 sequenced arthropod genomes will be an enormous help for our understanding of agriculturally beneficial insects, such as the honeybee, parasitic wasps, and the silkworm, but also of agricultural pests. We expect that this improved understanding will lead to improved pollination, silk production and to the development of new and environmentally safe ways of crop protection.

Acknowledgment We thank Anders Bo Ronnegaard Hansen for typing the manuscript and the Danish Research Agency, and Novo Nordisk Foundation for financial support.

References

Adams MD, Celniker SE, Holt RA et al (2000) The genomic sequence of *Drosophila melanogaster*. Science 287:2185–2195

Arensburger P, Megy K, Waterhouse RM et al (2010) Sequencing of *Culex quinquefasciatus* establishes a platform for mosquito comparative genomics. Science 330:86–88

Arthur W, Chipman AD (2005) The centipede *Strigamia maritima:* what it can tell us about the development and evolution of segmentation. Bioessays 27:653–660

Bai H, Zhu F, Shah K et al (2011) Large-scale RNAi screen of G protein-coupled receptors involved in larval growth, molting and metamorphosis in the red flour beetle. BMC Genomics 12:388

Bonasio R, Zhang G, Ye C et al (2010) Genomic comparison of the ants *Camponotus floridanus* and *Harpegnathos saltator*. Science 329:1068–1071

Cazzamali G, Grimmelikhuijzen CJP (2002) Molecular cloning and functional expression of the first insect FMRFamide receptor. Proc Natl Acad Sci USA 99:12073–12078

Clark AG, Eisen MB, Smith DR et al (2007) Evolution of genes and genomes on the *Drosophila* phylogeny. Nature 450:203–218

Colbourne JK, Pfrender ME, Gilbert D et al (2011) The ecoresponsive genome of *Daphnia pulex*. Science 331:555–561

Collin C, Hauser F, Krogh-Meyer P et al (2011) Identification of the *Drosophila* and *Tribolium* receptors for the recently discovered insect RYamide neuropeptides. Biochem Biophys Res Commun 412:578–583

Dircksen H, Neupert S, Predel R et al (2011) Genomics, transcriptomics and peptidomics of *Daphnia pulex* neuropeptides and protein hormones. J Proteome Res 10:4478–4504

Douzery EJ, Snell EA, Bapteste E et al (2004) The timing of eukaryotic evolution: does a relaxed molecular clock reconcile proteins and fossils? Proc Natl Acad Sci USA 101:15386–15391

Glenner H, Thomsen PF, Hebsgaard MB et al (2006) Evolution. The origin of insects. Science 314:1883–1884

Gordon KH, Waterhouse PM (2007) RNAi for insect-proof plants. Nat Biotechnol 25:1231–1232

Grimmelikhuijzen CJP, Cazzamali G, Williamson M et al (2009) Invertebrate neurohormone GPCRs. In: Squire L (ed) Encyclopedia of neuroscience, vol 5. Academic, Oxford, pp 205–212

Hansen KK, Stafflinger E, Schneider M et al (2010) Discovery of a novel insect neuropeptide signaling system closely related to the insect adipokinetic hormone and corazonin hormonal systems. J Biol Chem 285:10736–10747

Hansen KK, Hauser F, Williamson M et al (2011) The *Drosophila* genes CG14593 and CG30106 code for G-protein-coupled receptors specifically activated by the neuropeptides CCHamide-1 and CCHamide-2. Biochem Biophys Res Commun 404:184–189

Hauser F, Nothacker HP, Grimmelikhuijzen CJP (1997) Molecular cloning, genomic organization and developmental regulation of a novel receptor from *Drosophila melanogaster* structurally related to members of the thyroid-stimulating hormone, follicle-stimulating hormone, luteinizing hormone/choriogonadotropin receptor family from mammals. J Biol Chem 272:1002–1010

Hauser F, Sondergaard L, Grimmelikhuijzen CJP (1998) Molecular cloning, genomic organization and developmental regulation of a novel receptor from *Drosophila melanogaster* structurally related to gonadotropin-releasing hormone receptors from vertebrates. Biochem Biophys Res Commun 249:822–828

Hauser F, Cazzamali G, Williamson M et al (2006) A review of neurohormone GPCRs present in the fruitfly *Drosophila melanogaster* and the honey bee *Apis mellifera*. Prog Neurobiol 80:1–19

Hauser F, Cazzamali G, Williamson M et al (2008) A genome-wide inventory of neurohormone GPCRs in the red flour beetle *Tribolium castaneum*. Front Neuroendocrinol 29:142–165

Hauser F, Neupert S, Willamson M et al (2010) Genomics and peptidomics of neuropeptides and protein hormones present in the parasitic wasp *Nasonia vitripennis*. J Proteome Res 9:5296–5310

Holt RA, Subramanian GM, Halpern A et al (2002) The genome sequence of the malaria mosquito *Anopheles gambiae*. Science 298:129–149

Horodyski FM, Verlinden H, Filkin N et al (2011) Isolation and functional characterization of an allatotropin receptor from *Manduca sexta*. Insect Biochem Mol Biol 41:804–814

Kim YJ, Bartalska K, Audsley N et al (2010) MIPs are ancestral ligands for the sex peptide receptor. Proc Natl Acad Sci USA 107:6520–6525

Kirkness EF, Haas BJ, Sun W et al (2010) Genome sequences of the human body louse and its primary endosymbiont provide insights into the permanent parasitic lifestyle. Proc Natl Acad Sci USA 107:12168–12173

Lander ES, Linton LM, Birren B et al (2001) Initial sequencing and analysis of the human genome. Nature 409:860–921

Lenz C, Williamson M, Hansen GN et al (2001) Identification of four *Drosophila* allatostatins as the cognate ligands for the *Drosophila* orphan receptor DAR-2. Biochem Biophys Res Commun 286:1117–1122

Li B, Predel R, Neupert S et al (2008) Genomics, transcriptomics, and peptidomics of neuropeptides and protein hormones in the red flour beetle *Tribolium castaneum*. Genome Res 18:113–122

Ma L, Xu H, Zhu J et al (2011) Ras1(CA) overexpression in the posterior silk gland improves silk yield. Cell Res 21:934–943

Mita K, Kasahara M, Sasaki S et al (2004) The genome sequence of silkworm, *Bombyx mori*. DNA Res 11:27–35

Morse RA, Calderone NW (2000) The value of honey bee pollination in the United States. Bee Cult 128:1–15

Nassel DR, Winther AM (2010) *Drosophila* neuropeptides in regulation of physiology and behavior. Prog Neurobiol 92:42–104

Nene V, Wortman JR, Lawson D et al (2007) Genome sequence of *Aedes aegypti,* a major arbovirus vector. Science 316:1718–1723

Nusslein-Volhard C, Wieschaus E (1980) Mutations affecting segment number and polarity in *Drosophila*. Nature 287:795–801

Nygaard S, Zhang G, Schiott M et al (2011) The genome of the leaf-cutting ant *Acromyrmex echinatior* suggests key adaptations to advanced social life and fungus farming. Genome Res 21:1339–1348

Oerke EC, Dehne HW (2004) Safeguarding production-losses in major crops and the role of crop protection. Crop Prot 23:275–285

Omenetto FG, Kaplan DL (2010) New opportunities for an ancient material. Science 329:528–531

Richards S, Gibbs RA, Weinstock GM et al (2008) The genome of the model beetle and pest *Tribolium castaneum*. Nature 452:949–955

Richards S, Gibbs RA, Gerardo NM et al (2010) Genome sequence of the pea aphid *Acyrthosiphon pisum*. PLoS Biol 8:e1000313

Robinson GE, Hackett KJ, Purcell-Miramontes M et al (2011) Creating a buzz about insect genomes. Science 331:1386

Secher T, Lenz C, Cazzamali G et al (2001) Molecular cloning of a functional allatostatin gut/brain receptor and an allatostatin preprohormone from the silkworm *Bombyx mori*. J Biol Chem 276:47052–47060

Smith CR, Smith CD, Robertson HM et al (2011a) Draft genome of the red harvester ant *Pogonomyrmex barbatus*. Proc Natl Acad Sci USA 108:5667–5672

Smith CD, Zimin A, Holt C et al (2011b) Draft genome of the globally widespread and invasive Argentine ant *(Linepithema humile)*. Proc Natl Acad Sci USA 108:5673–5678

Stables J, Green A, Marshall F et al (1997) A bioluminescent assay for agonist activity at potentially any G-protein-coupled receptor. Anal Biochem 252:115–126

Stafflinger E, Hansen KK, Hauser F et al (2008) Cloning and identification of an oxytocin/vasopressin-like receptor and its ligand from insects. Proc Natl Acad Sci USA 105:3262–3267

Staubli F, Jorgensen TJD, Cazzamali G et al (2002) Molecular identification of the insect adipokinetic hormone receptors. Proc Natl Acad Sci USA 99:3446–3451

Suen G, Teiling C, Li L et al (2011) The genome sequence of the leaf-cutter ant *Atta cephalotes* reveals insights into its obligate symbiotic lifestyle. PLoS Genet 7:e1002007

Terenius O, Papanicolaou A, Garbutt JS et al (2011) RNA interference in Lepidoptera: an overview of successful and unsuccessful studies and implications for experimental design. J Insect Physiol 57:231–245

The *C. elegans* Sequencing Consortium (1998) Genome sequence of the nematode *C. elegans:* a platform for investigating biology. Science 282:2012–2018

Tomita M (2011) Transgenic silkworms that weave recombinant proteins into silk cocoons. Biotechnol Lett 33:645–654

Weinstock GM, Robinson GE, Gibbs RA et al (2006) Insights into social insects from the genome of the honey bee *Apis mellifera*. Nature 443:931–949

Werren JH, Richards S, Desjardins CA et al (2010) Functional and evolutionary insights from the genomes of three parasitoid *Nasonia* species. Science 327:343–348

Wurm Y, Wang J, Riba-Grognuz O et al (2011) The genome of the fire ant *Solenopsis invicta*. Proc Natl Acad Sci USA 108:5679–5684

Xia Q, Zhou Z, Lu C et al (2004) A draft sequence for the genome of the domesticated silkworm (*Bombyx mori*). Science 306:1937–1940

Yamanaka N, Hua YJ, Roller L et al (2010) *Bombyx* prothoracicostatic peptides activate the sex peptide receptor to regulate ecdysteroid biosynthesis. Proc Natl Acad Sci USA 107:2060–2065

Yapici N, Kim YJ, Ribeiro C et al (2008) A receptor that mediates the post-mating switch in *Drosophila* reproductive behavior. Nature 451:33–37

Chapter 9
RNA Interference and Its Potential for Developing New Control Methods Against Insect Pests

Murad Ghanim and Adi Kliot

1 Introduction

RNA interference (RNAi) is a highly specific and conserved mechanism, and refers to the specific cleavage of mRNA molecules leading to silencing or inactivation of gene expression. RNAi is one of the most important discoveries made in biological sciences in the last two decades. Although the hallmark manuscript describing this discovery in the model nematode *Caenorhabditis elegans* was published in 1998 (Fire et al. 1998), the first research that showed the way to these discoveries started in plant research. Three decades ago, Jorgensen and colleagues studied Anthocyanin biosynthesis in petunia plants and the role of the chalcone synthase (CS) enzyme in this pathway. They noticed that over-expressing transgenic CS in petunia for enhancing their violet color made them whiter. The expression of CS in these transgenic flowers was 50 times lower than its expression in normal flowers (Napoli et al. 1990). A research conducted by Fire, Mello and colleagues many years later found that delivering dsRNA into the nematode *C. elengans* induced specific and potent gene silencing 10–100 times more than delivering either sense or anti-sense molecules alone (Fire et al. 1998). It was puzzling that the complete antisense representing the delivered RNA molecules has never been found, to complete the hypothesis that dsRNA are involved in the inactivation of the endogenous mRNA. This led to the investigations by Baulcombe and colleagues to search for similar sequences of the antisense RNA. Their work yielded the discovery that there are such molecules, representing part of the delivered antisense RNA; however these molecules were much shorter than expected. They were the first to discover molecules of 25 nucleotides

M. Ghanim (✉) • A. Kliot
Department of Entomology, Institute of Plant Protection, Agricultural Research Organization, The Volcani Center, Bet Dagan, 50250, Israel
e-mail: ghanim@agri.gov.il

I. Ishaaya et al. (eds.), *Advanced Technologies for Managing Insect Pests*,
DOI 10.1007/978-94-007-4497-4_9, © Springer Science+Business Media Dordrecht 2012

long, which were proposed to activate the RNAi machinery in the cell (Hamilton and Baulcombe 1999). It was evident later that shorter RNA molecules, 21–23 nucleotides long, were the driving force that initiate the whole process (Zamore et al. 2000; Hammond et al. 2000). Despite these important discoveries, it was yet unclear how these molecules are generated and what drives the degradation of the target mRNA, however, it was clear that these processes involve enzymatic reactions and cleavage by nucleases. Two cleavage events occur in this process: the first is cleaving the dsRNA into small interfering RNAs (siRNAs), which then drive the second cleavage of the target mRNA. The first cleavage was found to occur by the activity of a type III RNase which was then termed "Dicer" (Bernstein et al. 2001). The target mRNA cleavage was found to be a result of the RNA-induced silencing complex (RISC) activity. RISC recognizes the target mRNA based on homology with the siRNA molecules. Characterizing the activity of the RISC complex was fostered after the discovery of the first Argonaute proteins, which are the main enzymes that cleave the target mRNA in the cell (Martinez et al. 2002). Further ongoing research that continues until today is focusing on the identification of other Argonaute proteins in other organisms, the specificity of these proteins, the recognition of dsRNA by Dicer, the generation of siRNA, and how siRNA molecules are uncoiled and driven by RISC to the target mRNA.

Another important discovery that made RNAi more potent in the nematode *C. elegans* compared with other organisms was the presence of the RNA dependent RNA polymerase (RdRp) (Sijen et al. 2001). This enzyme was first discovered in plants, and was found to be responsible for driving the amplification and the spread of siRNA molecules and the RNAi signal. In insects, however, no orthologs of this enzyme were found, and it is still unclear how the signal of RNAi is spread throughout the body (Tomoyasu et al. 2008).

2 *In Vitro* and *In Vivo* RNAi

Initiating the RNAi machinery requires that the dsRNA or the signal activating this machinery both be present in the same cell. Much of the important discoveries in understanding how RNAi is initiated and activated were made thanks to *in vitro* experiments made in cell lines that represent several organisms including humans (HeLa cell line) and insects (S2 cell lines from *D. melanogaster*). The use of cell lines was an important step for conducting large scale silencing experiments in which several hundreds to thousands of genes, sometimes representing whole genomes, were silenced and tested for the appearance of impaired phenotypes (Boutros and Ahringer 2008). The ability to introduce dsRNA directly into growing cells by mixing it with the media, then observing the resulting phenotypes, greatly advanced the discovery of new gene functions. In many cases, it was even possible to link functions and interactions between genes by inducing similar phenotypes (Friedman and Perrimon 2006). Although, *in vitro* experiments with cell lines were powerful in conducting large-scale RNAi screens, they were limited to phenotypes in a single cell, and in many cases silenced genes did not induce any phenotypes. It was therefore necessary to develop *in vivo* systems for gene silencing. Such technologies were developed in model species that could be stably transformed with foreign DNA, which could then be integrated into their genome. These species included *C. elegans*, *D. Melanogaster* and in recent years other dipterans and lipedopteran insect species. The most elegant system that was developed thus far uses the responsive element GAL4 attached to a generic promoter and the upstream activator sequence (UAS), which can target RNAi in any cell type of a whole organism, for which a relevant GAL4 driver line is available. Using this system, it is possible to conduct one-by-one or whole genome gene silencing experiments (Yapici et al. 2008; Kennerdell and Carthew 2000; Dietzl et al. 2007). The first non-model organism that was investigated using this system was the red flour beetle *Tribolium castaneum*, followed by the milkweed bug *Oncopeltus fasciatus* in which the function of Hox genes was studied (Hughes and Kaufman 2000). Segmentation and Hox genes, as well as other known genetic pathways in insects including the molting hormone ecdysone pathway, were the most studied genetic systems in model and non-model organisms, since they were very well-known from *D. melanogaster* research for the past 50 years. The function of many genes in these pathways were investigated in non-model insects including *T. Castaneum*, *Locusta migratoria* and *Periplaneta ameriana*, and it was generally confirmed that they have similar functions to those known in vertebrates, suggesting a similar ancestral mechanism that control these processes (Zhang et al. 2011; Bai and Palli 2010; Parthasarathy et al. 2010; Tan and Palli 2008; Pueyo et al. 2008; He et al. 2006; Bucher et al. 2002).

3 Delivery and Cellular Uptake of dsRNA

An important factor that limits the effectiveness of RNAi, in both *in vitro* and *in vivo* experiments, remains the accessibility of the dsRNA of siRNA molecules to the target cell, and the ability of these molecules to cross cell and tissue barriers. The RNAi machinery can be activated in a single cell and can remain restricted to the cell, a situation called cell autonomous RNAi (Meister and Tuschl 2004; Fire 2007; Jinek and Doudna 2009). In the non-cell autonomous RNAi, the signal for activating the machinery can be acquired from the environment, such as in unicellular organisms. The RNAi signal can also be acquired by a multicellular organism bearing tissues and complicated system, and in this case the machinery is called systemic RNAi. In systemic RNAi, the signal is transported from one cell to another. In multicellular organisms, several mechanisms were described and showed that molecules participating in the RNAi machinery are transported within and between tissues. These mechanisms were investigated based on previous knowledge on RNA transport within and between cells. The first mechanism involves the systemic RNAi defective mutants which were impaired in their ability to exhibit systemic RNAi. The genes SID-1 and SID-2 were identified and they seem to work together and are involved in importing dsRNA into the cell (Winston et al. 2002, 2007). In *D. Melanogaster*, long-term lasting RNAi was never observed; however, when the SID-1 protein from *C. elegans* was expressed in S2 *D. melanogaster* cell lines, these cells exhibited higher ability to import dsRNA into the cell. In the *D. melanogaster* genome, no SID-1 or SID-2 homologues were found, however, SID-1 homologues were identified in the genomes of other non-model insects such as *T. castaneum*, *Apis mellifera*, *Bombyx mori* and the pea aphid *Acyrthosiphon pisum* (Xu and Han 2008). Recent evidence suggested that the dsRNA uptake into S2 *D. melanogaster* cells involves receptor-mediated endocytosis and not SID-1 as was previously described. This assumption was based on the fact that inhibition of the endocytosis prevented the effects caused by dsRNA-activated RNAi machinery (Saleh et al. 2006; Ulvila et al. 2006). Endocytosis of dsRNA seems to be also an active mechanism in *C. elegans* (Saleh et al. 2006), which suggests that the receptor mediated endocytosis might be a general mechanism present in other organisms such as insect pests. Using FITC-labeled dsRNA, Saleh et al. (2006) showed that the dsRNA molecules were associated with vesicles, suggesting that the uptake of these molecules is specific and receptor-mediated. If the receptor-mediated endocytosis of dsRNA is general as has been described in several organisms, it is possible to orally introduce dsRNA, which might then be internalized by gut cells and be spread to other cells. This approach exhibits a great potential for targeting important genes that encode important traits with agricultural relevance, such as resistance to insecticides, virus transmission and interactions with plants.

A main factor that was found to be important in the spread of the RNAi signal and inducing the systemic silencing is the presence of the RNA dependent RNA polymerase (RdRP). RdRp was shown to amplify the signal and the spread of the siRNAs through the plasmodesmata between plant cells. RdRp orthologues were found in nematodes (Sijen et al. 2001), but were never found in insects (Jose and Hunter 2007; Richards and Tribolium Genome Sequencing Consortium 2008). This result suggests another mechanism that might be responsible for the spread of systemic RNAi in insects.

4 Methods for dsRNA Application and Potency in Inducing Silencing

RNAi is activated once dsRNA molecules are available in the cell. The availability of dsRNA molecules are thus a limiting factor for initiating the machinery. Artificial delivery of dsRNA into the organism or the cell can be achieved using several methods, and the speed of the interference depends mainly on the availability of the dsRNA. Indeed, experiments employing direct microinjection of dsRNA into *D. melanogaster* embryos, direct injection into the hemolymph, transformation with constructs bearing hairpin sequences, transfection of dsRNA into cell cultures using in vitro synthesized dsRNA or engineered viruses specific to infect the cells, were all efficient methods for obtaining fast and strong phenotypes after silencing. Several experiments done on cell cultures have shown that dsRNA has to reach the cell cytoplasm for triggering efficient silencing, and when this RNA is simply added in the medium, RNAi is not triggered (Beck and Strand 2003).

The most relevant method for dsRNA delivery, that has a practical application for developing insect pest control methods, is oral feeding. Several methods for delivering dsRNA by oral ingestion were developed, depending on the mode of food ingestion by the insect. In order to develop an efficient, long-lasting and practical method for oral delivery, it is pivotal to ensure constitutive expression of the dsRNA, and ensure its availability for ingestion by the insect in a cost-effective manner. In any feeding experiment with dsRNA, midgut cells are the first target in which the silencing may be activated. The question is whether the silencing in midgut cells is spread by systemic RNAi to other tissues and cell types. Several examples in the literature suggest efficient induction of RNAi in midgut cells, however only limited number of cases demonstrated that the signal is systematically spreading after dsRNA ingestion (Sivakumar et al. 2007; Whyard et al. 2009; Bautista et al. 2009; Tian et al. 2009; Zhu et al. 2011). Many insects were used in artificial feeding experiments on dsRNA for testing the potency of the method. For example, feeding the light brown apple moth *Epiphyas postvittana* with dsRNA could result in down regulating the expression of a carboxyesterases gene in the adult's gut. A similar treatment resulted in repressing the expression of an antennal gene. This was achieved by feeding the larvae with dsRNA, a treatment that resulted in efficient silencing of the gene in the adult stage, suggesting a persistence of the RNAi signal through different developmental stages to the adult (Turner et al. 2006). Feeding experiments with artificially synthesized dsRNA, or bacteria expressing hairpin sequences specifically targeting one gene in the diet, were also applied to other insects with varying levels of success. A recent example tested the potential of feeding with dsRNA that was *in vitro* synthesized or expressed in bacteria to manage different developmental stages of the Colorado potato beetle *Leptinotarsa decemlineata* (Say). The results demostrated the success in triggering the silencing of five target genes, and this silencing was accompanied by significant mortality and reduced body weight gain (Zhu et al. 2011).

It is generally believed that in order to obtain significant effects on the insect, high amounts of dsRNA in the diet are required. It was also shown that the sensitivity and the response depend on the developmental stage used for dsRNA feeding, and in some cases the RNAi signal was heritable and persisted from one generation to another. Specific gene silencing was sometime achieved for genes in certain tissues; however, the signal did not spread to other tissues. The levels of silencing depend in many cases on the length of the introduced dsRNA, longer sequences up to a certain limit will usually be more efficient than shorter sequences below 200 bp. Insects that were used in dsRNA feeding experiments include *Plutella xylostella*, *S. exigua*, *M. sexta*, *S. frugiperda*, *Ostrinia nubilalis*, *Epiphyas postvittana*, *Diatraea saccharalis*, *Leptinotarsa decemlineata* and *Trichoplusia ni*, with varying levels of silencing (Bautista et al. 2009; Tian et al. 2009; Whyard et al. 2009; Yang et al. 2010; Turner et al. 2006; Khajuria et al. 2010; Griebler et al. 2008; Zhu et al. 2011; Rodriguez-Cabrera et al. 2010).

An interesting result was obtained when the blood sucking bug *Rhodnius prolixus* (Hemiptera) was fed with dsRNA for silencing the salivary gland gene nitroporin 2 (NP2). Successful silencing was obtained both by feeding and by directly injecting the dsRNA into the hemolymph (Araujo et al. 2006). In both cases, repression in gene expression was obtained; however, microinjection was about twofolds more effective than feeding. This result demonstrates that feeding on dsRNA in the diet is an easy approach ensuring the delivery of dsRNA, however, it is not possible to control the amount of the ingested dsRNA by the insect, or the amount that reaches the target cells. Direct microinjection of the dsRNA in the hemolymph represents a more potent approach that seems to induce a stronger effect. The gut barrier seems to pose a filter effect that reduces the amount of dsRNA passing to target cells and tissues. Indeed, many examples in the literature showed that microinjection of dsRNA induced potent and consistent suppression of gene expression; however, this method remains a tool for research purposes and not for the development of control methods in the field. Additionally, microinjection is a tool that can be used with big insects, while small and soft-bodied insects such as whiteflies and mites are hard to inject. Many successful cases in which microinjection of dsRNA induced significant gene silencing effects were reported, most importantly with the red flour beetle *T. castaneum.* Microinjection works so efficient in this species that feeding experiments were never conducted. The first example for efficient silencing in this beetle was the successful complete suppression of the sensory bristle-forming gene *Tc-achaete-scute* that resulted in complete loss-of-bristle phenotype (Tomoyasu and Denell 2004). Heritable RNAi signal from one generation to another was also demonstrated in mothers and their progeny after injecting dsRNA that targeted several homeobox genes including *distalless*, *maxillopedia* and *proboscipedia* (Bucher et al. 2002). Many other insects were used to investigate the potency of RNAi after dsRNA microinjection into embryos. Such insects include, *B. mori* and several other lepidopteran species such as *Mamestra brassicae*, *Plodia interpunctella*, and *S. exigua* (Liu et al. 2008; Masumoto et al. 2009; Pan et al. 2009; Tomita and Kikuchi 2009; Fabrick et al. 2004; Tsuzuki et al. 2005).

5 Insect and Tissue Sensitivity to RNAi

While many reports showed efficient RNAi induction in various insect orders, it is obvious that not all insects have the same sensitivity to RNAi (Tomoyasu et al. 2008), and in many cases insects did not have any sensitivity. The level of sensitivity may well depend on many factors that need to be optimized. These factors include the dsRNA application method, concentration of the injected/fed dsRNA, length of the dsRNA, nucleotide composition, persistence of the silencing signal and other unknown factors or factors that were not investigated. Since not all these factors can be controlled or optimized, the number of insects sensitive to RNAi treatment might be higher than reported. While it seems that the most resistant insect to RNAi is *D. melanogaster*, where the dsRNA needs to cross several tissues and barriers (Tomoyasu et al. 2008), the nematode *C. elegans* remains the most sensitive organism that seems to exhibit both local and systemic RNAi properties (May and Plasterk 2005). The accessibility of dsRNA or siRNA molecules to cells and tissues and their systemic spread is the most important factor for RNAi efficiency. However, other factor such as degradation mechanisms in the insect and the response/expression of components of the RNAi machinery including Dicer and Argonaute, are other important factors that may determine the efficiency of the pathway. Permeability of the target tissue to dsRNA and siRNA is another important factor, and it has been shown that certain tissues are more potent in activating the RNAi machinery than others. For example silencing the lipophorin receptor was easier and faster in the fat body of *Blattella germanica* than in the ovary (Ciudad et al. 2007). In some cases, the expression levels of the RNAi genes are lower in certain tissues than others which may suggest lower efficiency of the whole RNAi machinery. As an example, the expression of Dicer and Argonaute, is lower in salivary gland tissues of the mosquito *Anopheles gambiae* if compared to other tissues, which may explain the lower sensitivity of salivary gland to RNAi (Boisson et al. 2006).

6 Successful Examples of RNAi with Agricultural Relevance

The core machinery of RNAi seems to be conserved across organisms and many reports have shown successful gene silencing in insect pests of agricultural importance. The high sensitivity of some agricultural pests to RNAi might be a first step in developing control methods based on this genetic approach, and avoiding chemical pesticides and their harmful effects. Unlike nematodes, the RNA dependent RNA polymerase (RdRp), which is suggested to enhance the interference and the spread of the RNAi signal was never found in insects. This observation suggests that RNAi in insects is likely to be less efficient than nematodes (Gatehouse 2008; Price and Gatehouse 2008; Gordon and Waterhouse 2007). Furthermore, among insect orders, there is a variation in the presence of key genes involved in triggering and spreading the RNAi signal. For example, lepidopterans lack homologues of the SID-1 gene, which is thought to be required for systemic RNAi.

The breakthrough using RNAi to control insect pests came when transgenic plants expressing insect-specific hairpin dsRNAs were consumed by those insect pests, leading to efficient silencing in the insect. In this work, the authors reported the efficient silencing of a V-type ATPase gene from the western corn rootworm *Diabrotica virgifera virgifera* when its respective dsRNA was expressed in transgenic corn plants. When the roots of these plants were consumed by the larvae, efficient silencing of the gene was observed (Baum et al. 2007). This gene was selected based on an initial screen in a cDNA library that identified 290 genes which were considered good targets for silencing. dsRNA was synthesized for each one of these genes, and their silencing was tested *in vitro* from artificial diet. Only 14 genes exhibited strong silencing effects and mortality of the larvae with low dsRNA concentrations. The dsRNA representing the V-type ATPase demonstrated rapid knockdown of the mRNA and triggered specific RNAi within 24 h post ingestion from the diet. To demonstrate the practical relevance of the work, transgenic corn expressing ATPase dsRNA were prepared and used for larvae feeding experiments. The results showed significant plant protection from the larvae compared with the control. It should be noted that the same dsRNA prepared for silencing the V-type ATPase genes *in vitro*, exhibited efficiency in silencing homologues genes in other coleopteran insects including the southern corn rootworm *Diabrotica undecimpunctata* howardii and the Colorado potato beetle *L. decemlineata*, however, the cotton boll weevil *Anthonomus grandis* Boheman showed insensitivity to the silencing. These experiments demonstrate the ability for cross-silencing between several gene targets or pyramiding multiple targets as was previously demonstrated in *D. melanogaster* (Schmid et al. 2002). The ability to target specific genes by designing specific dsRNA is also very promising in avoiding the harm of beneficial organisms such as natural enemies. To this end, bioinformatics tools are required for selecting species-specific sequences when preparing the dsRNA.

An additional successful example in creating plants protected from insect pest using RNAi was demonstrated by targeting a cytochrome P450 monooxygenase gene used by the cotton bollworm *Helicoverpa armigera* for detoxifying gossypol, a plant secondary metabolite from cotton (Mao et al. 2007). In this work, the authors identified over expression of the gene CYP6AE14 after exposure to gossypol. This over expression enables the bollworm to grow successfully on cotton. When dsRNA of this gene was expressed in Tobacco or Arabidopsis and the plants were used as a feeding source for the cotton bollworm, a significant reduction in the expression of this gene, and increased sensitivity of the larvae to consume gossypol from artificial diet, were documented (Mao et al. 2007).

The two examples presented above demonstrate that it is feasible to use RNAi for crop protection, and that this technology holds a great promise for developing environment friendly control methods against insect pests. The number of genes that can be targeted for each pest using RNAi is hypothetically high, since many gene functions are important in the biology and development of insect pests and thus are good candidates for gene silencing. How can good candidates be identified? One approach is performing large-scale gene silencing experiments that result in identifying essential genes whose inactivation leads to strong suppression of the pest development. An additional strategy for selecting essential genes is predicting their function based on homologues genes from model or genome-enabled organisms. Selecting candidate genes for silencing is an essential step that will determine the success of the whole process in making a resistant crop against any insect pests. The search for pest control methods that do not rely on chemicals will continue. Such methods exist for several insect pest groups and the best and most successful examples are *Bacillus thuringiensis* toxins used against lepidopterans and coleopterans. Similar toxins, however, are not available for insect pests like sap-sucking insects, such as whiteflies, aphids and others. For such insects, RNAi based strategies using plants seem promising, however since these insects are phloem feeders, it is required that the dsRNA be specifically expressed or transported into the phloem sieve elements.

7 RNAi in Hemipterans and the Whitefly *Bemisia tabaci*

The whitefly *Bemisia tabaci* is a cosmopolitan insect pest, and considered one of the most important plant virus vectors worldwide (Byrne and Bellows 1991). This insect strictly feeds on the phloem sap and secretes huge amounts of honeydew. The control of *B. tabaci* relies mainly on chemical insecticides, however this species is known for its exceptional ability to develop resistance to all major insecticides (Horowitz et al. 2011). Therefore, the control of *B. tabaci* and similar insect pests will reach a bottleneck in using chemical insecticides not only for their reduced efficiency and the resistance problems, but also because of their toxicity to humans and beneficial organisms and consumers' tendency to prefer unsprayed products. Several non-chemical control methods are used to control *B. tabaci*, including cultural, physical and biological actions, however these methods account for about 10% of damage reduction caused by *B. tabaci.* It is thus essential to develop other control methods against *B. tabaci*, whiteflies in general and other sap-sucking insect pests such as aphids, planthoppers and psyllids. RNAi holds a promise as one of these methods that has been recently studied for its potential to control sap sucking pests. Indeed, several groups reported successful silencing of gene transcripts in *B. tabaci* and other hemipterans. In the pea aphid *A. pisum* a specific salivary gland transcript termed C002 was targeted for silencing by injecting specific siRNA into the body cavity of the aphid. The silencing caused a dramatic decrease in the expression of the transcript over a 3-day period after injection, followed by rapid aphid mortality (Mutti et al. 2006). One year later, a protocol for gene silencing in *A. pisum* using injected dsRNA was developed. Two genes with different expression patterns were targeted and both showed maximal significant decrease in their expression 7 days post injection (Jaubert-Possamai et al. 2007).

The first successful implementation of RNAi in a practical approach for creating resistant plants against a hemipteran pest was recently reported. Three genes from the brown planthopper *Nilaparvata lugens* Stal, (the hexose transporter gene NlHT1, the carboxypeptidase gene Nlcar and the trypsin-like serine protease gene Nltry) were cloned and dsRNA constructs for transforming rice were prepared. Some of the expressed dsRNAs in the plants were processed to siRNA. When nymphs were fed on the rice plants expressing dsRNA, the levels of transcripts of the three genes were reduced, however, no mortality of the pest was observed (Zha et al. 2011).

We have previously shown that RNAi is active in *B. tabaci* (Ghanim et al. 2007), and by dsRNA injections into whole insects, we have demonstrated tissue specific gene silencing, by targeting several genes specifically expressed in salivary glands, midgut and whole insects. Injecting dsRNA that targeted the *Chickadee* homologue of *D. melanogaster* (a protein involved in the actin-based dynamics in the developing oocytes) from *B. tabaci* resulted in a strong phenotype in oocytes dissected from injected females with the dsRNA (Fig. 9.1) (Ghanim et al. 2007). Our approach showed that the RNAi machinery is active in *B. tabaci* and that injecting dsRNA into the body cavity can induce strong phenotype. This result holds promise for developing RNAi-based control methods for whiteflies. Along this line, we investigated

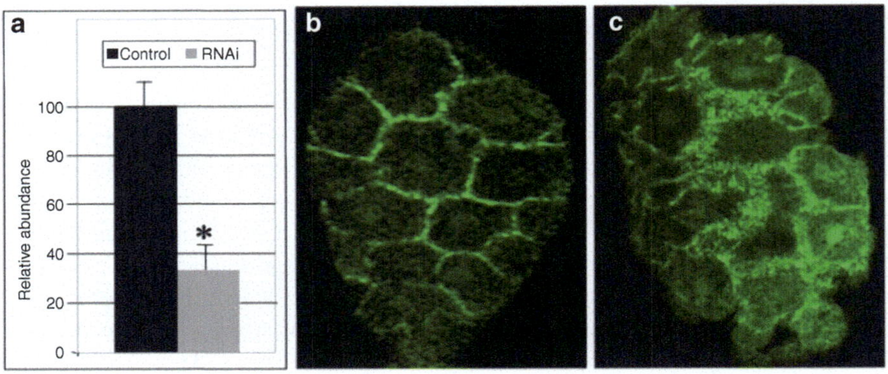

Fig. 9.1 Disruption of actin network in *B. tabaci* developing eggs following RNAi-silencing of the *chickadee* gene homologue. (**a**) Decrease in *chickadee* expression following injection of dsRNA as estimated by RT-PCR. Data shown are the mean ± SEM of three independent experiments. Asterisk refers to significant reduction in the expression of the *chickadee* gene. (**b**) Phalloidin-FITC staining of developing oocytes dissected from 2-days old whitefly female injected with GFP-dsRNA, showing the subcortical actin networks in the follicular cells membranes. (**c**) Phalloidin-FITC staining of developing oocytes dissected 18 h and injection with *chickadee* dsRNA, showing malformations in the follicular subcortical actin

the possibility of triggering RNAi by ingestion. dsRNA against the CYP6CM1 P450 monooxygenase gene was in vitro synthesized and labeled with dUTP-Cy5 to track the RNA after ingestion. The dsRNA was supplied in artificial diet containing 15% sucrose and 1 μg/μl labeled dsRNA, and *B. tabaci* was given a 24 h feeding period on this diet, after which the labeled dsRNA was monitored in adult females, guts and ovaries. The labeled dsRNA was observed in the midgut epithelial cells, hemolymph and developing oocytes, suggesting that it is able to penetrate the midgut barrier and reach the hemolymph and other tissues (Fig. 9.2). The expression of the CYP6CM1 gene was significantly lower in midguts dissected from females fed on dsRNA compared with females that fed on artificial diet only, as assessed by RT-PCR analysis (Fig. 9.2). Although these experiments are now being optimized, the results show that *in vitro* synthesized dsRNA molecules are able to cross the midgut and reach the hemolymph and several important organs where silencing of gene expression may lead to significant effects on the insect. These results further demonstrate the potential in developing plants expressing dsRNA for specific silencing in the insect. dsRNA expressed in the plant can be acquired and transported to several target sites and organs, which may increase the efficacy of this approach.

8 Conclusions and Future Perspective

Many recent studies showed that the damage caused by insect pests can be reduced or inhibited using RNAi methods, suggesting its potential for developing new and environment friendly genetic control methods. Whole genome sequences, which will

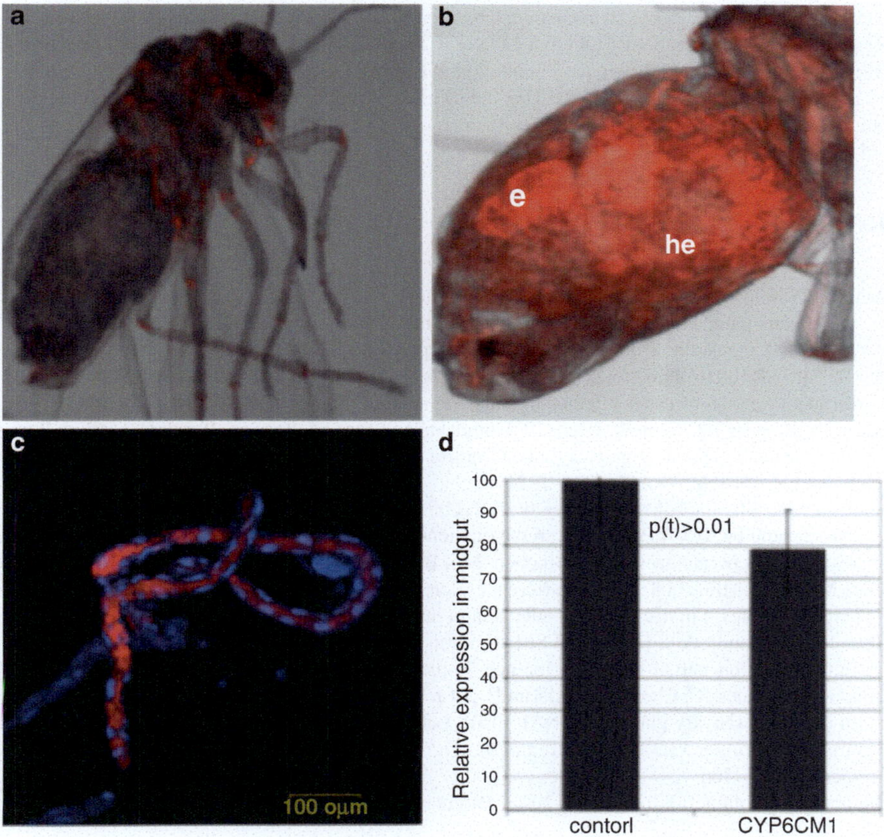

Fig. 9.2 *B. tabaci* CYP6CM1 Cy5-labeled dsRNA feeding from artificial diet. (**a**) control female fed on 15% sucrose only. (**b**) Females fed with 15% sucrose supplemented with Cy5-labeled dsRNA showing the labeled dsRNA in the hemolymph and developing eggs. (**c**) dissected midgut from females fed with 15% sucrose supplemented with Cy5-labeled dsRNA showing the dsRNA in cells and DAPI blue stain showing the nuclei of the cells. (**d**) significant reduction in CYP6CM1 expression following feeding on dsRNA for 24 h from artificial diet. *h* hemolymph, *e* developing eggs

be soon available for many insect species, including many economically important pests, combined with better understanding of the RNAi machinery will foster research and discoveries towards developing sustainable RNAi control methods. Genome sequences are pivotal for large-scale RNAi screens and generating candidate genes with important functions. Some big companies and enterprises already initiated large scale gene function discoveries for similar purposes. Many issues surrounding the use of RNAi for pest control need to be resolved. Those issues are mostly related to whether genetically modified plants will be used in food production and consumption, and the specificity of the introduced dsRNA molecules in controlling the target pest and avoiding non-target organisms such as humans and beneficial organisms.

Acknowledgements Research in Ghanim laboratory was supported by the Binational Agricultural Research and Development Fund BARD grant IS-4062-07, Binational Science Foundation BSF grant 2007045, Israel Science Foundation ISF grant 884/07 and the Chief Scientist of the Israeli Ministry of Agriculture grant 131-1433-09. This is contribution number 500/12 from ARO publications.

References

Araujo RN, Santos A, Pinto FS, Gontijo NF, Lehane MJ, Pereira MH (2006) RNA interference of the salivary gland nitrophorin 2 in the triatomine bug *Rhodnius prolixus* (Hemiptera: Reduviidae) by dsRNA ingestion or injection. Insect Biochem Mol Biol 36:683–693

Bai H, Palli SR (2010) Functional characterization of bursicon receptor and genome-wide analysis for identification of genes affected by bursicon receptor RNAi. Dev Biol 344:248–258

Baum JA, Bogaert T, Clinton W, Heck GR, Feldmann P, Ilagan O, Johnson S, Plaetinck G, Munyikwa T, Pleau M, Vaughn T, Roberts J (2007) Control of coleopteran insect pests through RNA interference. Nat Biotechnol 25:1322–1326

Bautista MA, Miyata T, Miura K, Tanaka T (2009) RNA interference-mediated knockdown of a cytochrome P450, CYP6BG1, from the diamondback moth, *Plutella xylostella*, reduces larval resistance to permethrin. Insect Biochem Mol Biol 39:38–46

Beck M, Strand MR (2003) RNA interference silences Microplitis demolitor bracovirus genes and implicates glc1.8 in disruption of adhesion in infected host cells. Virology 314:521–535

Bernstein E, Caudy AA, Hammond SM, Hannon GJ (2001) Role for a bidentate ribonuclease in the initiation step of RNA interference. Nature 409:363–366

Boisson B, Jacques JC, Choumet V, Martin E, Xu J, Vernick K, Bourgouin C (2006) Gene silencing in mosquito salivary glands by RNAi. FEBS Lett 580:1988–1992

Boutros M, Ahringer J (2008) The art and design of genetic screens: RNA interference. Nat Rev Genet 9:554–566

Bucher G, Scholten J, Klingler M (2002) Parental RNAi in Tribolium (Coleoptera). Curr Biol 12:R85–R86

Byrne DN, Bellows TS Jr (1991) Whitefly biology. Annu Rev Entomol 36:431–457

Ciudad L, Belles X, Piulachs MD (2007) Structural and RNAi characterization of the German cockroach lipophorin receptor, and the evolutionary relationships of lipoprotein receptors. BMC Mol Biol 8:53

Dietzl G, Chen D, Schnorrer F, Su KC, Barinova Y, Fellner M, Gasser B, Kinsey K, Oppel S, Scheiblauer S, Couto A, Marra V, Keleman K, Dickson BJ (2007) A genome-wide transgenic RNAi library for conditional gene inactivation in *Drosophila*. Nature 448:151–156

Fabrick JA, Kanost MR, Baker JE (2004) RNAi-induced silencing of embryonic tryptophan oxygenase in the pyralid moth, *Plodia interpunctella*. J Insect Sci 4:15

Fire AZ (2007) Gene silencing by double-stranded RNA (Nobel lecture). Angew Chem Int Ed 46:6967–6984

Fire A, Xu S, Montgomery MK, Kostas SA, Driver SE, Mello CC (1998) Potent and specific genetic interference by double-stranded RNA in *Caenorhabditis elegans*. Nature 391:806–811

Friedman A, Perrimon N (2006) A functional RNAi screen for regulators of receptor tyrosine kinase and ERK signaling. Nature 444:230–234

Gatehouse JA (2008) Biotechnological prospects for engineering insect-resistant plants. Plant Physiol 146:881–887

Ghanim M, Kontsedalov S, Czosnek H (2007) Tissue- specific gene silencing by RNA interference in the whitefly *Bemisia tabaci* (Gennadius). Insect Biochem Mol Biol 37:732–738

Gordon KHJ, Waterhouse PM (2007) RNAi for insect-proof plants. Nat Biotechnol 25: 1231–1232

Griebler M, Westerlund SA, Hoffmann KH, Meyering-Vos M (2008) RNA interference with the allatoregulating neuropeptide genes from the fall armyworm *Spodoptera frugiperda* and its effects on the JH titer in the hemolymph. J Insect Physiol 54:997–1007

Hamilton AJ, Baulcombe DC (1999) A species of small antisense RNA in posttranscriptional gene silencing in plants. Science 286:950–952

Hammond SM, Bernstein E, Beach D, Hannon GJ (2000) An RNA-directed nuclease mediates posttranscriptional gene silencing in *Drosophila* cells. Nature 404:293–296

He ZB, Cao YQ, Yin YP, Wang ZK, Chen B, Peng GX, Xia YX (2006) Role of *hunchback* in segment patterning of *Locusta migratoria manilensis* revealed by parental RNAi. Dev Growth Differ 48:439–445

Horowitz AR, Antignus Y, Gerling D (2011) Management of *Bemisia tabaci* whiteflies. In: Thompson WMO (ed) The whitefly, *Bemisia tabaci* (Homoptera: Aleyrodidae) interaction with Geminivirus-infected host plants. Springer, Dordrecht, pp 293–322

Hughes CL, Kaufman TC (2000) RNAi analysis of *Deformed*, *proboscipedia* and *Sex combs reduced* in the milkweed bug *Oncopeltus fasciatus*: novel roles for Hox genes in the hemipteran head. Development 127:3683–3694

Jaubert-Possamai S, Trionnaire GL, Bonhomme J, Christ phides GK, Rispe C, Tagu D (2007) Gene knockdown by RNAi in the pea aphid *Acyrthosiphon pisum*. BMC Biotechnol 7:63

Jinek M, Doudna JA (2009) A three-dimensional view of the molecular machinery of RNA interference. Nature 457:405–412

Jose AM, Hunter CP (2007) Transport of sequence-specific RNA interference information between cells. Annu Rev Genet 41:305–330

Kennerdell JR, Carthew RW (2000) Heritable gene silencing in *Drosophila* using double-stranded RNA. Nat Biotechnol 18:896–898

Khajuria C, Buschman LL, Chen MS, Muthukrishnan S, Zhu KY (2010) A gut specific chitinase gene essential for regulation of chitin content of peritrophic matrix and growth of *Ostrinia nubilalis* larvae. Insect Biochem Mol Biol 40:621–629

Liu W, Yang F, Jia S, Miao X, Huang Y (2008) Cloning and characterization of Bmrunt from the silkworm *Bombyx mori* during embryonic development. Arch Insect Biochem Physiol 69:47–59

Mao YB, Cai WJ, Wang JW, Hong GJ, Tao XY, Wang LJ, Huang YP, Chen XY (2007) Silencing a cotton bollworm P450 monooxygenase gene by plant-mediated RNAi impairs larval tolerance of gossypol. Nat Biotechnol 25:1307–1313

Martinez J, Patkaniowska A, Urlaub H, Luhrmann R, Tuschl T (2002) Single-stranded antisense siRNAs guide target RNA cleavage in RNAi. Cell 110:563–574

Masumoto M, Yaginuma T, Niimi T (2009) Functional analysis of Ultrabithorax in the silkworm, *Bombyx mori*, using RNAi. Dev Genes Evol 219:437–444

May RC, Plasterk RH (2005) RNA interference spreading in *C. elegans*. Methods Enzymol 392:308–315

Meister G, Tuschl T (2004) Mechanisms of gene silencing by double-stranded RNA. Nature 431:343–349

Mutti NS, Park Y, Reese JC, Reeck GR (2006) RNAi knockdown of a salivary transcript leading to lethality in the pea aphid, *Acyrthosiphon pisum*. J Insect Sci 6:38

Napoli C, Lemieux C, Jorgensen R (1990) Introduction of a chimeric chalcone synthase gene into petunia results in reversible co-suppression of homologous genes in trans. Plant Cell 2:279–289

Pan MH, Wang XY, Chai CL, Zhang CD, Lu C, Xiang ZH (2009) Identification and function of Abdominal-A in the silkworm, *Bombyx mori*. Insect Mol Biol 18:155–160

Parthasarathy R, Sheng Z, Sun Z, Palli SR (2010) Ecdysteroid regulation of ovarian growth and oocyte maturation in the red flour beetle, *Tribolium castaneum*. Insect Biochem Mol Biol 40:429–439

Price DR, Gatehouse JA (2008) RNAi-mediated crop protection against insects. Trends Biotechnol 26:393–400

Pueyo JI, Lanfear R, Couso JP (2008) Ancestral Notch-mediated segmentation revealed in the cockroach *Periplaneta americana*. Proc Natl Acad Sci USA 105:16614–16619

Richards S, Tribolium Genome Sequencing Consortium (2008) The genome of the model beetle and pest *Tribolium castaneum*. Nature 452:949–955

Rodriguez-Cabrera L, Trujillo-Bacallao D, Borrás-Hidalgo O, Wright DJ, Ayra-Pardo C (2010) RNAi-mediated knockdown of a *Spodoptera frugiperda* trypsinlike serine-protease gene reduces susceptibility to a *Bacillus thuringiensis* Cry1Ca1 protoxin. Environ Microbiol 12:2894–2903

Saleh MC, van Rij RP, Hekele A, Gillis A, Foley E, O'Farrell PH, Andino R (2006) The endocytic pathway mediates cell entry of dsRNA to induce RNAi silencing. Nat Cell Biol 8:793–802

Schmid A, Schindelholz B, Zinn K (2002) Combinatorial RNAi: a method for evaluating the functions of gene families in *Drosophila*. Trends Neurosci 25:71–74

Sijen T, Fleenor J, Simmer F, Thijssen KL, Parrish S, Timmons L, Plasterk RHA, Fire A (2001) On the role of RNA amplification in dsRNA-triggered gene silencing. Cell 107:465–476

Sivakumar S, Rajagopal R, Venkatesh GR, Srivastava A, Bhatmagar RK (2007) Knockdown of aminopeptidase-N from *Helicoverpa armigera* larvae and in transfected Sf21 cells by RNA interference reveals its functional interaction with *Bacillus thuringiensis* insecticidal protein Cry1Ac. J Biol Chem 282:7312–7319

Tan A, Palli SR (2008) Edysone receptor isoforms play distinct roles in controlling molting and metamorphosis in the red flour beetle, *Tribolium castaneum*. Mol Cell Endocrinol 291:42–49

Tian H, Peng H, Yao Q, Chen H, Xie Q, Tang B, Zhang W (2009) Developmental control of a lepidopteran pest *Spodoptera exigua* by ingestion of bacteria expressing dsRNA of a non-midgut gene. PLoS One 4:e6225

Tomita S, Kikuchi A (2009) Abd-B suppresses lepidopteran proleg development in posterior abdomen. Dev Biol 328:403–409

Tomoyasu Y, Denell RE (2004) Larval RNAi in Tribolium (Coleoptera) for analyzing adult development. Dev Genes Evol 214:575–578

Tomoyasu Y, Miller SC, Tomita S, Schoppmeier M, Grossmann D, Bucher G (2008) Exploring systemic RNA interference in insects: a genome-wide survey for RNAi genes in Tribolium. Genome Biol 9:R10

Tsuzuki S, Sekiguchi S, Kamimura M, Kiuchi M, Hayakawa Y (2005) A cytokine secreted from the suboesophageal body is essential for morphogenesis of the insect head. Mech Dev 122:189–197

Turner CT, Davy MW, MacDiarmid RM, Plummer KM, Birch NP, Newcomb RD (2006) RNA interference in the light brown apple moth, *Epiphyas postvittana* (Walker) induced by double-stranded RNA feeding. Insect Mol Biol 15:383–391

Ulvila J, Parikka M, Kleino A, Sormunen R, Ezekowitz RA, Kocks C, Rämet M (2006) Double-stranded RNA is internalized by scavenger receptor-mediated endocytosis in *Drosophila* S2 cells. J Biol Chem 281:14370–14375

Whyard S, Singh AD, Wong S (2009) Ingested double-stranded RNAs can act as species-specific insecticides. Insect Biochem Mol Biol 39:824–832

Winston WM, Molodowitch C, Hunter CP (2002) Systemic RNAi in *C. elegans* requires the putative transmembrane protein SID-1. Science 295:2456–2459

Winston WM, Sutherlin M, Wright AJ, Feinberg EH, Hunter CP (2007) *Caenorhabditis elegans* SID-2 is required for environmental RNA interference. Proc Natl Acad Sci USA 104: 10565–10570

Xu W, Han Z (2008) Cloning and phylogenetic analysis of sid-1- like genes from aphids. J Insect Sci 8:30

Yang Y, Zhu YC, Ottea J, Husseneder C, Leonard BR, Abel C, Huang F (2010) Molecular characterization and RNA interference of three midgut aminopeptidase N isozymes from Bacillus thuringiensis-susceptible and -resistant strains of sugarcane borer, *Diatraea saccharalis*. Insect Biochem Mol Biol 40:592–603

Yapici N, Kim YJ, Ribeiro C, Dickson BJ (2008) A receptor that mediates the postmating switch in *Drosophila* reproductive behavior. Nature 451:33–37

Zamore PD, Tuschl T, Sharp PA, Bartel DP (2000) RNAi: double-stranded RNA directs the ATPdependent cleavage of mRNA at 21 to 23 nucleotide intervals. Cell 101:25–33

Zha W, Peng X, Chen R, Du B, Zhu L, He G (2011) Knockdown of midgut genes by dsRNA-transgenic plant-mediated RNA interference in the hemipteran insect *Nilaparvata lugens*. PLoS One 6:e20504

Zhang Z, Xu J, Sheng Z, Sui Y, Palli SR (2011) Steroid receptor co-activator is required for juvenile hormone signal transduction through a bHLH-PAS transcription factor, methoprene tolerant. J Biol Chem 286:8437–8447

Zhu F, Xu J, Palli R, Ferguson J, Palli SR (2011) Ingested RNA interference for managing the populations of the Colorado potato beetle, *Leptinotarsa decemlineata*. Pest Manag Sci 67:175–182

Chapter 10
Comparative Aspects of Cry Toxin Usage in Insect Control

András Székács and Béla Darvas

1 Introduction

Bacillus thuringiensis Berliner microorganisms are aerobe, Gram positive, endospore-forming insect pathogenic bacteria, discovered by Ishiwata in 1901 from silkworm (*Bombyx mori* L.) and described by Berliner in 1915 (Hilbeck and Schmidt 2006; Roh et al. 2007). *B. thuringiensis* strains are ubiquitously present in our environment, as soil-borne bacteria and as insect larval pathogens. The identification of the protein composition in the parasporal bodies of numerous *B. thuringiensis* strains and the discovery of the unique physico-chemical features and biological specificity of the protein toxins (termed Cry toxins) has led to several landmark events in pest control practices.

Initially *B. thuringiensis* products (spores and δ-endotoxins) were applied (Table 10.1), but later, with the discovery of several *Bt* strains and their genetic investigations, two or more strains of *Bt* subspecies were mixed together in a way that facilitates the formation of combination of *cry* genes settled on plasmids. *Bt* strains EG2348, EG2349 and EG2371 (Ecogen, Inc.) were created through a process called transconjugation, a phenomenon known to occur in nature and considered analogous to hybridization in higher organisms. In the next step two or more subspecies (serological or pathological variants) of *Bt* were mixed together. Thus, *Bt* strains EG7826, EG7841 (Ecogen, Inc.) and GC-91 (AGC Ltd.) have also been produced through this process. These novel strains may be considered modified *Bt* strains, although the term hybrid *Bt* strain (gene exchange within a species) would be more accurate.

A. Székács (✉) • B. Darvas
Department of Ecotoxicology and Environmental Analysis, Plant Protection Institute, Hungarian Academy of Sciences, Budapest, Hungary
e-mail: a.szekacs@cfri.hu; drdarvas.bela@chello.hu

I. Ishaaya et al. (eds.), *Advanced Technologies for Managing Insect Pests*,
DOI 10.1007/978-94-007-4497-4_10, © Springer Science+Business Media Dordrecht 2012

Table 10.1 Application forms of *Bacillus thuringiensis*

Subspecies	Strain	Cry toxin types	Selected trade names/codes	Type
aizawai	SA-2	Cry1	XenTari	NAT
aizawai	MYX833	Cry1C	M/C	CC
israelensis	SA-3	Cry4, Cry1	Acrobe, Aquabac, Bacticide, Bactilarvae, Bactoculis, Bakthane, Gnatrol, Larvatrol, Prehatch, Sentry, Teknar, VectoBac, Vectobar, Vectocid	NAT
japonensis	buibui	Cry8Ga1	M-Press	NAT
kurstaki		Cry1, Cry2	Agrobac, Bactec, Bactuside, Baritone, Baturad, Biobit, Biolep, BioPas, Collapse, Cordalene, Costar, Delfin, Dipel, Foray, Forwarbit, Halt, Insectobiol, Javelin, Lipel	NAT
kurstaki	EG2348	Cry1Aa, Cry1Ac, Cry2A	Condor, Ecotech Pro, Rapax, Wormox	NAT
kurstaki	EG2349	Cry1	Bollgard[a]	NAT
kurstaki	EG2371	Cry1	Cutlass[a], Ecotech Bio	NAT
kurstaki/*P. fluorescens*	MYX7275	Cry1A	MVP	CC
kurstaki/*P. fluorescens*	MYX104	Cry1Ac	Guardjet, M-Peril, MVP II,	CC
kurstaki x *aizawai*	GC-91	Cry1Ac, Cry1C	Agree, Design, Turex	HYB
kurstaki x *aizawai*	EG7826	Cry1Ac, Cry1F	Lepinox	HYB
kurstaki x *aizawai*	EG7841	Cry1Ac	Crymax	HYB
kurstaki x *aizawai*/*P. fluorescens*	MYX300	Cry1Ac, Cry1C	Mattch	CC
kurstaki x *tenebrionis* x *kumamotoensis*	EG2424	Cry1, Cry3	Foil[a], Jackpot	HYB
kurstaki x *tenebrionis* x *kumamotoensis*	EG7673	Cry3Aa, Cry3Bb	Raven	HYB
tenebrionis (= *morrisoni, san diego*)	SA-10	Cry3, Cry1	M-One[a], Novodor	NAT
tenebrionis/*P. florescens*		Cry3A	M-Trak[a]	CC

Comments: *NAT* – native *B. thuringiensis*, *HYB* – hybrid *B. thuringiensis*, *CC* – δ-endotoxin (CellCap)
[a]withdrawn from the market

The bioencapsulation and delivery system CellCap is a proprietary technology of Mycogen Co. for enhancing field persistence. The *cry* gene(s) coding for the desired endotoxin(s) is (are) isolated from *Bt* strains and transferred into a *Pseudomonas fluorescens* Migula host isolated from the phylloplane. In the production of CellCap preparations, genetically modified (GM) *P. fluorescens* cells are cultured in large-scale fermentors. Unlike *Bt* cells, which undergo lysis at the end of the fermentation cycle, the *P. fluorescens* cell walls remain intact. *P. fluorescens* cells are then killed in the fermentor before harvest using a proprietary physico-chemical procedure. This process also fixes the cell wall by cross-linking its components, creating a stable, dead cell biocapsule that encapsulates and protects the Cry toxins. Thus, the active component of any CellCap product contains no living cells: it rather consists of the selected Cry toxin(s) encapsulated within a dead cell biocapsule. The best-known GM *P. fluorescens* strains, containing different *cry* genes, are MYX104, MXY300, MYX833 and MYX7275 (Mycogen Co.) (Table 10.1).

Beside the introduction and broadening applications of *Bt*-bioinsecticides as means of environmentally friendly insect control, another *Bt*-based molecular biological application of increasing significance has been the development of insect resistant GM plants by the insertion of *cry* transgenes into the plant genome, the expression of which being responsible for the biosynthesis of Cry toxin proteins. This enables the resultant transgenic *Bt* plants to produce these microbial proteins and gain protection against sensitive insect pests through a mechanism similar to the pathogenicity of *B. thuringiensis* strains. As the molecular basis of the insect pathogenicity of the two approaches are identical, these agricultural practices are often considered equivalent. The low environmental and ecological impact of *Bt*-based bioinsecticides (Darvas and Polgár 1998) is observably so low that these formulated preparations gained acceptance even in ecological (organic) agriculture, the practice of which completely rejects the use of synthetic pesticides. In turn, *Bt* plants have also been proposed to be as environmentally safe as *Bt* insecticides. Nonetheless, although very closely related to each other in their biochemical mode of action, there are fundamental differences between these two insect control practices.

2 *Bt* Bioinsecticides

Bacillus thuringiensis strains applied in agricultural or hygienic (mosquito larva) treatment practices characteristically form parasporal bodies consisting of δ-endotoxins during sporulation. Certain varieties also contain parasporin, a recently described toxin causing cellular toxicity on tumor cells (Crickmore et al. 2009). In addition, several exotoxins (α-, β-, M-, etc.) and Vip (vegetative insecticidal protein) toxins may also be formed at the end of the vegetative stage of the bacteria, if food sources are limited for further vegetative periods (Bravo et al. 2007; Crickmore et al. 2009). Strains producing α-exotoxin (lecitinase C) and β-exotoxin (thermostable adenine nucleotide inhibiting RNA-polymerase) have been banned due to severe side-effects (mutagenicity and teratogenicity) of the latter.

The first insecticide containing *B. thuringiensis* was introduced in France in 1938 under the trade name of Sporeine and contained the *B. thuringiensis* subsp. *thuringiensis* pathotype. The HD-1 strain of *B. thuringiensis* subsp. *kurstaki*, isolated by Dulmage in 1970, has been found to be two orders of magnitude more active against agricultural pests (van Frankenhuyzen 1993), allowing broad application of *Bt* bioinsecticides. One of the major *Bt*-bioinsecticides is Dipel, developed from the HD-1 strain by Abbott Laboratories (1992).

2.1 Bt-Based Toxins: Structure and Classification

The δ-endotoxin proteins are further divided into two main groups, the pore-forming Cry (crystalline) and Cyt (cytolytic) toxins. Cry toxins are structurally related three-domain proteins consisting of an α-helix (domain 1), participating in the insertion into membranes, and two β-sheets (domains 2 and 3), taking part in the binding to the lectin receptors (Schnepf et al. 1998; Bravo et al. 2007). In Cyt toxins two α-helices surround a β-sheet, forming a simple α-β domain (Li et al. 1996). Cry toxins bind to special midgut receptors (Schnepf et al. 1998), while Cyt toxins form pores on the cell membrane through direct interaction with membrane lipids (Promdonkoy and Ellar 2003).

Bacillus thuringiensis strains used to be classified into 69 serotypes and 13 subgroups based on the H antigens of their flagellae and certain biochemical characteristics as there also exist strains without flagellae (van Frankenhuyzen 1993; Lecadet et al. 1999). These large toxin groups can be further divided by toxin structure (Cry1Aa, Cry1Ab, Cry1Ac, Cry1B, etc.). Most of the *Bt* strains produce several toxin types, for example *B. thuringiensis* subsp. *kurstaki* HD-1 produces Cry1Aa, Cry1Ab, Cry1Ac, Cry2A and Cry2B toxins (Arvidson et al. 1989; Lisansky et al. 1997). As the classification of several new toxins was problematic in this setup, a new system has been introduced on the basis of primary protein structure (amino acid sequence) similarities. Thus, the so far described 179 Cry and 9 Cyt toxins have been reclassified into 55 (Cry1 – Cry55) and 2 (Cyt1 – Cyt2) main toxin types, with several subtypes in each (e.g., Cry1Aa, Cry1Ba) (Crickmore et al. 1998, 2009). In addition, several toxins of structures different from the above are also known, including Bin (binary), Mtx (mosquitocidal) and Vip type toxins (Bravo and Soberón 2008). Cyt toxins not utilized in plant protection have cytolytic and hemolytic activity, and exert effects mainly in larvae of Diptera, or synergize the effect of other Cry toxins (Bravo et al. 2007; Gómez et al. 2007).

2.2 Mode of Action of Cry Toxins

Cry toxins exert *per os* type activity, and are divided in groups that each affect individuals within the same insect order. On this basis, Cry toxins can be sorted

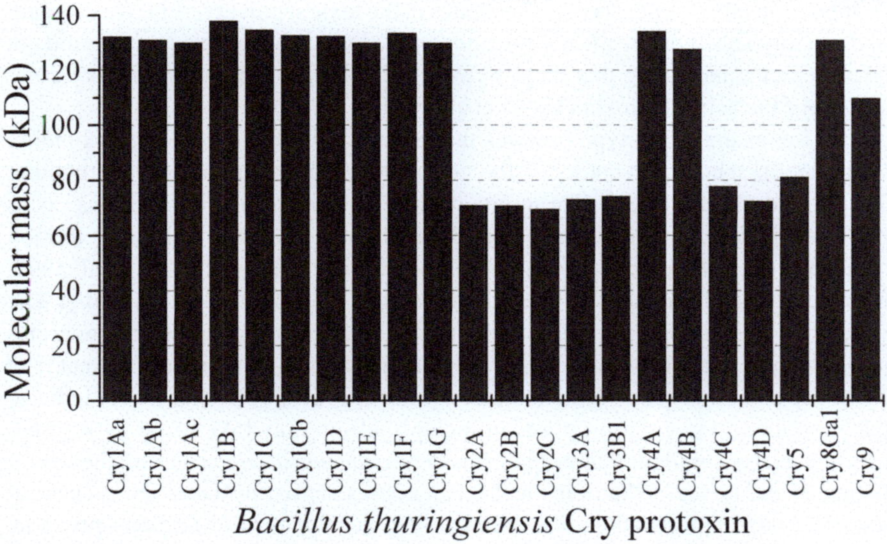

Fig. 10.1 Molecular mass of Cry protoxins

into five groups: exerting effects on Cry1 – mostly lepidopteran (*aizawai, kurstaki*, etc.), Cry2 – lepidopteran and dipteran (*kurstaki*), Cry3 – coleopteran (*tenebrionis, kumamotoensis*, etc.), Cry4 – Diptera (*israelensis*) larvae, Cry8 – coleopteran (*japonensis*), Cry9 – lepidopteran (*tolworthi*) specific toxins (Crickmore et al. 2009; van Frankenhuyzen 2009).

As for the mode of action of the most important Cry toxins, their effects lead, in several steps, to the lysis of the cells in the midgut epithelium. The Cry toxins, as formed in the *B. thuringiensis* bacteria, are proteins of 70–140 kDa molecular mass (Fig. 10.1), stabilized by disulfide bonds, and therefore, hard to decompose. This form is termed protoxin. The C-terminal domain in the characteristic three-domain structure of Cry1 protoxins is essential for exerting toxicity, and is believed to play a role in the formation of the bacterial parasporal bodies and to maintain the unique solubility properties of the protoxin crystals (de Maagd et al. 2003). Moreover, these C-terminal domains occur highly conserved, showing high (>90%) homology among Cry1 toxins. The N-terminal region of the toxic part of the crystalline toxins is more variable, showing 40–90% homology. The N-terminal region of the protoxin is markedly hydrophobic, while the C-terminal domain is dominantly hydrophilic. Proteases (trypsin, chymotrypsin, etc.) in the insect midgut cleave these protoxins to 55–65 kDa size activated toxins. The process occurs at high pH (10–11). Cry1A protoxins contain 16 cystein moieties (12 of which are conserved), and upon proteolytic removal of the first 28 amino acids and the C-terminal part of the protoxin, there remain no cystein in the remaining trypsin-resistant activated toxin. Of the 34 lysine moieties, only 3 remain in the trypsin-cleaved activated toxin, and even these and the arginine moieties present must be buried in the protein structure (causing resistance to further hydrolysis by trypsin). Chemical modification of approximately 12 of the tyrosine moieties in the toxic segment resulted in decrease in the cytolytic activity, while modification of the lysine and cystein moieties did not affect toxicity, indicating that these tyrosine moieties are located on the molecular surface of the activated toxin (Visser et al. 1993).

Cry toxins are lectin type proteins that undergo oligomerization upon binding to the lectin-specific receptors of the cell membranes in the midgut epithelium. The oligomer forms irreversible insertion into the lipid membrane, and thus opens pores in the cell membrane, induce colloid-osmotic swelling, disturbing the ion balance of the cell and causing its lysis (Knowles and Ellar 1987; Gill et al. 1992; Knowles 1994). Peristalsis of the gut stops, and the insect ceases feeding. The vegetative body of *B. thuringiensis* enters the larval coleoma through the microinjury (Schnepf et al. 1998), but any other microrganisms living in the gut tract may cause sepsis at that stage. Recent studies (Mason et al. 2011; Graf 2011) indicate that mortality upon creating pores in the epithelium may take place by septicemia caused by a midgut microorganism, *Enterococcus faecalis* entering the hemolymph of the larvae. Whatever mechanism causes sepsis, it makes the successful use of Cry toxin preparations and even the later development of *Bt* plants possible, as the lethal effect is triggered alone by the Cry toxin protein (Broderick et al. 2006), produced either by a microorganism or a plant expressing a *cry* transgene.

Collapse of the ion balance is sufficient for mortal paralysis in certain insect species (van Frankenhuyzen 1993). Sensitivity of insect species to given Cry toxins and consequently efficacy of different toxins on insect species varies.

2.3 Analysis of Cry Toxins

Quantitative detectability of the active ingredient content of biocides is essential both for technological control and environmental risk assessment. As the biological activity of *Bt* bioinsecticides is not necessarily directly proportional to the detectable content of given Cry toxins, this remains an unsolved problem for *Bt* preparations. Cry toxins are proteins, therefore, methods for their biochemical analysis possibly include high performance liquid chromatography (HPLC), gel or capillary electrophoresis, and immunoanalytical techniques. HPLC methods are mostly not suitable due to the instability and highly adsorptive character of the protein during separation. Therefore, electrophoretic, immunological and bioassay test methods are being used to identify and quantify Cry proteins (Hickle and Fitch 1990). The most commonly used immunoanalytical formats are lateral flow devices and 96-well microplate-based enzyme-linked immunosorbent assay (ELISA) (Grothaus et al. 2006). ELISAs are rapid and cost-effective methods in Cry toxin analysis, and numerous ELISAs have been developed and made commercially available for Cry endotoxins. Other immunoanalytical and specific receptor binding assay techniques exist: the rocket immunoelectrophoretic precipitation assay (Winkler et al. 1971), microsphere-based immunoassays (Ermolli et al. 2006a, b; Fantozzi et al. 2007), sensors (e.g., surface plasmon resonance biosensor, Okumura et al. 2001) and immunomagnetic electrochemical sensor (Volpe et al. 2006) have been developed for Cry toxins. Among these numerous analytical methods, ELISAs remain the methods of choice for their versatile applicability (Hickle and Fitch 1990; Hammock et al. 1991), and the use of ELISAs for Cry1Ab monitoring in *Bt* bioinsecticides and *Bt* plants has been reported extensively (Grothaus et al. 2006; Ermolli et al. 2006a; Palm et al. 1994; Adamczyk et al. 2001; Xie and Shu 2001; Zwahlen et al. 2003; Douville et al. 2001; Harwood et al. 2005; Székács et al. 2005, 2010a, b, 2012; Baumgarte and Tebbe 2005; Margarit et al. 2006; Nguyen and Jehle 2007; Crespo et al. 2008; Chen et al. 2009). It is important to emphasize that all commercially available ELISA methods have been developed using bacterial protoxins (see later), and therefore, they are directly applicable to bacterial preparations only.

Bt formulations are complex mixtures containing large amounts of damaged spores, intact δ-endotoxin crystals, residual amounts of fermentation medium and bacterial cell wall debris. The insecticidal efficacy of *Bt* formulations was characterized by specifying their bacterial spore content, yet it did not necessarily correlate with toxicity on insects. Attempts have also been made to standardize the endotoxin content of these preparations (Crespo et al. 2008), but did not succeed for different Cry toxin compositions due to varying actual fermentation conditions. Instead, solely biological activity requirements have been accepted, expressing the toxic efficacy on insects in International Units (IU) (van Frankenhuyzen 1993). A great boost occurred in the 1980s with the broadening use of *B. thuringiensis* subsp. *israelensis* preparations active on Diptera (mosquitoes and black flies) (Goldberg and Margalit 1977; de Barjac 1978; Federici et al. 1990) and Cry3-based preparations active on coleopteran (Gelernter 2004; Oppert et al. 2011). Yet *Bt* bioinsecticides of the highest importance remained the lepidopteran-specific preparations containing Cry1 and Cry2 type *Bt*-based endotoxins (e.g., Dipel).

The Cry endotoxin content of Dipel has been described inconsistently. Approximately 20–30% of the dry cell mass is constituted by Cry endotoxin crystals (Baum and Malvar 1995; Schnepf et al. 1998), nearly 80% of which is Cry1A (a, b, c) and approximately 20% is Cry2 (A, B) (Abbott Laboratories 1992). Nonetheless, actual δ-endotoxin concentrations in commercial and experimental formulations, determined by ion exchange chromatography, were found to be much lower, 0.3–1.7% δ-endotoxin (Bernhard and Utz 1993). Consequently, endotoxin content of distributed preparations varies extensively (U.S. EPA/OPP Pesticide-Related Database Queries; NPIRS National Pesticides Information Retrieval System; http://ppis.ceris.purdue.edu/).

It is of outstanding importance to consider the actual form of the Cry endotoxins within the *Bt* bioinsecticides, such as Dipel. The crystalline structure of these toxins is stabilized by disulfide bonds, and the crystal structure varies across toxin proteins. Due to the numerous disulfide bonds per protein molecule, the crystals are quite stable to solubilization at neutral pH. As a result, only a small minority of the toxin protein is immediately bioavailable (solubilizable), the vast majority of the crystal mass being only bioaccessible (temporarily non-bioavailable). Thus, just the solubilizable portion of the toxin content is analyzed promptly by ELISA; the bioaccessible part can be detected by ELISA only upon decomposition of the disulfide bonds stabilizing the crystal structure. Lastly, a minor part of the entire toxin content in the parasporal bodies is non-bioavailable: this amount of endotoxin is decomposed during the process of breaking the crystal structure, and therefore, cannot exert its potential biological activity by direct bioavailability or bioaccessibility.

The endotoxin composition in *Bt* bioinsecticides is illustrated on Fig. 10.2 using as an example the Cry1Ab endotoxin in Dipel. Dipel contains protoxins that require enzymatic activation by a hydrolytic process. Cry1Ab protoxins of molecular mass of 131 kDa form bipyramidal crystals stabilized by a maximum of 16 disulfide bonds per molecule: 14 of the cysteins are found at the C-terminal of the protein (Huber et al. 1981; Bietlot et al. 1990; Vazquez-Padron et al. 2004). These crystals are soluble only at high pH (Hickle and Fitch 1990) or in the presence of reducing agents of the disulfide bonds (e.g., mercaptoethanol or dithiothreitol). Solubilized Cry1Ab protoxin molecules undergo enzymatic cleavage in the alkaline medium of the insect midgut, and their cleavage by peptidases (such as trypsin) produces an activated toxin of approximately 63–65 kDa molecular mass, which is resistant to further hydrolysis (Chestukhina et al. 1982; Choma et al. 1990; Schnepf et al. 1998; Oppert 1999; Hilbeck 2001; Douville et al. 2001).

The nominal concentration of a common formulation of Dipel is 3.2%, meaning that the preparation contains 32 mg/g bacterial protein. While the biological efficacy was consistent, actual protein and Cry1Ab/Cry1Ac toxin concentrations of Dipel were found to be highly variable depending on the product batches (possibly differing from each other in actual fermentation conditions). Thus, total protein concentrations in various batches (obtained in separate product packages and in different years) ranged between 22.4 ± 2.2 and 51.4 ± 5.8 mg/g according to the bicinchoninic acid (BCA) method (Smith et al. 1985). Cry1Ab/Cry1Ac toxin content detectable by ELISA, however, was much more variable and depended on product batch and sample preparation conditions. The bacterial preparation extracted with neutral buffer (pH 7) resulted in detected Cry1Ab/Cry1Ac concentrations between 4.8 ± 0.6 and 60.2 ± 3.7 μg/g, with an average of 20.6 μg/g. This concentration is considered as the bioavailable Cry1Ab/Cry1Ac toxin content of the preparation. Dissolving the crystal structure under alkaline conditions results in increasing immobilization of the bioaccessible toxin stock and in higher detectable toxin concentration. Optimal solubilization was achieved at pH 10 (better than at pH 12), which could be further improved by the use of 50 mM mercaptoethanol or dithiothreitol in the extraction buffer reaching Cry1Ab/Cry1Ac concentrations of 84.5 ± 6.9 μg/g and 8.16 ± 0.87 mg/g without and with solubilizing agent, respectively. From these determinations, per hectare dosages of bioavailable and bioaccessible Cry1Ab/Cry1Ac toxin (20.6 mg/ha and 0.085–8.16 g/ha, respectively) can be easily calculated from the registered dose of 1 kg/ha for Dipel.

Fig. 10.2 (continued).*Bt* plants (e.g., *MON 810* maize) contain a single, truncated form of the Cry1Ab protein (approximately 91 kDa molecular mass), termed preactivated toxin, expressed at various concentrations in the plant tissues. Entering the insect digestive tract, this preactivated Cry1Ab toxin undergoes enzymatic cleavage resulting in the same hydrolysis-resistant core, activated Cry1Ab toxin. (The graphical representation of Cry1Ab toxin forms is adopted from Séralini 2010)

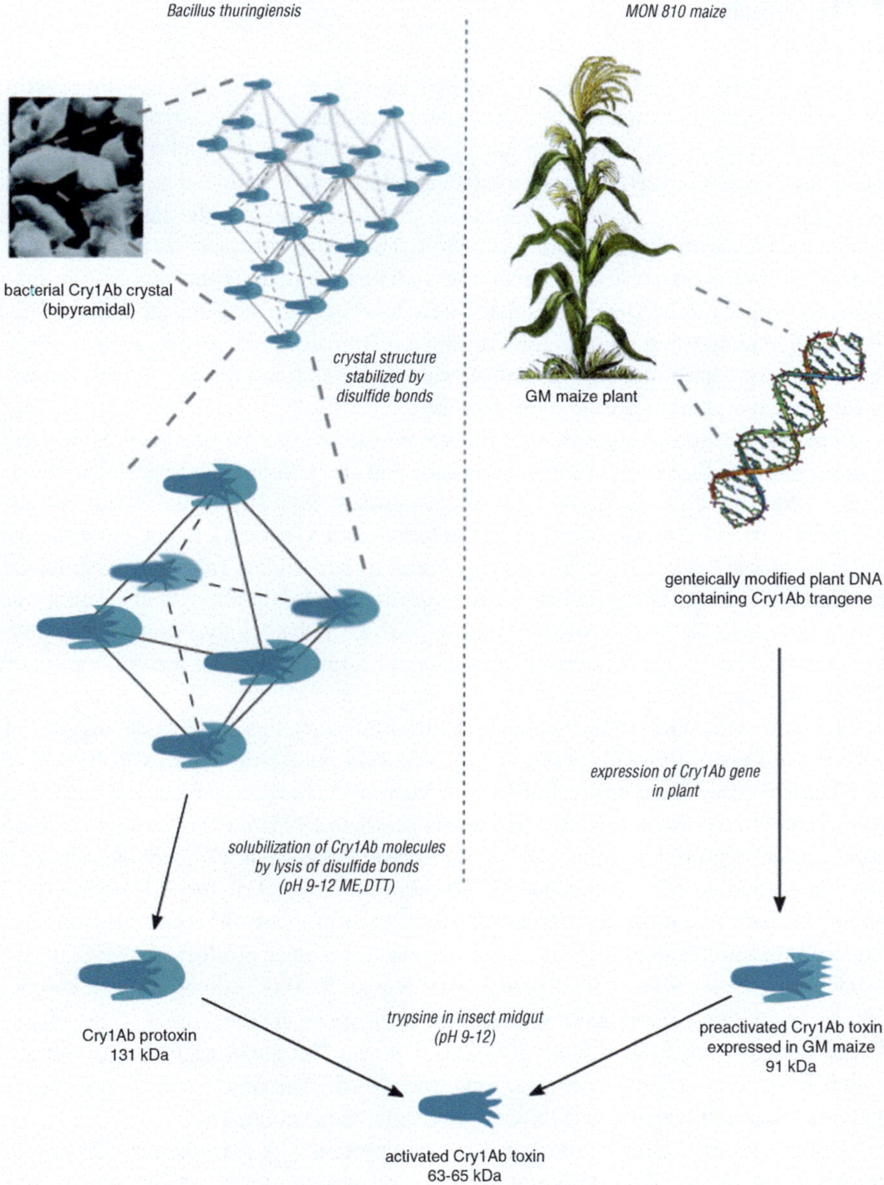

Fig. 10.2 Schematic representation of the various forms and activation of Cry1Ab protein in bacterial *Bt* preparations (*left*) and *Bt* crops (*right*). *Bt* bioinsecticides (e.g., Dipel) contain, among several other Cry protoxins, Cry1Ab protoxin molecules (131 kDa molecular mass) in bipyramidal crystals stabilized by disulfide bonds. Upon cleavage of the crystal structure by high pH or disulfide-reducing agents (e.g., mercaptoethanol, dithiothreitol), Cry1Ab protoxin molecules are solubilized. The solubilized Cry1Ab protoxin content undergoes enzymatic activation in the insect midgut to form the activated Cry1Ab toxin (63–65 kDa molecular mass) responsible for insecticidal action.

3 *Bt* Plants

At present there are 130 single or stacked event (i.e., producing one or several insecticide substances and/or tolerating one or more herbicide active ingredients) GM plant varieties under registration in the European Union, and a great part of these have been modified for pest resistance. It has been mentioned as an advantage of *Bt* plants that they provide continuous protection against the target pest and related species with similar modes of action. The active substance is not subject to certain environmental effects (direct UV radiation and rain) that could possibly lower its efficacy. It has been considered as a disadvantage, however, that the pollen of *Bt* maize containing the *cry* gene originated from *B. thuringiensis* may fertilize the flowers of traditional varieties of the same species (intraspecific hybridization) or their relatives (interspecific hybridization).

Bt plants registered or considered for registration within the European Union are cotton, maize and soybean (12 genetic events, and one additional withdrawn earlier). These GM plants, produced by three multinational firms (Monsanto Corporation, Pioneer Hi-Bred/Dow Agrosciences LLC/DuPont and Syngenta), target lepidopteran and coleopteran pests, and may be resistant to certain herbicides (Table 10.2). Although the genetic event *MON 810* has been registered for cultivation in the European Union, nine Member States (Austria, Hungary, Greece, Poland, Italy, France, Germany, Luxemburg and Bulgaria) announced national moratoria against the cultivation of this maize variety group.

Bt potato and additional *Bt* maize and cotton varieties have been registered outside the European Union. For example, the GM potato variety Russet Burbank/NewLeaf of Monsanto, producing Cry3A toxin and resistant against potato beetle was planted in the United States and Canada between 1995 and 2001. However, GM potato never captured a large share of the market because an efficient and cheaper insecticide against the potato beetle was commercialized at the same time, and public pressure on food processors led them to stop using this type of crop (US National Research Council 2010). StarLink maize varieties producing Cry9C toxin, registered only for animal feed between 1999 and 2000 (Castle et al. 2006), were rapidly withdrawn upon their widespread occurrence in the human food chain. These varieties have never been introduced in the European registration system, which is substantially more precautious than North America, partly due to agro-technological and legistative differences between these continents.

Bt plants produce large amounts of Cry toxin protein (Cry toxin/ha), and this toxin encapsulated in the plant cells remain for long periods in the environment. The effects of the toxin on the arthropods involved in the decomposition of the stubble- and on soil-microbial populations deserve further attention. Numerous current studies reveal low environmental impacts (US National Research Council 2010), yet the exact combination of all processes involved is not yet known. Pollen containing Cry toxin drifting off the fields may modify the habitat quality of the area and its borders, and therefore, can cause risks to rare and nationally protected butterflies. Butterflies living on nettle (*Urtica dioica*) and *Rubus* spp. at the perimeters or on

Table 10.2 Main single and stacked genetic events for lepidopteran and coleopteran resistance authorized in the European Union

Crop	Owner of the variety	Cry toxin types	Genetic event (Trade names)	Authorization Type	Authorization Stage	Type of genetic modification
Cotton	Monsanto	Cry1Ac	*MON 531* (Bollgard)	FF	R	Lepidopteran-resistant (*nptII*)
		Cry1Ac, Cry2Ab	*MON 15985* (Bollgard II)	FF	R	Lepidopteran-resistant (*nptII*)
Maize	Monsanto	Cry1Ab	*MON 810* (MaizeGard, YieldGard)	FF, IP, C	R	Lepidopteran-resistant
		Cry3Bb1	*MON 863* (MaxGard)	FF, IP	R	Coleopteran-resistant
		Cry3Bb1	*MON 88017* (YieldGard VT RW)	FF, IP	D	Coleopteran-resistant + herbicide-tolerant (*glyphosate*)
		Cry1A.105, Cry2Ab2	*MON 89034* (YieldGard VT Pro)	FF, IP	D	Lepidopteran-resistant
	Pioneer Hi-Bred/Dow/DuPont	Cry1F	*DAS-1507* (Herculex I)	FF	D	Lepidopteran-resistant
		Cry34Ab1, Cry35Ab1	*DAS-59122-7* (Herculex RW)	FF	D	Coleopteran-resistant + herbicide-tolerant (*glufosinate*)
	Syngenta	Cry1Ab	*SYN-BT011-1*	FF	P	Lepidopteran-resistant + herbicide-tolerant (*glufosinate*)
		Cry3A	*SYN-IR604* (Agrisure RW)	FF, IP	D	Coleopteran-resistant
		Vip3Aa19e	*SYN-IR162-4*	FF, IP	D	Lepidopteran-resistant
Soybean	Monsanto	Cry3A	*MON 87701*	FF, IP	D	Lepidopteran-resistant

Comments: *FF* – food and feed, *IP* – import and processing, *C* – cultivation, *D* – documentation submitted, *P* – product, *R* – authorization under renewal, *nptII* – containing antibiotics (kanamycin) resistance gene thorn apple (*Datura stramonium*) at the first 50 m of corn fields are at high risk of exposure, in particular, larvae of the peacock butterfly (*Nymphalis io*) in Central Europe (Darvas et al. 2004; Lauber et al. 2010; Lauber 2011).

Rapid insect resistance development has been observed with products producing a single Cry toxin (Tabashnik et al. 2008; Bagla 2010; Storer et al. 2010; Gassmann et al. 2011). To delay the development of such pest resistance, so-called "pyramid" *Bt* crops, producing two or more Cry toxins active against the same pest are being developed. Such pyramid strategy is best applicable, when selection for resistance to one of the toxins applied does not cause cross-resistance to the other(s) (Zhao et al. 2005; Gassmann et al. 2009; Tabashnik et al. 2009b). Nonetheless, although the evolution of pest resistance to pyramid *Bt* crops is slower, resistance development is driven by the same evolutionary processes as for single *Bt* Cry toxin varieties (Ives et al. 2011). *Bt* maize varieties producing toxins other than Cry, such as Vip3Aa19e (*SYN-IR162-4*) without cross-resistance with Cry1 toxins may also provide a solution to the problem of Cry1 toxin resistant strains of lepidopteran pests.

3.1 Coexistence of Bt Plants with Conventional Varieties

Probably one of the practically most important risks of GM crops is the possibility of gene flow. The reason is that the transgene is contained in and transmitted by the reproductive organs of the GM plants. Therefore, in the course of the cultivation of GM varieties, especially in the wind- (e.g., maize) and insect-pollinated (e.g., canola) plant species, the escape of the transgene with pollen (biological gene flow) or with seed/reproductive organs mixed with conventional varieties (physical gene flow) cannot be prevented, jeopardizing natural biodiversity, as well as traditional and organic farming. At present, gene flow is more problematic for glyphosate tolerant GM crops (for a review, see Székács and Darvas 2012), but also applies to *Bt* crops. Due to such possibilities of biological and physical gene flow, even the most severe coexistence law can only provide short-term solutions. In the interest of the elimination of gene flow it is necessary for the pollen not to contain the transgene or at least not in an operational state (Heszky 2011a).

Data on maize indicate that a distance larger than 25 and 250 m is needed to keep admixture below the European Union labeling threshold of 0.9% (that used to be the limit of detection of the early PCR techniques) and for 0.1% threshold (as favored by organic farming organizations), respectively. Oilseed rape represents a more complex issue in Europe, because apart from pollen flow, persistence of volunteers in arable fields and their perimeters, as well as interspecific hybridization with wild relatives also play a role (van de Wiel and Lotz 2006).

Different levels of GM traces in seeds (ranging from 0.01% to 0.5%) are considered in the European Union for quantifying the final adventitious GM presence in crops, especially in seed production. Three solutions are frequently mentioned in order to reduce gene flow: (*i*) the isolation distances between GM and non-GM fields; (*ii*) sowing a non-GM maize buffer strip around GM fields; and finally (*iii*) using GM varieties with different flowering duration compared with non-GM

Varieties. The latter is highly effective but is dependent on meteorological conditions and is hampered by associated yield losses (Messean et al. 2006).

The viewpoint in organic production is that the product cannot be declared organic if it contains any detectable GMO content (in other words tolerance for GMO content is 0%). In case of 0.5% GMO content in traditional maize seed, the calculated amount of individual GM plants that will emerge with a crop density of 70,000 plants per hectare is 350 individual GM plants. Projected to 1.2 million hectares (corresponding to of maize production of Hungary, the second biggest maize seed producer in European Union), the amount of emerged GM maize represents 420 million individual GM plants, equivalent to 6,000 ha of pure GM maize cultivation size. Considering 0.1% GMO content in the seed, the corresponding result is 60 GM plants per hectare and 70 million GM maize plants, representing 1,000 ha of pure GM maize cultivation size (Heszky 2011b).

The direction of biological gene flow can be diverse, which is unfortunately rarely discussed: (*i*) Gene flow *via* pollen can occur from GM to traditional Varieties. This is regulated by the current coexistence law. Gene flow, however, can occur in the reverse direction as well: conventional variety may also contaminate GM varieties, which may cause biological risk e.g., in the case of glyphosate resistance; (*ii*) Gene flow *via* pollen can also occur from GM varieties to other GM varieties, which is not regulated by the current coexistence law; (*iii*) From grass, trees, open pollinated fruit species, etc. gene flow *via* pollen may occur from GM varieties to wild ecotypes of the same plant species and also in the reverse direction (Heszky 2011b).

Intraspecific hybridization among cultivars is a major issue among foreign pollinated plants, especially in seed production, where pollen competition does not occur (detasseled or male sterile plants). The necessary isolation distance may reach 1,000 m in maize seed production if the tolerance limit is approximately 0.1% of foreign seed. Maize cultivars may be subdivided into three means groups by flowering time: early, normal and late pollination cultivars (see FAO numbering). The occurrence of silk or maturing of female flowers of maize follows the tasseling pattern. Pollen emission lasts 10–14 days in general for hybrid species, although it may last even twice as long in the external five rows, where individual plants may be at different developmental stages, well-known as border effect. Cross-pollination may occur only among cultivars of similar flowering time. In case of cross-pollination, however, the *cry1* gene transferred *via* pollen produces Cry1 toxin in the seed of originally non-modified plants already in the same year, as seen by RT-PCR and ELISA techniques. Therefore, although measures ensuring co-existence of GM and non-GM crops have been devised for many countries in Europe, Asia, and America, long term co-existence of a conventional a GM variety with same flowering time in case of cross-pollination is an ecological nonsense.

3.2 Compatibility of Bt Plants with Integrated Pest Management

A clear advantage of *Bt* crops is their specific mode of action. The broad range of *cry* genes provides the possibility of their applicability against various insect orders.

The utility of traditional formulated *Bt* products is well documented (Ravensberg 2011). Although sprayable *Bt* preparations with oral activity (e.g., *kurstaki* – Dipel; *israelensis* – VectoBac) are well applicable, certain unfavorable features have been revealed: the distribution of Cry toxin is uneven on the plant surface to be protected, the spray does not provide protection against pests that feed inside the plant; the toxin is decomposed upon UV irradiation; and the preparation is washed off from plants by rain (Roh et al. 2007).

In contrast, *Bt* plants provide relatively continuous protection against target pests and related species. This, however, means that the truncated Cry toxin is synthesized by the *Bt* plant continuously, regulated by the gene construct introduced into the plant and by the own genetic program of the plant, independently from the actual occurrence and population dynamics of the insects, causing an extensive presence of the Cry toxins in plants. In this context, *Bt* plants do not comply with the principles of integrated pest management (IPM), as the occurrence of the toxin is not limited to the duration of the possible damage by the pest, and does not implement any threshold value to the acceptable damage level. Although *Bt* crops can bring significant advantages under given climatic conditions, including reduction in use of broad-spectrum insecticides that is one of the primary goal of IPM (Cannon 2000; Romeis et al. 2008; US National Research Council 2010), there is opposition regarding the acceptance of *Bt* crops in IPM. Regardless of how mild there transgenic protein may be considered, the use of these crops currently cannot fulfill the main ecological principle of IPM that any protection step against any given pest is justified only if pest damage exceeds a critical threshold level.

Moreover, the Cry toxin varieties expressed in *Bt* plants are not necessarily the same as those in the corresponding *Bt* bioinsecticides. For example, maize varieties in the *MON 810* variety group produce a single preactivated Cry1Ab toxin of approximately 91 kDa molecular mass (Fig. 10.2), a truncated form of the bacterial Cry1Ab protoxin that undergoes enzymatic cleavage in the insect midgut, resulting in the same hydrolyzed, 63–65 kDa active toxin as Dipel (Hilbeck 2001; Székács et al. 2010a). Besides the obvious biochemical consequences, this fact of not identical Cry1Ab active ingredient in *Bt* bioinsecticides and *Bt* plants has connotations of utmost importance in pesticide/crop registration and in analysis of the active ingredient content.

As for registration issues, on the basis of the above, *MON 810* maize produces an active substance that is not a registered bioinsecticide ingredient. The Cry1Ab active ingredient of *MON 810* maize varieties is preactivated Cry1Ab toxin (91 kDa), yet toxicology studies in the registration documentation have been carried out with either bacterial protoxin (one of the active ingredients of Dipel, 131 kDa) or with the active toxin (63–65 kDa). This may be considered by some as a formal issue, yet a rather important one, as no pesticide active ingredient, regardless how similar it is to a registered one, can be exempt from individual registration and (eco)toxicological evaluation. This is well known in the registration of pesticides or pharmaceuticals, where complete toxicological evaluation is required for authorization of an active ingredient even if it differs structurally only slightly (e.g. in a single substituent) from a registered active ingredient. The same should also apply to insecticidally active proteins, such as Cry toxins.

It is often claimed that the registration of *Bt* crops has gone through an extensive review process that has been deemed thorough in more than 23 countries, where these crops are used. Such evaluations, however, are mostly administrative, based on the documentation of the genetic event supplied by the owner and corresponding data from the scientific literature. Moreover, in the scope of registration of new *Bt* crops in the European Union, event based documentation are submitted to a selected Member State called the rapporteur country of the given event. In case of maize, rapporteur countries include France (*MON 810*, *SYN-BT011*), Germany (*MON 863*, *SYN-IR162*), Spain (*DAS-1507*), the Czech Republic (*MON 88017*), the Netherlands (*MON 89034*, *DAS-59122*) and the United Kingdom (*SYN-IR604*). It is hardly justifiable, why decisions about cultivated *Bt* maize varieties are made in countries of slight importance in European maize production, and why not in leading maize producers such as Hungary or Italy, beside France.

Toxicological assessment of the variety documentations is not unambiguous, either, as seen for example in the case of maize variety *MON 863*. In 2002, Monsanto Company submitted an application to the German authorities to import *MON 863* maize into the European Union. The submission contained a 13-week rat feeding study, performed by a third company (Covance Labs), but statistically analyzed by Monsanto (Lemen et al. 2002; Hammond et al. 2006). Based on the results EFSA's experts (2004) stated, "The results of 90-day sub-chronic rodent studies do not indicate adverse effects from consumption of *MON 863* and *MON 810* and the Panel concluded that there are no concerns over their safety." A Court of Appeal action in Germany in June 2005 allowed public access to all the raw data from this 13-week rat-feeding study, on the basis of which Séralini et al. (2007) performed an independent analysis, and arrived to a conclusion that *MON 863* consumption affected the two main organs of detoxification: liver (in case of females) and kidney (in case of males). It appears that the statistical methods used by Monsanto were not sufficiently detailed to see disruptions in biochemical parameters, in order to evidence possible pathological signs. The EFSA GMO Panel re-evaluated the statistical methods (EFSA 2007), and stated that the observed differences in test parameters were not indicatives of adverse effects, and the new statistical analysis had not raised toxicologically relevant issues. In parallel, Doull et al. (2007) also came to the same conclusion regarding *MON 863*, and studying a Cry1Ab toxin producing maize variety in a three-generation feeding study, Kiliç and Akay (2008) found no statistically significant differences in relative organ weights of rats, except for minimal histopathological changes in liver and kidney. Changes in creatinine, total protein and globulin levels were also determined in biochemical analysis.

Further statistical analysis done by de Vendômois et al. (2009), however, clearly revealed for three events (*MON 810 – cry1Ab* gene, *MON 863 – cry3Bb* gene, *MON 603 – cp4-epsps* gene) new, sex- and often dose-dependent side-effects upon consumption. Effects were mostly associated with the kidney and liver, although varying among the three events. Further effects were also noticed in the heart, adrenal glands, spleen and the haematopoietic system. Upon criticism received from Monsanto (2010); de Vendômois et al. (2010) summarized the debated alimentary chronic risks, and suggested they may come from unpredictable insertional mutagenesis or metabolic effects, or from new pesticide residues. Therefore, as chronic health effects including cancerous, hormonal, reproductive, nervous or immune diseases are increasing worldwide, gender differences and the non-linear dose- or time-related effects should be particularly considered in toxicology, mainly in attempts to reveal hormone-dependent diseases and first signs of toxicities (Séralini et al. 2009). Yet, although certain 90-day feeding tests were performed, longer studies are very rare. Upon worldwide commercialization of GM crops, especially stacked events, the standard toxicological evaluation is even more seriously inadequate as the so-called "cocktail effects" are not taken into consideration.

An additional example of the controversies in the toxicological evaluations is the mammalian toxicological assessment in the application for renewal of authorization of *MON 810* maize (European Food Safety Authority 2009b). The assessment refers to a 90-day rat feeding study with grain of *MON 810* maize (Hammond et al. 2006), in which the overall health, body weight, food consumption, clinical pathology parameters (hematology, blood chemistry, urinalysis), organ weights, and gross and microscopic appearance of tissues were found to be comparable between groups fed with diets containing *MON 810* maize and conventional maize varieties. In contrast to this conclusion, principal component analysis clearly revealed sex- and often dose-dependent new side effects linked with consumption of GM maize of genetic events *MON 810* (containing Cry1Ab toxin), *MON 863* (containing Cry3Bb1 toxin) and *NK603* (glyphosate tolerant). Effects were mostly associated with the dietary detoxifying organs (de Vendômois et al. 2009). Moreover, Séralini et al. (2011) found the results of Hammond et al. (2006) highly controversial, claiming that 90-day tests are insufficient in length to evaluate chronic toxicity, and the hepatorenal toxicity signs observed may indicate the onset of chronic diseases. Similar toxicological considerations apply for stacked trait GM plants with both Roundup (glyphosate) tolerance and transgenic insecticidal Cry toxin-based insect resistance. Current findings indicate that Cry1 toxins cause cell death at high concentrations (above 100 ppm), while reduce caspase 3/7 activation induced by Roundup on a human embryonic kidney cell line (Mesnage et al. 2011), indicating that Cry toxins are not inert on non-target human cells and can interact with the side-effects of glyphosate.

As for the analytical consequences of the differing Cry1Ab toxin proteins in *Bt* bioinsecticides and *Bt* crops, the ELISA systems devised against the bacterial protoxins are improper to be used in the analytical sense to directly apply to the quantitative determination of plant-expressed toxins. It is obvious to concede that antibodies generated about the protoxin will show different (lower) affinity to the truncated forms of the toxin. Thus, ELISA systems against Cry1Ab/Cry1Ac bacterial protoxins cannot be applied without correction to the measurement of plant toxin levels by using analytical standards of the protoxin protein. A simple possibility to overcome this problem is the use of plant-produced toxins as analytical standards, however, the preactivated Cry1Ab protein produced by *MON 810* maize is commercially not available. A more elaborate approach has been the determination of the cross-reactivity (CR) of the ELISA systems with the activated toxin. On the basis of established enzymatic activation protocols (Lilley et al. 1980; Lambert et al. 1996;

Shao et al. 1998; Miranda et al. 2001; Mohan and Gujar 2003), this has been achieved for Cry1Ab (Székács et al. 2010a), and two commercial ELISA systems have been shown to have CR to the active toxin of 0.41–0.56. It has to be emphasized, that all values reported in the scientific literature, obtained by protoxin-based immunoassays, are subject to correction with such CR values. In other words, reported toxin concentration values underestimate the actual levels if they do not take the lower CR of the plant expressed Cry1Ab toxin with the antibodies into consideration. To overcome this problem, ELISA systems against the plant-expressed toxin(s) should be developed and made available, or at least the plant-expressed toxin(s) should be made readily available as a protein standard.

3.3 Biomass and Expressed Cry1Ab Toxin Levels in Bt Maize

Maize biomass production is affected by a number of parameters, the most important of which being the plant variety, agrotechnologies and weather conditions. Cultivating various *Bt*, near-isogenic and commercial maize varieties in the period between 2001 and 2011 at the Ecological Experimental Station of the Plant Protection Institute, Hungarian Academy of Sciences (Julianna-major, Nagykovácsi, Hungary) (Székács et al. 2005, 2010a, b; Takács et al. 2011), the overall maize (FAO number near 400) biomass produced by various varieties ranged between 70.9 and 96.2 t/ha. The mass proportion of each plant organ was determined in each year for each variety, and was found consistent among different years and varieties with the highest mass being the cobs (measured with all husk leaves and pistils), 31.9–42.6% of the entire biomass, followed by the foliage 21.1–32.0%, and stalk 21.7–23.0% (Fig. 10.3).

Cry1Ab toxin content was determined for all *Bt* maize varieties, with the CR between activated toxin and protoxin (Székács et al. 2010a) and the uneven distribution within organs (Székács et al. 2010b) considered. Thus, in 2001 Cry1Ab toxin concentration was demonstrated to be 9.6–17.2, 2.3–5.3 and 1.4 μg/g in the leaves, roots and stalk of *MON 810* maize DK-440 BTY, showing seasonal fluctuation (Székács et al. 2010a) with maximal toxin content in the leaves in the vegetative five-leaf (V5) phenological stage. The per hectare production of plant-expressed Cry1Ab toxin was calculated, and was found to range between 147 and 456 g of Cry1Ab toxin/ha during the 2001–2011 period. Calculating the Cry1Ab toxin distribution among plant tissues (Fig. 10.3) revealed important observations: (*i*) The vast majority of the Cry1Ab toxin is found in the foliage (69.0–72.1%) (Székács et al. 2010a). (*ii*) Toxin content in the seeds represents a minor portion (1.9–6.8%). (*iii*) *MON 810* maize expresses the Cry1Ab toxin at moderate concentrations in the root as well, and as a result, a significant proportion (7.7–9.7%) of the overall produced toxin quantity is found in the roots and inevitably remain, along with other plant parts, in the stubble.

These toxin levels may further be elevated by soil fertilization, the use of long maturation maize varieties and the use of stacked genetic events. Soil quality, especially

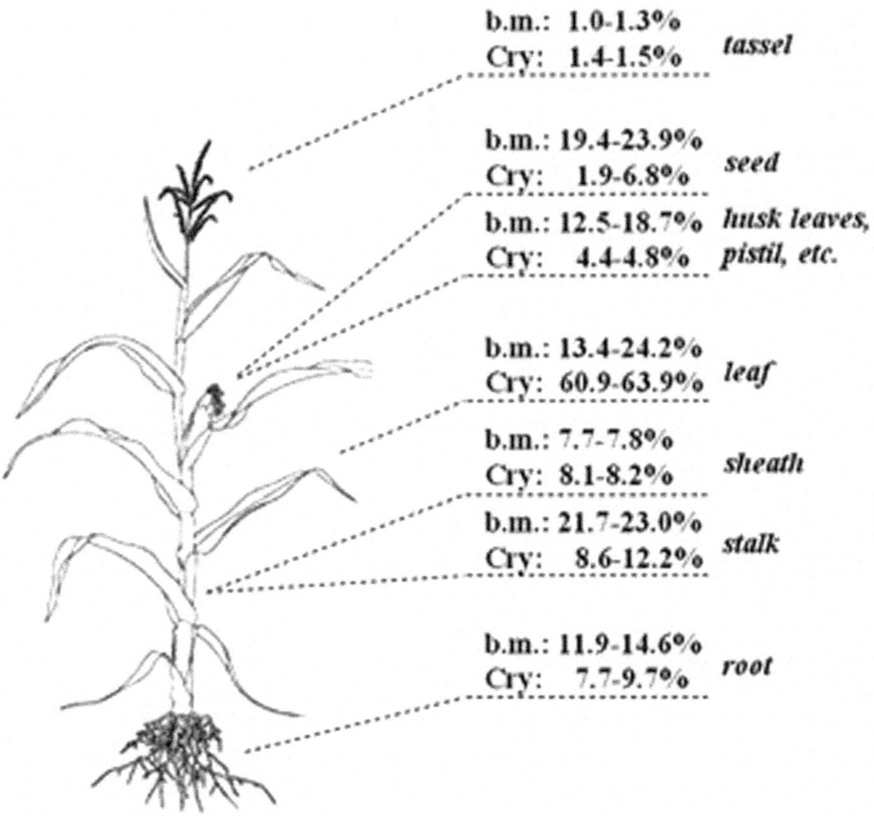

Fig. 10.3 Average biomass (b.m.) and Cry1Ab toxin production (Cry) by *MON 810* maize varieties. Meanwhile the largest biomass is *represented by the seed, leaf and stem* (approximately equal to each other), the largest proportion of the transgenic Cry1Ab toxin content is found in the leaves

the use of N-fertilizers was shown to exert a strong influence on Cry1Ab toxin expression by increasing biomass production by up to 80–85% (Bruns and Abel 2003; Ma and Subedi 2005) and consequently Cry1Ab toxin levels in *MON 810* by up to 3.7-fold (Bruns and Abel 2004). Thus, in the study of Bruns and Abel (2003, 2004) examining the effect of N-fertilization on Cry1Ab toxin production in the leaves of two *MON 810* maize varieties (Pioneer 33V08 Bt és DK-626 Bt) at V2 phenological stage and in the husk leaves of three varieties (AgriGold 6729 Bt, Pioneer 33V08 Bt, DeKalb DBT 418 Bt) at R3 phenological stage, an 1.4–1.5-fold increase was seen in Cry1Ab concentration in the leaves at 224–336 kg/ha N-fertilizer (NH_4NO_3) application, and over threefold increase in the husk leaves at nearly 300 kg/ha N-fertilization in two varieties (AgriGold 6729 Bt, Pioneer 33V08 Bt) and none in the third one (DeKalb DBT-418 Bt). The absolute amount of Cry1Ab toxin produced is further increased by the higher biomass of the maize varieties due to N-fertilization. In our experience fertilization increased the biomass produced by

DK-440 BTY in pot experiments up to 1.5–2.0-fold. Fertilization, therefore, may multiply Cry1Ab toxin production both through boosting Cry1Ab biosynthesis (along with other physiological biosynthesis processes in photosynthesizing tissues) and through increased biomass production. Cry1Ab toxin production varies among *Bt* maize varieties of different genetic events (Fearing et al. 1997; Baumgarte and Tebbe 2005; Nguyen and Jehle 2007; Székács et al. 2010a). Therefore, the use of long maturation (high biomass) maize variety of high toxin production (e.g., DK-818) may also result in such double enhancement effect: a *Bt* variety of DK-818 produced Cry1Ab toxin in the leaves at concentrations about 80% higher than in certain short maturation varieties (e.g., DK-440 BTY), while resulting in increased biomass production by 1.9–3.2-fold, to 180 t/ha. In that case, the overall quantity of produced Cry1Ab could reach as high as 1,930 g/ha. Apparently, the combination of individual genetic events in stacked GM crops may have an effect on the expression of the given transgenes, as a twofold increase has been observed in Cry1Ab toxin levels in stacked event maize varieties than in single event *MON 810* cultivars (European Food Safety Authority 2005). Moreover, the active ingredient is not subject to such strong environmental effects (UV radiation, rain) than in spray application that would lower their efficacy.

Yield increases are often mentioned as major advantages of *Bt* crops (Betz et al. 2000), particularly when pest pressure is high, but such yield advantages are related to avoided loss by pest damage, not the growth or production capacity of the maize variety. In the case of Cry1-expressing maize varieties, achievable yields depend on damage by the European corn borer, and therefore, as indicated in the literature (Ma and Subedi 2005) and in national distinctness, uniformity and stability (DUS) tests (e.g., in Hungary), maize yields remain unchanged in areas, where this pest is insignificant (Füsti Molnár 2011; Darvas et al. 2011).

3.4 Utility and Duration of Cry1Ab Toxin Production in Plant Tissues

In contrast to cause-directed crop protection involving spraying pesticide applications, GM plants express Cry toxin also during periods, when it is not necessary for pest control and also in plant tissues, where it is not needed. It has also been mentioned as a benefit that *Bt* crops reduce the need for broad spectrum insecticides, however, this does not apply to all climatic conditions. For example significant occurrence of the European corn borer in Hungary is rare (once in every 10 years), therefore, farmers do not even protect their crops against this pest. In such regions, cost efficacy cannot be justified for the very same reason. In addition, Cry1Ab toxin does not exert any effect on other maize pests such as aphids and mites or soil inhabiting coleopteran species. Moreover, *MON 810* maize resistant to European corn borer produces significant amounts of Cry1Ab toxin in its roots (approximately one quarter of the level produced in the leaves, where European corn borer feeds only in its L1 stage), while this pest does not damage the root at all. The roots release Cry1Ab into the soil *via* their exudate (Icoz and Stotzky 2008b) during the entire vegetation period, and the toxin may be taken up by plants cultivated subsequently (Icoz et al. 2009). Although more studies are needed about such exudation of Cry toxins into the soil, numerous studies indicated little or no effect on soil organisms (Blackwood and Buyer 2004; Griffiths et al. 2006; Cortet et al. 2007; Icoz et al. 2008; US National Research Council 2010; Zeilinger et al. 2010; Tan et al. 2010), while others found low but significant effects of *Bt* maize on microbial community structure in soil (Turrini et al. 2004; Castaldini et al. 2005; Oliveira et al. 2008). In their extensive evaluation, Icoz and Stotzky (2008b) concluded that the effects of *Bt* maize on the soil biota are transient, but possible long-term impacts cannot be excluded.

The decrease in the incidence of corn ear in infestation by *Fusarium* species co-occurring with damage by larvae is considered a benefit (Munkvold 2003; Clements et al. 2003; Folcher et al. 2010). Yet, visual signs of fungal infection do not necessarily correlate with mycotoxin content, the composition of which reflects to proportions of various *Fusarium* species. Seed infection may occur not only by surface injuries (chewing by insects), but also by infestation through the pistil, characteristic to certain *Fusarium* species (Darvas et al. 2011). Thus, the level of given mycotoxins (e.g., zearalenone) does not correlate with the rate of larval damage (Folcher et al. 2009). In the case of *Bt* maize (*MON 810* and *SYN-EV176* variety groups), only a limited decrease of certain mycotoxins has been verified (Papst et al. 2005).

Seeds of *MON 810* maize produce significantly lower amounts of Cry1Ab toxin than the leaves (approximately one-tenth) (Székács et al. 2010a). Therefore, this maize variety cannot offer an optimal method to control corn ear damage by the larvae of the cotton bollworm (*Helicoverpa armigera* Hübner) and the European corn borer (Darvas et al. 2011). The high Cry1Ab toxin content in the foliage also raises the question of safe use of *MON 810* maize as silage. At least, 60% of the initial Cry1Ab protein concentration remained in the fresh silage (Kamota et al. 2011). Cry1Ab content in silage exhibited no clear-cut pattern of decrease over the time of 4 months. Thus, average Cry1Ab toxin content was detected to be 1.88 ± 0.71 μg/g *MON 810* silage (Rauschen and Schuphan 2006). In case of lactating dairy cows, there was no difference in ruminal degradability, determined separately for maize silage and grain of *MON 810*. Nutritional value and production efficacy for *MON 810* maize silage was similar to its near isogenic line (Donkin et al. 2003). No transgenic DNA from maize containing stacked (2GM) *cry1Ab* and *mepsps* genes, or Cry1Ab protein produced in the plant were detected in milk from cows fed with GM corn silage (Calsamiglia et al. 2007). A long-term study over 25 months was conducted to evaluate the effects of *MON 810* maize (silage, kernels and whole-crop cobs) on performance of lactating dairy cows. Cows fed with *MON 810* maize were exposed daily to Cry1Ab protein intake of 6.0 and 6.1 mg in the first and second lactation of the trial, respectively. There were no consistent effects of feeding with *MON 810* or its isogenic line on milk composition and body condition (Steinke et al. 2010). Health studies on mammalian model as rats (Hammond et al. 2006) met strong criticism (de Vendômois et al. 2009; Séralini et al. 2011) as mentioned earlier.

As seen above, *Bt* plants produce high amounts of Cry1Ab toxin per hectare – depending on the proportion of the vegetative plant parts in the biomass. The alarmingly high toxin amount in the foliage raises severe concerns regarding its utility for feed/silage or even leaving foliage in the stubble. In contrast, toxin content is of lesser concern in the seeds, maybe that is the reason, besides high insect specificity of the toxin, why seed consumption issues of *MON 810* maize have not emerged. Toxin expression in the roots and pollen, where it is not needed for plant protection purposes and causes unwanted exposure and ecotoxicological consequences prompts further development of *Bt* crops towards tissue-specific gene expression systems. The time of degradation of Cry toxins in the soil have been shown to depend on several factors, including climatic conditions (temperature, soil water content), soil characteristics (pH, composition) and soil microbial life. Cry toxins have been shown to be rapidly degraded microbially in soil (Clark et al. 2005). The plant-expressed toxin, however, being protected against decomposition in the plant cells, may persist in the soil (Baumgarte and Tebbe 2005; Icoz et al. 2008). There are numerous studies indicating no persistence or environmental effects of Cry toxins (Hopkins and Gregorich 2003; Pagel-Wieder et al. 2007; Icoz and Stotzky 2008a), others however, are less comforting. Results indicate 1–8% of the toxin content in stubble can be detected 1 year later, upon harvest. This can still be a substantial amount in the case of large vegetative mass varieties, compared to toxin amounts released with the bioinsecticide Dipel. Toxin accumulation and biological effects observed on insects show a pattern dependent on soil type (Tapp and Stotzky 1998). Moreover, Cry1 toxin persistence is a function of the entire amount of Cry toxin produced, its distribution among plant tissues, production and decomposition dynamics, as well as additional abiotic factors (Zwahlen et al. 2003). Certain collembolan species show a tendency to avoid maize debris containing Cry1Ab toxin in stubble as compared to the isogenic variety (Bakonyi et al. 2006, 2011), and their fecundity is decreased on this food type, that may partially explain low decomposition rate of a *MON 810* maize variety (DK-440 BTY) in stubble.

3.5 Secondary Effects Through Cry1 Toxin Containing Pollen and Stubble

Non-target organisms may get exposed through feeding to Cry toxins produced by *Bt* crops. The main routes of exposure include drifted pollen settled on food plants for phytophagous species, Cry toxin consuming pray and host for predators and parasitoids, plant debris and residues for decomposing organisms, flower and pollen for pollinating insects, and community connections for symbiontic organisms. Although Cry toxins (including Cry1Ab) have been claimed to pose no risk to various non-target organisms in numerous studies due to low toxicity or lack of exposure (Romeis et al. 2006), possible affectedness of non-target organisms related to the target pest is outstandingly problematic. Pollination by Cry1 producing crops present a possible risk in the pollination period to the habitats of larvae of protected and rare lepidopteran insects living in the weedy perimeters of cultivation sites, if pollen containing Cry1 toxin is settled to these habitats and endures there for longer periods.

The analysis of pollen shedding revealed that a 300–600 pollen/cm^2 density is not rare on the top leaves of maize, but exceptional outside the maize fields (Pleasants et al. 2001; Darvas et al. 2004; Li et al. 2005). Ample intervals appear at the edges of maize fields for deposited pollen, depending on the pollen yield of the hybrid, the prevailing wind direction and pressure, and leaf surface characteristics of the plants on which maize pollen is deposited (Pleasants et al. 2001; Darvas et al. 2004). *Bt* maize pollen is ingested by non-target organisms that consume weeds emerging on maize fields. In a laboratory assay, larvae of *Danaus plexippus* (L.), reared on milkweed leaves dusted with pollen from *Bt* maize (N4640-*Bt*), consumed less plant material, grew slower and suffered higher mortality than larvae reared on leaves dusted with maize pollen without Cry1 toxin (Losey et al. 1999). These results were later questioned, and the impact of *Bt* maize pollen from commercial hybrids was suggested negligible on monarch butterfly populations (Sears et al. 2001) due to low levels of exposure. In subsequent experiments, increased mortality, delayed development of lepidopteran larvae and decreased larval, pupal and adult weight were revealed in experiments performed with pollen containing Cry1 toxin (*MON 810* or *SYN-Bt11*: ~90 ng Cry1Ab/g pollen) using densities typical at maize field adjacencies (Jesse and Obrycki 2000; Dively et al. 2004; Anderson et al. 2005).

The majority of lepidopteran adults visit agricultural areas for feeding and oviposition, so the diversity of flowering plants in the neighboring environments severely affects the butterfly community. Only extended monitoring can exactly reflect the effects of the environmental changes (Lang 2004).

There are several protected butterfly species in Europe, especially in the Pannonian Biogeographic Region, where the number of protected species is more than 200 (European Environment Agency 2002; Darvas et al. 2004). In this region larvae of *Nymphalis urticae* (L.), *Nymphalis io* (L.), *Nymphalis c-album* (L.), *Vanessa atalanta* (L.) feeding on *Urtica dioica* L.; *Argynnis niobe* (L.), *Argynnis pandora* (Dennis et Schiffermüller), *Brenthis ino* (Rottemburg), *Spialia sertorius* (Hoffmannsegg) feeding on *Rubus* spp.; and *Acherontia atropos* (L.) feeding on *Datura stramonium* L. may be affected principally (Darvas et al. 2004; Lauber et al. 2010; Lauber 2011). *Urtica dioica* and *Rubus* spp. are common at the perimeters and *D. stramonium* is frequent weed at the 50 m in the maize field.

In Germany, toxicity of *Bt* maize pollen to *Papilio machaon* L. was also studied. First instar larvae were exposed to different pollen densities applied to leaf disks of *Pastinaca sativa* L. for 48 h. The LD_{50} with regard to larvae surviving to adulthood was ~14 pollen grains of *SYN-EV176* consumed by first instar larvae (Lang and Vojtech 2006).

A mathematical model analyzed exposure of larvae of some non-target species: for example *N. io* and *V. atalanta* in four European countries. A dose-mortality relationship was integrated with a dose-distance relationship to estimate mortality both within maize field and at varying distances from the field edge. Perry et al. (2010) concluded the estimated environmental impact was low. Lang et al. (2011) found that the incomplete and uncertain input data cause a higher uncertainty than indicated by Perry et al. (2010), and the possibility that the effects might be worse than predicted. Moreover, Perry et al. (2010) assumed larvae of *V. atalanta*, and *N. io* equally susceptible to Cry1Ab. They cited Darvas et al. (2004) as a reference for such equitoxicity, even though the cited paper contains no data about species sensitivity. In contrary, we reported LC_{50} of Dipel being 15.14 ppm to *V. atalanta* and 4.39 ppm to *N. io* first instar larvae (for example for the EFSA GMO Panel in Parma at June 11, 2008 – see Rodics et al. 2011; Lauber 2011). In another version of the mathematical model extended to non-target effects of Cry1F toxin in Bt maize pollen (Perry et al. 2012), therefore, the sensitivity of non-target insects is considered purely on a theoretical basis. The value of a mathematical model rests on the basis of the certainty of its input data (the biological effect in this case), which is highly questionable for the given model. Moreover, no acceptable mortality threshold applies in environmental risk assessment regarding protected species. In other world only the lack of exposure can limit the risk of a toxic substance to protected species, but any rate of mortality is unacceptable if the species is exposed to the substance. Pollen drifted from maize fields modifies habitat characteristics of protected species, which is not allowed by the Habitat Directive of the European Union (European Council 1992).

A frequently mentioned justification of pest control on the basis of plant-expressed toxin is the argument that broad-spectrum insecticides have more severe toxic effects on non-target organisms (Romeis et al. 2006). This may be true for various crops, climatic conditions and pests, but not for *MON 810* maize, the Pannonian Biogeographic Region (e.g., Hungary) and the European corn borer, which is not a regular pest in Hungary, and therefore, there is no reasons to use chemical or agro-biotechnological protection against it.

The large, globular particles of maize pollen settle in a relatively rapid course. Nearly 80% of it is settled within 6 m, but a small proportion may reach as far as several hundred meters. Pollen containing Cry toxin may reach natural aquatic habitats as well, where it may travel long distances without decomposition of the toxin. Rosi-Marshall et al. (2007) found that pollen containing Cry1 toxin along with other plant debris may enter aquatic ecosystems, where may impede the development of given trichopteran species and increase their mortality. Chambers et al. (2010) detected developmental delay of the development, but not the abundance or biomass of a trichopteran species (*Lepidostoma liba* Ross). Bøhn et al. (2008, 2010) demonstrated inhibitory effects of plant debris containing Cry1 toxin on the development and reproduction of the great water flea (*Daphnia magna* Straus), an indicator organism for water quality. None of these species belong to the group of known sensitivity to Cry1 toxins. Cry1Ab toxin protected from decomposition in the plant tissue could be detected from pollen transported by surface waters as late as 6 months after harvest (Tank et al. 2010).

The amount of Cry toxin containing pollen drifting from the maize field can be substantially reduced, yet not eliminated, by a refugee zone (border lines) with the isogenic line. Moreover, such refugee zone cannot solve the problem around canals across and temporarily water-covered spots on cultivation sites. The required size of the critical zone is proportional with the expressed Cry toxin content in the pollen, characteristic to the crop variety. A satisfactory solution to this problem would be if the *cry* gene would not be present in the pollen.

The Pannonian Biogeographic Region i.e., the Carpathian basin is of outstanding importance in Europe in respect of conservation of biodiversity. Hungary represent 81% of this region. This justifies the definition of strict specifications for co-existence of GM organisms in the biogeographic zone. Legal regulations fully complying with requirements on the conservation of natural resources and the precautionary principle are required.

Environmental risk assessment of agrochemical or agrobiotechnological substances is based on the identification and estimation of negative effects and a subsequent evaluation of real exposures in agricultural practice. Secondary effects on non-target species are often tested experimentally on model organisms to describe the potential effects. Decision-makers at different authorities and boards (e.g., EPA, FDA, USDA, EFSA) attempt to balance experimental results and offer a general solution on the basis of special findings. In turn, several debates emerge. Regulatory frameworks should advocate the tiered approach to assess possible non-target effects. According to Romeis et al. (2006), in risk assessment of *Bt* plants on non-target organisms, early tier (i.e., laboratory) tests are conducted to determine whether an organism is susceptible to the Cry toxin under worst case conditions. The main problem is that laboratory conditions usually do not represent worst cases, and the agent with selectivity at order level may often exert indirect tritrophic effect through sensitive parasitoids or predators of an insensitive pest. In their comments, Andow et al. (2006) argued that several of the proposed conclusions and recommendations are restrictive and premature. It is essential for a suitable environmental risk assessment to include direct and indirect effects on natural enemies, which may not be resolved in a mechanistic decision procedure. Lang et al. (2007) emphasized that laboratory settings with ample food supply and favorable climatic circumstances ensure that experimental animals are in a good condition, provide an advantageous status to cope with exposure to Cry toxins. In worst case scenarios, however, additional stressors such as low temperature, rain, food shortage, or especially parasites and diseases are likely to exacerbate the effect. For example, *N. io* larval populations are regularly reduced by an endemic pathogen (cypovirus 2) and certain parasitoids (e.g., *Sturmia bella* (Meigen), Tachinidae and *Microgaster subcompleta*, Nees, Ichneumonoidea and *Pteromalus puparum* L., Pteromalidae) in the Pannonian Biogeographic Region. These controlling agents may divide a single *N. io* population into different susceptible and tolerant subpopulations, modulating the effect of an additional pathogenic factor such as food containing Cry1Ab toxin (Lang et al. 2007; Lauber 2011).

3.6 Pest Resistance to Cry Toxins

Sublethal effects (that possibly occur with spray applications upon the wash-out effect of natural precipitation or with *Bt* crops with low gene expression levels or insufficient exposure of the pest) may contribute to the occurrence of Cry resistance or cross-resistance. This phenomenon is attributed to two main factors, sublethal effects caused by the transgenic Cry1Ab toxin on pest sub-populations and Cry toxin composition of the *Bt* crop (namely a single toxin, preactivated Cry1Ab toxin). Sublethal effects may contribute to the occurrence of Cry resistance due to low gene expression levels and thus insufficient exposure of the pest. In laboratory experiments, the model species, Indian meal moth (*Plodia interpunctella* Hübner) was found to develop resistance already in the 10th generation (Darvas 2011), indicating rapid obsoletion of *Bt* maize varieties. What makes this problem even more troublesome is that *P. interpunctella* larvae resistant to *MON 810* show tolerance also to Dipel. Such cross-resistance means that development of resistance to Cry1 toxins, may lead to loss of applicability of both *Bt* crops and conventional *Bt* preparations.

The rapid onset of resistance development is explained by the fact that *MON 810* maize contains a single Cry toxin (preactivated Cry1Ab) only. Microbial *Bt*-bioinsecticides, in contrast, contain several Cry toxins, and therefore, resistance development is more hindered. Although the mode of action of Cry toxins is similar, their pathways do differ from each other at least in the receptor protein in the insect midgut. The severity of the resistance problem is well indicated by the fact that the occurrence of resistance has been an accentuated and critical issue in environmental risk assessment (beside non-target effects and toxin loads on the environment) within the re-registration of the *MON 810* variety group in the European Union (European Food Safety Authority 2009a).

During 2005–2006, field-evolved Cry1 toxin resistance has been documented on three noctuid species: *Spodoptera frugiperda* (J. E. Smith) to Cry1F toxin in Puerto Rico, *Busseola fusca* (Fuller) to Cry1Ab toxin in *Bt* maize in South Africa, and *Helicoverpa zea* (Boddie) to Cry1Ac and Cry2Ab toxins in *Bt* cotton in the Southeastern United States (Tabashnik et al. 2008, 2009a). In 2001, Cry1Ab-resistant individuals of *O. nubilalis* were identified from a field collection from Kandiyohi, Minnesota, based on increased survival at a diagnostic Cry1Ab concentration. The resistant strain exhibited later over 800-fold resistance to Cry1Ab. Resistance was primarily autosomal, and was controlled by more than one locus or multiple alleles at one locus (Crespo et al. 2009).

The resistance management approach, often termed as "high-dose refuge strategy" works best if the dose of the toxin ingested by insects on *Bt* plants is high enough to kill all or nearly all of the aforementioned hybrid progeny (Gould 1998). Meihls et al. (2008) reported rapid resistance development without refuges, and slower or no occurrence of resistance with refuges in the case of *Diabrotica* species. Variety owners suggest the use of isogenic maize (10–20% proportion) in the fields of *Bt* maize to sustain susceptible pest populations. This purpose can be served by the isogenic border line sown in order to avoid pollen drift from the *Bt* maize field.

The refuge theory has several flaws in the practical sense: (*i*) the pest is being "bred" on substantial areas, bringing pest damage to a constant level, which is practically nonsense; (*ii*) in refuge zones where the *MON 810* x non-GM variety hybrids are frequent, different seeds in a cob result in survivorship for Lepidopteran cob pest and help the selection of a Cry1 resistant subpopulation; (*iii*) the developmental time is substantially different for survivor insects in the *MON 810* field and in the refuge zone (insects may develop twice as slow on *Bt* maize than on the isogenic line). This situation results in a modified time-table for the subsequent lepidopteran generations in the given year, lowering the chance for the summer generations of *O. nubilalis* or *H. armigera* to mate. Susceptible insects could survive in the overwintering population (first generation), but the presently popular stalk crusher technology destroys *O. nubilalis* larvae in the stalks.

Bt maize varieties producing Cry3 toxin were developed against *Diabrotica* species. Expression of the *cry3* gene is usually poor, and in consequence a small portion of the *Diabrotica* spp. larvae may survive. This is an ideal setup for selection of a Cry3 resistant subpopulation. Gassmann et al. (2011) reported field evolved resistance: Western corn rootworm displayed significantly higher survival on Cry3Bb1 maize in Iowa in 2009, *MON 863* maize variety having been commercialized since 2003. No significant correlation was found among populations for survival on Cry34/35Ab1 (*DAS-59122*) and Cry3Bb1 (*MON 853*, *MON 88017*) maize, suggesting a lack of cross-resistance between these Cry3 toxins.

The occurrence of cross-resistance depends on the specificity of the toxin-receptor interaction. At least four Cry receptors have been identified in larvae of diamondback moth (*Plutella xylostella* L.): there are (also) separate receptors for Cry1Aa, Cry1Ba and Cry1Ca toxins, while Cry1Aa, Cry1Ab, Cry1Ac, Cry1Fa and Cry1Ja toxins distributively bind to the fourth receptor (Ferré and Van Rie 2002). However, this explains only partially the strong cross-resistance of Cry1C-resistant *P. xylostella* larvae to CrylAb, CrylAc and Cry1F toxins (CrylAc and Cry1F found in WideStrike). Low of mediocre cross-resistance was seen with Cry1Aa and Cry9C toxins (the latter in StarLink). Cross-resistance did not occur with Cry1Bb, CrylJa and Cry2A toxins (the latter in Bollgard II or YieldGard VT Pro) (Liu et al. 2001). Due to various biochemical mechanisms behind resistance, cross-resistance against toxins with different receptors (e.g., Cry1Ac and Cry2Aa) may occur (Jurat-Fuentes et al. 2003).

As a result, stacked event crops expressing several toxins in parallel, appearing in increasing numbers nowadays in registration and cultivation, may limit the problem of Cry resistance, if prudent resistance studies were included in their development. The overall amount of Cry toxins produced by these crops is consequently higher, as the given Cry toxins (i.e., several Cry toxins used in resistance management or *Bt* crop variety groups producing Cry toxins specific to both the corn borer and corn rootworm) each must be produced above sublethal doses.

4 Conclusions

Based on the above, *Bt*-based bioinsecticides and crops cannot be considered by far as equivalent technologies. Their application differs as *Bt* bioinsecticides allow singular applications, while *Bt* crops exert a continuous production of the Cry toxin. This results in higher environmental doses of the plant-expressed toxin(s) than in the case of the *Bt* bioinsecticide. For example a single treatment of Dipel bioinsecticide at the registered dosage (1 kg/ha) contains 4.8–60.2 mg/ha (average 20.6 mg/ha) of bioavailable Cry1Ab toxin, while the amount of bioaccessible amount of Cry1Ab toxin is 0.085–8.16 g/ha. In contrast, the production of plant-expressed Cry1Ab toxin was found to be 147–456 g Cry1Ab toxin/ha, representing 18–56 treatments with Dipel (on the basis of its maximally detected bioaccessible Cry1Ab toxin content, 8.16 g/ha). The level of plant-expressed Cry1Ab toxin can be further elevated by soil fertilization (2.3–6.8-fold) and the use of long maturation maize varieties (2.5–5.8-fold), representing, in worst case scenarios, in 625–1,930 treatments with Dipel. Moreover, it has to be mentioned that stacked genetic events may further elevate toxin production (twofold). These ratios are even higher if lower bioaccessible Cry1Ab protoxin content biopesticides or bioavailable Cry1Ab toxin contents are considered.

Beside toxin ratios, another characteristic difference is that while *Bt* bioinsecticides are composed of several crystalline toxins, single genetic event *Bt* crops express only a single toxin molecule. This has severe consequences in resistance development, which may be alleviated, yet not eliminated by the use of "pyramid" *Bt* event varieties, expressing several Cry toxins acting on the same insect order, as the evolutionary driving force remain the same. The active ingredient of *Bt* bioinsecticides are bacterial protoxins stabilized in crystalline form and requiring enzymatic activation, while *Bt* plants (e.g., *MON 810*) express a truncated form of the protoxin, so-called preactivated toxin. This has severe consequences in product registration, as the active ingredient toxin in the *Bt* crop is not the registered active substance of the corresponding *Bt* bioinsecticide, and the required toxicology studies have been carried out not with the plant-expressed preactivated toxin, but with the bacterial protoxin or the enzyme-activated active toxin. Moreover, commercial ELISA systems utilizing antibodies against the bacterial protoxin and analytical standards of that protoxin consistently underdetect actual toxin content in *Bt* plants due to their lower cross-reactivities to the plant-expressed preactivated toxin. As a result, all reported results obtained by protoxin-based ELISAs, including manufacturer documentation, are subject to correction. And finally, although *Bt* crops have been widely advocated to be included in integrated pest management (IPM) practices or even in ecological agriculture, *Bt* crops cannot fulfill the main ecological principle of IPM that any protection measures should be timed only to the period(s) when pest damage exceeds the critical level, and therefore, regardless how environmentally mild their active ingredient is, do not comply with IPM.

References

Adamczyk JJ Jr, Adam LC, Hardee DD (2001) Field efficacy and seasonal expression profiles for terminal leaves of single and double *Bacillus thuringiensis* toxin cotton genotypes. J Econ Entomol 94:1589–1593

Anderson PL, Hellmich RL, Prasifka JR, Lewis LC (2005) Effects on fitness and behavior of monarch butterfly larvae exposed to a combination of Cry1Ab-expressing corn anthers and pollen. Environ Entomol 34:944–952

Andow DA, Lövei GL, Arpaia S (2006) Ecological risk assessment for *Bt* crops. Nat Biotechnol 24:749–751
Arvidson H, Dunn PE, Strnad S, Aronson AI (1989) Specificity of *Bacillus thuringiensis* for lepidopteran larvae: factors involved *in vivo* and in the structure of a purified protoxin. Mol Microbiol 3:1533–1543
Bagla P (2010) Hardy cotton-munching pests are latest blow to GM crops. Science 327:1439
Bakonyi G, Szira F, Kiss I, Villányi I, Seres A, Székács A (2006) Preference tests with collembolas on isogenic and *Bt*-maize. Eur J Soil Biol 42:S132–S135
Bakonyi G, Dolezsai A, Mátrai N, Székács A (2011) Long-term effects of *Bt*-maize (*MON 810*) consumption on the Collembolan *Folsomia candida*, over multiple generations: a laboratory study. Insects 2:243–252
Baum JA, Malvar T (1995) Regulation of insecticidal crystal protein production in *Bacillus thuringiensis*. Mol Microbiol 18:1–12
Baumgarte S, Tebbe CC (2005) Field studies on the environmental fate of the Cry1Ab *Bt*-toxin produced by transgenic maize (MON810) and its effect on bacterial communities in the maize rhizosphere. Mol Ecol 14:2539–2551
Bernhard K, Utz R (1993) Production of *Bacillus thuringiensis* insecticides for experimental and commercial uses. In: Entwistle PF, Cory JS, Bailey MJ, Higgs S (eds) *Bacillus thuringiensis*, an environmental biopesticide: theory and practice. Wiley, New York, pp 255–267
Betz FS, Hammond BG, Fuchs RL (2000) Safety and advantages of *Bacillus thuringiensis*-protected plants to control insect pests. Regul Toxicol Pharmacol 32:156–173
Bietlot HPL, Vishnubhatla I, Carey PR, Pozsgay M, Kaplan H (1990) Characterization of the cysteine residues and disulphide linkages in the protein crystal of *B. thuringiensis*. J Biochem 267:309–315
Blackwood CB, Buyer JS (2004) Soil microbial communities associated with *Bt* and non-*Bt* corn in three soils. J Environ Qual 33:832–836
Bøhn T, Primicerio R, Hessen DO, Traavik T (2008) Reduced fitness of *Daphnia magna* fed a *Bt*-transgenic maize variety. Arch Environ Contam Toxicol 55:584–592
Bøhn T, Traavik T, Primicerio R (2010) Demographic responses of *Daphnia magna* fed transgenic *Bt*-maize. Ecotoxicology 19:419–430
Bravo A, Soberón M (2008) How to cope with insect resistance to *Bt* toxins? Trends Biotechnol 26:573–579
Bravo A, Gill SS, Soberón M (2007) Mode of action of *Bacillus thuringiensis* Cry and Cyt toxins and their potential for insect control. Toxicon 49:423–435
Broderick NA, Raffa KF, Handelsman J (2006) Midgut bacteria required for *Bacillus thuringiensis* insecticidal activity. Proc Natl Acad Sci USA 103:15196–15199
Bruns HA, Abel CA (2003) Nitrogen fertility effects on *Bt* delta-endotoxin and nitrogen concentrations of maize during-early growth. Agron J 95:207–211
Bruns HA, Abel CA (2004) Effects of nitrogen fertility on *Bt* endotoxin levels in *Bt* hybrid maize. In: 4th proceedings of the international crop science congress. Poster 3.8. The Regional Institute Ltd, Gosford, Australia. (http://www.cropscience.org.au/icsc2004/poster/3/8/453_brunsha.htm)
Calsamiglia S, Hernandez B, Hartnell GF, Phipps R (2007) Effects of corn silage derived from a genetically modified variety containing two transgenes on feed intake, milk production, and composition, and the absence of detectable transgenic deoxyribonucleic acid in milk in Holstein dairy cows. J Dairy Sci 90:4718–4723
Cannon RJC (2000) *Bt* transgenic crops: risks and benefits. Integr Pest Manag Rev 5:151–173
Castaldini M, Turrini A, Sbrana C, Benedetti A, Marchionni M, Mocali S, Fabiani A, Landi S, Santomassimo F, Pietrangeli B, Nuti MP, Miclaus N, Giovannetti M (2005) Impact of *Bt* corn on rhizospheric and soil eubacterial communities and on beneficial mycorrhizal symbiosis in experimental microcosms. Appl Environ Microbiol 71:6719–6729
Castle LA, Wu G, McElroy D (2006) Agricultural input traits: past, present and future. Curr Opin Biotechol 17:105–112

Chambers CP, Whiles MR, Rosi-Marshall EJ, Tank JL, Royer TV, Griffiths NA, Evans-White MA, Stojak AR (2010) Responses of stream macroinvertebrates to *Bt* maize leaf detritus. Ecol Appl 20:1949–1960

Chen M, Ye G, Liu Z, Fang Q, Hu C, Peng Y, Shelton AM (2009) Analysis of Cry1Ab toxin bioaccumulation in a food chain of *Bt* rice, an herbivore and a predator. Ecotoxicology 18:230–238

Chestukhina GG, Kostina LI, Mikhailova AL, Tyurin SA, Klepikova FS, Stepanov VM (1982) Crystal-forming proteins of *Bacillus thuringiensis*. Arch Microbiol 132:159–162

Choma CT, Surewicz WK, Carey PR, Pozsgay M, Kaplan H (1990) Unusual proteolysis of the protoxin and toxin from *Bacillus thuringiensis*. J Protein Chem 9:87–94

Clark BW, Phillips TA, Coats JR (2005) Environmental fate and effects of *Bacillus thuringiensis* (*Bt*) proteins from transgenic crops: a review. J Agric Food Chem 53:4643–4653

Clements MJ, Campbell KW, Maragos CM, Pilcher C, Headrick JM, Pataky JK, White DG (2003) Influence of Cry1Ab protein and hybrid genotype on fumonisin contamination and *Fusarium* ear rot of corn. Crop Sci 43:1283–1293

Cortet J, Griffiths BS, Bohanec M, Demsar D, Andersen MN, Caul S, Birch ANE, Pernin C, Tabone E, de Vaufleury A, Ke X, Krogh PH (2007) Evaluation of effects of transgenic *Bt* maize on microarthropods in a European multi-site experiment. Pedobiologia 51:207–250

Crespo ALB, Spencer TA, Nekl E, Pusztai-Carey M, Moar WJ, Siegfried BD (2008) Comparison and validation of methods to quantify Cry1Ab toxin from *Bacillus thuringiensis* for standardization of insect bioassays. Appl Environ Microbiol 74:130–135

Crespo ALB, Spencer TA, Alves AP, Hellmich RL, Blankenship EE, Magalhäesa LC, Siegfried BD (2009) On-plant survival and inheritance of resistance to Cry1Ab toxin from *Bacillus thuringiensis* in a field-derived strain of European corn borer, *Ostrinia nubilalis*. Pest Manag Sci 65:1071–1081

Crickmore N, Zeigler DR, Feitelson J, Schnepf E, Van Rie J, Lereclus D, Baum J, Dean DH (1998) Revision of the nomenclature for the *Bacillus thuringiensis* pesticidal crystal proteins. Microbiol Mol Biol Rev 62:807–813

Crickmore N, Zeigler DR, Schnepf E, Van Rie J, Lereclus D, Baum J, Bravo A, Dean DH (2009) *Bacillus thuringiensis* toxin nomenclature. (http://www.lifesci.sussex.ac.uk/home/Neil_Crickmore/Bt/)

Darvas B (2011) GM plants and resistance – resistance-management. In: Darvas B, Székács A (eds) Hungarian background on views of 1st generation genetically modified plants. Agricultural Committee of the Hungarian Parliament, Budapest, Hungary, pp 140–141. (http://www.kormany.hu/download/2/9d/20000/GenetEM.pdf)

Darvas B, Polgár LA (1998) Novel type insecticides: specificity and effects on non-target organisms. In: Ishaaya I, Degheele D (eds) Insecticides with novel modes of action, mechanism and application. Springer, Berlin, pp 188–259

Darvas B, Csóti A, Gharib A, Peregovits L, Ronkay L, Lauber É, Polgár AL (2004) Some data to the risk analysis of *Bt*-corn pollen and protected Lepidoptera species in Hungary. Növényvédelem 40:441–449 (in Hungarian)

Darvas B, Bánáti H, Takács E, Lauber É, Szécsi Á, Székács A (2011) Relationships of *Helicoverpa armigera*, *Ostrinia nubilalis* and *Fusarium verticillioides* on *MON 810* maize. Insects 2:1–11

de Barjac H (1978) Une nouvelle variété de *Bacillus thuringiensis* très toxique pour les moustiques: *Bacillus thuringiensis* subsp. *israelensis* serotype 14. C R Acad Sci Paris D 286:797–800

de Maagd RA, Bravo A, Berry C, Crickmore N, Schnepf HE (2003) Structure, diversity, and evolution of protein toxins from spore-forming entomopathogenic bacteria. Annu Rev Genet 37:409–433

de Vendômois JS, Roullier F, Cellier D, Séralini G-E (2009) A comparison of the effects of three GM corn varieties on mammalian health. Int J Biol Sci 5(7):706–726

de Vendômois JS, CellierD VC, Clair E, Mesnage R, Séralini G-E (2010) Debate on GMOs health risks after statistical findings in regulatory tests. Int J Biol Sci 6:590–598

Dively GP, Rose R, Sears MK, Hellmich RL, Stanley-Horn DE, Calvin DD, Russo JM, Anderson PL (2004) Effects on monarch butterfly larvae (Lepidoptera: Danaidae) after continuous exposure to Cry1Ab-expressing corn during anthesis. Environ Entomol 33:1116–1125

Donkin SS, Velez JC, Totten AK, Stanisiewski EP, Hartnell GF (2003) Effects of feeding silage and grain from glyphosate-tolerant or insect-protected corn hybrids on feed intake, ruminal digestion, and milk production in dairy cattle. J Dairy Sci 86:1780–1788

Doull J, Gaylor D, Greim HA, Lovell DP, Lynch B, Munro IC (2007) Report of an expert panel on the reanalysis by Séralini et al. (2007) of a 90-day study conducted by Monsanto in support of the safety of a genetically modified corn variety (MON 863). Food Chem Toxicol 45:2073–2085

Douville M, Gagné F, Masson L, McKay J, Blaise C (2001) Tracking the source of *Bacillus thuringiensis* Cry1Ab endotoxin in the environment. Biochem Syst Ecol 33:219–232

Ermolli M, Fantozzi A, Marini M, Scotti D, Balla B, Hoffmann S, Querci M, Paoletti C, Van den Eede G (2006a) Food safety: screening tests used to detect and quantify GMO proteins. Accredit Qual Assur 11:55–57

Ermolli M, Prospero A, Balla B, Querci M, Mazzeo A, Van Den Eede G (2006b) Development of an innovative immunoassay for CP4EPSPS and Cry1AB genetically modified protein detection and quantification. Food Addit Contam 23:876–882

European Council (1992) Directive 92/43/EEC of 21 May 1992 on the conservation of natural habitats and of wild fauna and flora. OJ L 206, 22.7.1992, pp 1–7

European Environment Agency (2002) Europe's biodiversity – biogeographical regions and seas. In: Pinborg U, Larsson TJ (eds) European Environment Agency, Copenhagen, Denmark. (http://www.eea.europa.eu/publications/report_2002_0524_154909)

European Food Safety Authority (2004) Opinion of the scientific panel on genetically modified organisms on a request from the Commission related to the notification for the placing on the market of insect protected genetically modified maize MON 863 and MON 863 x MON 810, for import and processing, under Part C of Directive 2001/18/EC from Monsanto. EFSA J 49:1–25

European Food Safety Authority (2005) Opinion of the scientific panel on genetically modified organisms on an application (Reference EFSAGMO-DE-2004-03) for the placing on the market of insect-protected genetically modified maize MON 863 x MON 810, for food and feed use, under Regulation (EC) No 1829/2003 from Monsanto. EFSA J 252:1–23

European Food Safety Authority (2007) Statement of the scientific panel on genetically modified organisms on the analysis of data from a 90-day rat feeding study with MON 863 maize. doi:10.2903/j.efsa.2007.753. (http://www.efsa.europa.eu/en/efsajournal/pub/753.htm)

European Food Safety Authority (2009a) Technical meeting between EFSA GMO panel environmental experts and environmental experts from Member States (May 26, 2009). European Food Safety Authority, Parma, Italy. (http://www.efsa.europa.eu/en/events/event/gmo090526.htm)

European Food Safety Authority (2009b) Applications (EFSA-GMO-RX-MON810) for renewal of authorisation for the continued marketing of (1) existing food and food ingredients produced from genetically modified insect resistant maize MON810; (2) feed consisting of and/or containing maize MON810, including the use of seed for cultivation; and of (3) food and feed additives, and feed materials produced from maize MON810, all under Regulation (EC) No 1829/2003 from Monsanto. EFSA J 1149:1–85

Fantozzi A, Ermolli M, Marini M, Scotti D, Balla B, Querci M, Langrell SR, Van den Eede G (2007) First application of a microsphere-based immunoassay to the detection of genetically modified organisms (GMOs): quantification of Cry1Ab protein in genetically modified maize. J Agric Food Chem 55:1071–1076

Fearing PL, Brown D, Vlachos D, Meghji M, Privalle L (1997) Quantitative analysis of CryIA(b) expression in *Bt* maize plants, tissues, and silage and stability of expression over successive generations. Mol Breed 3(3):169–176

Federici BA, Lüthy P, Ibarra JE (1990) Parasporal body of *Bacillus thuringiensis israelensis*. Structure, protein composition, and toxicity. In: de Barjac H, Sutherland DJ (eds) Bacterial control of mosquitoes and black flies: biochemistry, genetics and applications of *Bacillus thuringiensis israelensis*. Rutgers University Press, New Brunswick, pp 16–44

Ferré J, Van Rie J (2002) Biochemistry and genetics of insect resistance to *Bacillus thuringiensis*. Annu Rev Entomol 47:501–533

Folcher L, Jarry M, Weissenberger A, Gérault F, Eychenne N, Delos M, Regnault-Roger C (2009) Comparative activity of agrochemical treatments on mycotoxin levels with regard to corn borers and *Fusarium mycoflora* in maize (*Zea mays* L.) fields. Crop Prot 28:302–308

Folcher L, Delos M, Marengue E, Jarry M, Weissenberger A, Eychenne N, Regnault-Roger C (2010) Lower mycotoxin levels in *Bt* maize grain. Agron Sustain Dev 30:711–719

Füsti Molnár G (2011) Results of the Hungarian variety evaluation of genetically modified varieties awaiting government authorization. In: Darvas B, Székács A (eds) Hungarian background on views of 1st generation genetically modified plants. Agricultural Committee of the Hungarian Parliament, Budapest, Hungary, pp 125–128. (http://www.kormany.hu/download/2/9d/20000/GenetEM.pdf)

Gassmann A, Carrière Y, Tabashnik BE (2009) Fitness costs of insect resistance to *Bacillus thuringiensis*. Annu Rev Entomol 54:147–163

Gassmann AJ, Petzold-Maxwell JL, Keweshan RS, Dunbar MW (2011) Field-evolved resistance to *Bt* maize by western corn rootworm. PLoS One 6(7):e22629. doi:10.1371/journal.pone.0022629

Gelernter W (2004) The rise and fall of *Bacillus thuringiensis tenebrionis*. Phytoparasitica 32:321–324

Gill SS, Cowles EA, Pietrantonio PV (1992) The mode of action of *Bacillus thuringiensis* endotoxins. Annu Rev Entomol 37:615–634

Goldberg LJ, Margalit J (1977) A bacterial spore demonstrating rapid larvicidal activity against *Anopheles sergentii*, *Uranotaenia unguiculata*, *Culex univittatus*, *Aedes aegypti* and *Culex pipiens*. Mosq News 37:355–358

Gómez I, Pardo-López L, Munoz-Garay C, Fernandez LE, Pérez C, Sánchez J, Soberón M, Bravo A (2007) Role of receptor interaction in the mode of action of insecticidal Cry and Cyt toxins produced by *Bacillus thuringiensis*. Peptides 28:169–173

Gould F (1998) Sustainability of transgenic insecticidal cultivars: integrating pest genetics and ecology. Annu Rev Entomol 43:701–726

Graf J (2011) Shifting paradigm on *Bacillus thuringiensis* toxin and a natural model for *Enterococcus faecalis* Septicemia. mBio 2. doi:10.1128/mBio.00161-11

Griffiths BS, Caul S, Thompson J, Birch AN, Scrimgeour C, Cortet J, Foggo A, Hackett CA, Krogh PH (2006) Soil microbial and faunal community responses to *Bt* maize and insecticide in two soils. J Environ Qual 35:734–741

Grothaus GD, Bandla M, Currier T, Giroux R, Jenkins GR, Lipp M, Shan G, Stave JW, Pantella V (2006) Immunoassay as an analytical tool in agricultural biotechnology. J AOAC Int 89: 913–928

Hammock BD, Gee SJ, Harrison RO, Jung F, Goodrow M, Li Q-X, Lucas AD, Székács A, Sundaram KMS (1991) Immunochemical technology in environmental analysis: addressing critical problems. In: Van Emon J, Mumma RO (eds) Immunochemical methods for environmental analysis, vol 442, ACS Symp Ser. American Chemical Society, Washington, DC, pp 112–139

Hammond BG, Dudek R, Lemen JK, Nemeth MA (2006) Results of a 90-day safety assurance study with rats fed grain from corn borer-protected corn. Food Chem Toxicol 44:1092–1099

Harwood JD, Wallin WG, Obrycki JJ (2005) Uptake of *Bt* endotoxins by nontarget herbivores and higher order arthropod predators: molecular evidence from a transgenic corn agroecosystem. Mol Ecol 14:2815–2823

Heszky L (2011a) Scientific problems associated with the cultivation of transgenic (GM) crops. In: Darvas B, Székács A (eds) Hungarian background on views of 1st generation genetically modified plants). Agricultural Committee of the Hungarian Parliament, Budapest, Hungary, pp 130–135. (http://www.kormany.hu/download/2/9d/20000/GenetEM.pdf)

Heszky L (2011b) Coexistence is professionally unacceptable, practically unaccomplishable. In: Darvas B, Székács A (eds) Hungarian background on views of 1st generation genetically modified plants). Agricultural Committee of the Hungarian Parliament, Budapest, Hungary, pp 119–122. (http://www.kormany.hu/download/2/9d/20000/GenetEM.pdf)

Hickle LA, Fitch WL (1990) Analytical chemistry of *Bacillus thuringiensis*. An overview. In: Hickle LA, Fitch WL (eds) Analytical chemistry of *Bacillus thuringiensis*, vol 432, ACS Symp Ser. American Chemical Society, Washington, DC, pp 1–8

Hilbeck A (2001) Implications of transgenic, insecticidal plants for insect and plant biodiversity. Perspect Plant Ecol Evol Syst 4:43–61

Hilbeck A, Schmidt JEU (2006) Another view on *Bt* proteins – How specific are they and what else might they do? Biopesticides Int 2:1–50

Hopkins DW, Gregorich EG (2003) Detection and decay of the *Bt* endotoxin in soil from a field trial with genetically modified maize. Eur J Soil Sci 54:793–800

Huber HE, Luthy P, Rudolf H-R, Cordier J-L (1981) The subunits of the parasporal crystal of *Bacillus thuringiensis*: size linkage and toxicity. Arch Microbiol 129:14–18

Icoz I, Stotzky G (2008a) Cry3Bb1 protein from *Bacillus thuringiensis* in root exudates and biomass of transgenic corn does not persist in soil. Transgenic Res 17:609–620

Icoz I, Stotzky G (2008b) Fate and effects of insect-resistant *Bt* crops in soil ecosystems. Soil Biol Biochem 40:559–586

Icoz I, Saxena D, Andow DA, Zwahlen C, Stotzky G (2008) Microbial populations and enzyme activities in soil *in situ* under transgenic corn expressing Cry proteins from *Bacillus thuringiensis*. J Environ Qual 37:647–662

Icoz I, Andow D, Zwahlen C, Stotzky G (2009) Is the Cry1Ab protein from *Bacillus thuringiensis* (*Bt*) taken up by plants from soils previously planted with *Bt* corn and by carrot from hydroponic culture? Bull Environ Contam Toxicol 83:48–58

Ives AR, Glaum PR, Ziebarth NL, Andow DA (2011) The evolution of resistance to two-toxin pyramid transgenic crops. Ecol Appl 21:503–515

Jesse LCH, Obrycki JJ (2000) Field deposition of *Bt* transgenic corn pollen: lethal effects on the monarch butterfly. Oecologia 125:241–248

Jurat-Fuentes JL, Gould FL, Adang MJ (2003) Dual resistance to *Bacillus thuringiensis* Cry1Ac and Cry2Aa toxins in *Heliothis virescens* suggests multiple mechanisms of resistance. Appl Environ Microbiol 69:5898–5906

Kamota A, Muchaonyerwa P, Mnkeni PNS (2011) Effects of ensiling of *Bacillus thuringiensis* (*Bt*) maize (MON810) on degradation of the crystal 1Ab (Cry1Ab) protein and compositional quality of silage. Afr J Biotechnol 10(76):17484–17489

Kiliç A, Akay MT (2008) A three generation study with genetically modified *Bt* corn in rats: biochemical and histopathological investigation. Food Chem Toxicol 46:1164–1170

Knowles BH (1994) Mechanism of action of *Bacillus thuringiensis* insecticidal delta-endotoxins. Adv Insect Physiol 24:275–308

Knowles BH, Ellar DJ (1987) Colloid-osmotic lysis is a general feature of the mechanism of action of *Bacillus thuringiensis* δ-endotoxins with different insect specificity. Biochim Biophys Acta 924:509–518

Laboratories A (1992) *Bt* products manual. Abbott, North Chicago

Lambert B, Buysse L, Decock C, Jansens S, Piens C, Saey B, Seurinck J, van Audenhove K, van Rie J, van Vliet A, Peferoen M (1996) A *Bacillus thuringiensis* insecticidal crystal protein with a high activity against members of the family Noctuidae. Appl Environ Microbiol 62:80–86

Lang A (2004) Monitoring the impact of *Bt* maize on butterflies in the field: estimation of required sample sizes. Environ Biosafety Res 3:55–66

Lang A, Vojtech E (2006) The effects of pollen consumption of transgenic *Bt* maize on the common swallowtail, *Papilio machaon* L. (Lepidoptera, Papilionidae). Basic Appl Ecol 7:296–306

Lang A, Lauber É, Darvas B (2007) Early tier tests are not sufficient for GMO risk assessment. Nat Biotechnol 25:35–36

Lang A, Brunzel S, Dolek M, Otto M, Theißen B (2011) Modelling in the light of uncertainty of key parameters: a call to exercise caution in field predictions of *Bt*-maize effects. Proc R Soc B 278:980–981

Lauber É (2011) The Cry1-toxin content of MON 810 and the affectivity of its pollen on Hungarian protected lepidopteran species (Supervisor: Darvas B.) PhD dissertation, Corvinus University of Budapest, Budapest, pp 1–102 (in Hungarian)

Lauber É, Peregovits L, Ronkay L, Csóti A, Székács A, Darvas B (2010) Protected lepidopteran larvae and Cry1Ab toxin exposure by *Bt* maize pollen in the Pannonian Region. In: 9th European congress of entomology, programme and book of abstracts. Hungarian Entomological Society, Budapest, Hungary, p 205

Lecadet MM, Frachon E, Dumanoir VC, Ripouteau H, Hamon S, Laurent P, Thiéry I (1999) Updating the H-antigen classification of *Bacillus thuringiensis*. J Appl Microbiol 86:660–672

Lemen JK, Hammond BG, Riordan SG, Jiang C, Nemeth M (2002) 13-Week dietary subchronic comparison study with MON 863 corn in rats preceded by a 1-week baseline food consumption determination with PMI certified rodent diet #5002. Monsanto Co. No: MSL-18175. (http://www.greenpeace.de/fileadmin/gpd/user_upload/themen/gentechnik/Monsanto_Rattenfuetterungsstudie.pdf)

Li J, Koni PA, Ellar DJ (1996) Structure of the mosquitocidal δ-endotoxin CytB from *Bacillus thuringiensis* sp. *kyushuensis* and implications for membrane pore formation. J Mol Biol 257:129–152

Li W, Wu K, Wang X, Wang G, Guo Y (2005) Impact of pollen grains from *Bt* transgenic corn on the growth and development of Chinese tussah silkworm, *Antheraea pernyi* (Lepidoptera: Saturniidae). Environ Entomol 34:922–928

Lilley M, Ruffell RN, Somerville HJ (1980) Purification of the insecticidal toxin in crystals of *Bacillus thuringiensis*. J Gen Microbiol 118:1–11

Lisansky SG, Coombs J, Dale T, Frederick R (1997) Biopesticides – markets, technology, registration and IPR companies. CPL Scientific Information Services Ltd., Newbury

Liu Y-B, Tabashnik B, Meyer SK, Crickmore N (2001) Cross-resistance and stability of resistance to *Bacillus thuringiensis* toxin Cry1C in diamondback moth. Appl Environ Microbiol 67: 3216–3219

Losey JE, Rayor LS, Carter ME (1999) Transgenic pollen harms monarch larvae. Nature 399:214

Ma BL, Subedi KD (2005) Development, yield, grain moisture and nitrogen uptake of *Bt* corn hybrids and their conventional near-isolines. Field Crop Res 93:199–211

Margarit E, Reggiardo MI, Vallejos RH, Permingeat HR (2006) Detection of *BT* transgenic maize in foodstuffs. Food Res Int 39:250–255

Mason KL, Stepien TA, Blum JE, Holt JF, Labbe NH, Rush JS, Raffa KF, Handelsman J (2011) From commensal to pathogen: translocation of *Enterococcus faecalis* from the midgut to the hemocoel of *Manduca sexta*. mBio 2. doi:10.1128/mBio.00065-11

Meihls LN, Higdon ML, Siegfried BD, Miller NJ, Sappington TW, Ellersieck MR, Spencer TA, Hibbard BA (2008) Increased survival of western corn rootworm on transgenic corn within three generations of on-plant greenhouse selection. Proc Natl Acad Sci USA 105:19177–19182

Mesnage R, Clair E, Gress S, Then C, Székács A, Séralini G-E (2011) Cytotoxicity on human cells of Cry1Ab and Cry1Ac *Bt* insecticidal toxins alone or with a glyphosate-based herbicide. J Appl Toxicol. doi:10.1002/jat.2712

Messean A, Angevin F, Gómez-Barbero M, Menrad K, Rodríguez-Cerezo E (2006) New case studies on the coexistence of GM and non-GM crops in European agriculture. European Commission Joint Research Centre Institute for Prospective Technological Studies. Technical Report EUR 22102 EN, pp 1–112

Miranda R, Zamudio FZ, Bravo A (2001) Processing of Cry1Ab [delta]-endotoxin from *Bacillus thuringiensis* by *Manduca sexta* and *Spodoptera frugiperda* midgut proteases: role in protoxin activation and toxin inactivation. Insect Biochem Mol Biol 31:1155–1163

Mohan M, Gujar GT (2003) Characterization and comparison of midgut proteases of *Bacillus thuringiensis* susceptible and resistant diamondback moth (Plutellidae: Lepidoptera). J Invertebr Pathol 82:1–11

Monsanto (2010) Monsanto response: de Vendômois et al. 2009 (A comparison of the effects of three GM corn varieties on mammalian health) regarding: MON 863, MON 810 and NK603. Monsanto Scientific Affairs. (http://www.monsanto.com/newsviews/Documents/SpirouxdeVendimois.pdf)

Munkvold GP (2003) Cultural and genetic approaches to managing mycotoxins in maize. Annu Rev Phytopathol 41:99–116

Nguyen TH, Jehle JA (2007) Quantitative analysis of the seasonal and tissue-specific expression of Cry1Ab in transgenic maize *MON 810*. J Plant Dis Prot 114:82–87

Okumura S, Akao T, Mizuki E, Ohba M, Inouye K (2001) Screening of the *Bacillus thuringiensis* Cry1Ac δ-endotoxin on the artificial phospholipid monolayer incorporated with brush border membrane vesicles of *Plutella xylostella* by optical biosensor technology. J Biochem Biophys Methods 47:177–188

Oliveira AP, Pampulha ME, Bennett JP (2008) A two-year field study with transgenic *Bacillus thuringiensis* maize: effects on soil microorganisms. Sci Total Environ 405:351–357

Oppert B (1999) Protease interactions with *Bacillus thuringiensis* insecticidal toxins. Arch Insect Biochem Physiol 42:1–12

Oppert BS, Morgan TD, Kramer KJ (2011) Efficacy of *Bacillus thuringiensis* Cry3Aa protoxin and protease inhibitors toward coleopteran storage pests. Pest Manag Sci 67:568–573

Pagel-Wieder S, Niemeyer J, Fischer WR, Gessler F (2007) Effects of physical and chemical properties of soils on adsorption of the insecticidal protein (Cry1Ab) from *Bacillus thuringiensis* at Cry1Ab protein concentrations relevant for experimental field sites. Soil Biol Biochem 39:3034–3042

Palm CJ, Donegan K, Harris D, Seidler RJ (1994) Quantification in soil of *Bacillus thuringiensis* var. *kurstaki* δ-endotoxin from transgenic plants. Mol Ecol 3:145–151

Papst C, Utz HF, Melchinger AE, Eder J, Magg T, Klein D, Bohn M (2005) Mycotoxins produced by *Fusarium* spp. in isogenic, *Bt vs.* non-*Bt* maize hybrids under European corn borer pressure. Agron J 97:219–224

Perry JN, Devos Y, Arpaia S, Bartsch D, Ehlert C, Gathmann A, Hails RS, Hendriksen NB, Kiss J, Messéan A, Mestdagh S, Neemann G, Nuti M, Sweet JB, Tebbe CC (2012) Estimating the effects of Cry1F *Bt*-maize on non-target Lepidoptera using a mathematical model of exposure. J Appl Ecol 49:29–37

Perry JN, Devos Y, Arpaia S, Bartsch D, Gathmann A, Hails RS, Kiss J, Lheureux K, Manachini B, Mestdagh S, Neemann G, Ortego F, Schiemann J, Sweet JB (2010) A mathematical model of exposure of nontarget Lepidoptera to *Bt*-maize pollen expressing Cry1Ab within Europe. Proc R Soc B 277:1417–1425

Pleasants JM, Hellmich RL, Dively GP, Sears MK, Stanley-Horn DE, Mattila HR, Foster JE, Clark P, Jones GD (2001) Corn pollen deposition on milkweeds in and near cornfields. Proc Natl Acad Sci USA 98:11919–11924

Promdonkoy B, Ellar DJ (2003) Investigation of the pore-forming mechanism of a cytolytic δ-endotoxin from *Bacillus thuringiensis*. Biochem J 374:255–259

Rauschen S, Schuphan I (2006) Fate of the Cry1Ab protein from *Bt*-maize MON810 silage in biogas production facilities. J Agric Food Chem 54(3):879–883

Ravensberg WJ (2011) Critical factors in the successful commercialization of microbial pest control products. Prog Biol Control 10:295–356

Rodics K, Homoki H, Bakonyi G, Darvas B, Székács A (2011) The hereafter of Hungarian scientific lectures for EFSA GMO Panel (Parma, June 11, 2008). In: Darvas B, Székács A (eds) Hungarian background on views of 1st generation genetically modified plants. Agricultural Committee of the Hungarian Parliament, Budapest, Hungary, pp 155–169. (http://www.kormany.hu/download/2/9d/20000/GenetEM.pdf)

Roh JY, Choi JY, Li MS, Jin BR, Je YH (2007) *Bacillus thuringiensis* as a specific, safe, and effective tool for insect pest control. J Microbiol Biotechnol 17:547–559

Romeis J, Meissle M, Bigler F (2006) Transgenic crops expressing *Bacillus thuringiensis* toxins and biological control. Nat Biotechnol 24:63–71

Romeis J, Shelton A, Kennedy GG (eds) (2008) Integration of insect-resistant genetically modified crops within IPM programs. Springer, Dordrecht

Rosi-Marshall EJ, Tank JL, Royer TV, Whiles MR, Evans-White M, Chamber C, Griffiths NA, Pokelsek J, Stephen ML (2007) Toxins in transgenic crop byproducts may affect headwater stream ecosystems. Proc Natl Acad Sci USA 104:16204–16208

Schnepf E, Crickmore N, Van Rie J, Lereclus D, Baum J, Feitelson J, Zeigler DR, Dean DH (1998) *Bacillus thuringiensis* and its pesticidal crystal proteins. Microbiol Mol Biol Rev 62:775–806

Sears MK, Hellmich RL, Stanley-Horn DE, Oberhauser KS, Pleasants JM, Mattila HR, Siegfried BD, Dively GP (2001) Impact of *Bt* corn pollen on monarch butterfly populations: a risk assessment. Proc Natl Acad Sci USA 98:11937–11942

Séralini G-E (2010) Ces OGM qui changent le monde. Flammarion, Paris

Séralini G-E, Cellier D, de Vendômois JP (2007) New analysis of a rat feeding study with a genetically modified maize reveals signs of hepatorenal toxicity. Arch Environ Contam Toxicol 45:2073–2085

Séralini G-E, de Vendômois JS, Cellier D, Sultan C, Buiatti M, GallagherL AM, Dronamraju KR (2009) How subchronic and chronic health effects can be neglected for GMOs, pesticides or chemicals. Int J Biol Sci 5:438–443

Séralini G-E, Mesnage R, Clair E, Gress S, de Vendômois JS, Cellier D (2011) Genetically modified crops safety assessments: present limits and possible improvements. Environ Sci Eur 23(10):1–10

Shao Z, Cui Y, Liu X, Yi H, Ji J, Yu Z (1998) Processing of delta-endotoxin of *Bacillus thuringiensis* subsp. *kurstaki* HD-1 in *Heliothis armigera* midgut juice and the effects of protease inhibitors. J Invertebr Pathol 72:73–81

Smith PK, Krohn RI, Hermanson GT, Mallia AK, Gartner FH, Provenzano MD, Fujimoto EK, Goeke NM, Olson BJ, Klenk DC (1985) Measurement of protein using bicinchoninic acid. Anal Biochem 150:76–85

Steinke K, Guertler P, Paul V, Wiedemann S, Ettle T, Albrecht C, Meyer HHD, Spiekers H, Schwarz FJ (2010) Effects of long-term feeding of genetically modified corn (event MON810) on the performance of lactating dairy cows. J Anim Physiol Anim Nutr 94(5):185–193

Storer NP, Babcock JM, Schlenz M, Meade T, Thompson GD, Bing JW, Huckaba RM (2010) Discovery and characterization of field resistance to *Bt* maize: *Spodoptera frugiperda* (Lepidoptera: Noctuidae) in Puerto Rico. J Econ Entomol 103:1031–1038

Székács A, Darvas B (2012) Forty years with glyphosate. In: Hasaneen MNAE-G (ed) Herbicides – properties, synthesis and control of weeds. InTech, Rijeka, Croatia. pp 247–284. (http://www.intechopen.com/articles/show/title/forty-years-with-glyphosate)

Székács A, Juracsek J, Polgár LA, Darvas B (2005) Levels of expressed Cry1Ab toxin in genetically modified corn DK-440-BTY (YieldGard) and stubble. FEBS J 272(Suppl 1):508

Székács A, Lauber É, Juracsek J, Darvas B (2010a) Cry1Ab toxin production of *MON 810* transgenic maize. Environ Toxicol Chem 29:182–190

Székács A, Lauber É, Takács E, Darvas B (2010b) Detection of Cry1Ab toxin in the leaves of *MON 810* transgenic maize. Anal Bioanal Chem 396:2203–2211

Székács A, Weiss G, Quist D, Takács E, Darvas B, Meier M, Swain T, Hilbeck A (2012) Inter-laboratory comparison of Cry1Ab toxin quantification in *MON 810* maize by enzyme-immunoassay. Food Agric Immunol 23:99–121

Tabashnik BE, Gassmann AJ, Crowder DW, Carrière Y (2008) Insect resistance to *Bt* crops: evidence versus theory. Nat Biotechnol 26:199–202

Tabashnik BE, van Rensburg JBJ, Carriére Y (2009a) Field-evolved insect resistance to *Bt* crops: definition, theory, and data. J Econ Entomol 102:2011–2025

Tabashnik BE, Unnithan GC, Masson L, Crowder DW, Li X, Carrière Y (2009b) Asymmetrical cross-resistance between *Bacillus thuringiensis* toxins Cry1Ac and Cry2Ab in pink bollworm. Proc Natl Acad Sci USA 106:11889–11894

Takács E, Fónagy A, Juracsek J, Kugler N, Székács A (2011) Characterisation of tritrophic effects of *DAS-59122-7* maize on seven-spotted ladybird (*Coccinella septempunctata*) feeding on the bird cherry-oat aphid (*Rhopalosiphum padi*). IOBC/WPRS Bull 73:121–134

Tan FX, Wang JW, Feng YJ, Chi GL, Kong HL, Qiu HF, Wei SL (2010) *Bt* corn plants and their straw have no apparent impact on soil microbial communities. Plant Soil 329:349–364

Tank JL, Rosi-Marshall EJ, Royer TV, Whiles MR, Griffiths NA, Frauendorf TC, Treering DJ (2010) Occurrence of maize detritus and a transgenic insecticidal protein (Cry1Ab) within the stream network of an agricultural landscape. Proc Natl Acad Sci USA 107:17645–17650

Tapp H, Stotzky G (1998) Persistence of the insecticidal toxin from *Bacillus thuringiensis* subsp. *kurstaki* in soil. Soil Biol Biochem 30:471–476

Turrini A, Sbrana C, Nuti MP, Pietrangeli B, Giovannetti M (2004) Development of a model system to assess the impact of genetically modified corn and aubergine plants on arbuscular mycorrhizal fungi. Plant Soil 266:69–75
US National Research Council (2010) The impact of genetically engineered crops on farm sustainability in the United States. National Academies Press, Washington, DC
van de Wiel CCM, Lotz LAP (2006) Coexistence of genetically modified with unmodified crops. NJAS 54(1):17–35
van Frankenhuyzen K (1993) The challenge of *Bacillus thuringiensis*. In: Entwistle PF, Cory JS, Bailey MJ, Higgs S (eds) *Bacillus thuringiensis*, an environmental biopesticide: theory and practice. Wiley, Chichester, pp 1–36
van Frankenhuyzen K (2009) Insecticidal activity of *Bacillus thuringiensis* insecticidal proteins. J Invertebr Pathol 101:1–16
Vazquez-Padron RI, de la Riva G, Agüero G, Silva Y, Pham SM, Soberón M, Bravo A, Aïtouche A (2004) Cryptic endotoxic nature of *Bacillus thuringiensis* Cry1Ab insecticidal crystal protein. FEBS Lett 570:30–36
Visser B, Bosch D, Honée G (1993) Domain-function studies of *Bacillus thuringiensis* crystal proteins: a genetic approach. In: Entwistle PF, Cory JS, Bailey MJ, Higgs S (eds) *Bacillus thuringiensis*, an environmental biopesticide: theory and practice. Wiley, New York, pp 71–88
Volpe G, Ammid NH, Moscone D, Occhigrossi L, Palleschi G (2006) Development of an immunomagnetic electrochemical sensor for detection of BT-CRY1AB/CRY1AC proteins in genetically modified corn samples. Anal Lett 39:1599–1609
Winkler VW, Hansen GD, Yoder JM (1971) Immunochemical analysis of parasporal crystal digests of *Bacillus thuringiensis* as an index of insecticidal activity. J Invertebr Pathol 18:378–382
Xie X, Shu Q (2001) Studies on rapid quantitative analysis of *Bt* toxin by using Envirologix kits in transgenic rice. Sci Agric Sinic 34:465–468
Zeilinger AR, Andow DA, Zwahlen C, Stotzky G (2010) Earthworm populations in a northern U.S. cornbelt soil are not affected by long-term cultivation of *Bt* maize expressing Cry1Ab and Cry3Bb1 proteins. Soil Biol Biochem 42:1284–1292
Zhao J-Z, Cao J, Collins HL, Bates SL, Roush RT, Earle ED, Shelton AM (2005) Concurrent use of transgenic plants expressing a single and two *Bacillus thuringiensis* genes speeds insect adaptation to pyramided plants. Proc Natl Acad Sci USA 102:8426–8430
Zwahlen C, Hilbeck A, Gugerli P, Nentwig W (2003) Degradation of the Cry1Ab protein within transgenic *Bacillus thuringiensis* corn tissue in the field. Mol Ecol 12:765–775

Chapter 11
Plant Natural Products for Pest Management: The Magic of Mixtures

Yasmin Akhtar and Murray B. Isman

1 Introduction

As pesticides, certain natural products can be suitable alternatives to synthetic pesticides owing to their generally reduced negative impacts on humans, beneficial insects and the environment. Higher plants constitute a diverse source of highly bioactive agents that include some that have contributed significantly to the successful use of natural products and analogues for crop protection (Isman 2004). However, some readily biodegradable synthetic and semi-synthetic products in pest management have also been considered as green pesticides (Koul 2008). Many of the currently used synthetic insecticides are relatively toxic to non-target organisms and often deleterious to human and animal health. Many also pollute soils and water, due to their slow breakdown. Extensive use of insecticides has led to the rapid evolution of resistance in many insects to several classes of insecticides. As a result, there has been an ongoing search for new and safer products. The concept of using mixtures as effective tools for crop protection comes from the fact that insects encounter complex mixtures of nutrients and plant secondary compounds, when feeding on host plants (Bernays and Chapman 2001). There is now ample evidence demonstrating that responses of insect gustatory receptors are greatly affected by interactions between chemicals, including chemicals that alone may not be stimulating to any of the neurons within a sensillum (Schoonhoven et al. 1992). Plants usually present defenses as a suite of compounds, not as individual ones, and there is evidence that minor constituents may act as synergists, enhancing the effect of the major constituents through a variety of mechanisms. Synergistic effects of complex mixtures are thought to be important in natural plant defense against herbivores.

Y. Akhtar • M.B. Isman (✉)
Faculty of Land and Food Systems, University of British Columbia,
2357 Main Mall, Suite 248, Vancouver, BC V6T 1Z4, Canada
e-mail: yasmin.akhtar@ubc.ca; murray.isman@ubc.ca

I. Ishaaya et al. (eds.), *Advanced Technologies for Managing Insect Pests*,
DOI 10.1007/978-94-007-4497-4_11,

We have demonstrated that complex plant essential oils are often considerably more efficacious than the pure compounds isolated from them. Examples include oil of anise (from which *trans*-anethole is derived), various citrus and mint oils (Akhtar et al. 2012), rosemary oil (Miresmailli et al. 2006) and *Litsea* oils (Jiang et al. 2009).

Understanding the role and contribution of each constituent to the overall activity of an essential oil can facilitate the creation of artificial blends that optimize their efficacy against different pests. Identifying synergistic compounds within complex mixtures may allow for the development of more effective control agents as well as the use of smaller absolute amounts in the mixture to achieve satisfactory levels of efficacy. The main objective of this chapter is to look at the prospects of mixtures as effective, environmentally sound and cost effective insect control agents. In this context, we discuss antifeedant effects, repellent effects and toxicity of mixtures.

2 Antifeedant Effects of Mixtures

Antifeedants have been described as substances that deter feeding in insects. Natural defenses of plants are usually based on mixtures of deterrents. Schoonhoven (1982) stated that "plants never defend themselves with a monocomponent system".

Constituents of a mixture may exhibit different types of interactions including synergistic, antagonistic or additive based on their structures. Synergy or antagonism can occur when sensory neurons respond to the mixture in a way that exceeds or become less than the summed responses to the components (Kang and Caprio 1991).

Such synergy has been demonstrated for binary mixtures of various known monterpenoid antifeedants. Binary mixtures of linalool with 1,8-cineole, linalool with terpineol or thymol, or 1,8-cineole with terpineol or thymol were more deterrent than the individual compounds alone (Singh et al. 2009) against third instar *Chilo partellus* larvae. Similarly, synergy was reported for other binary mixtures; *trans*-anethole and thymol, thymol and citronellal, and α-terpineol and citronellal (Hummelbrunner and Isman 2001) against *Spodoptera litura*. These results suggest that the combined effect of binary mixtures of pure allelochemicals is greater than that of individual compounds in most cases.

Extracts of *Melia toosendan* containing 60–75% toosendanin showed greater growth inhibition and antifeedant effects against the variegated cutworm, *Peridroma saucia*, than toosendanin alone (Chen et al. 1995). This resulted presumably from synergistic effects of the minor constituents, as the isolated analogs were no more active than toosendanin (Isman et al. 1996).

Synergy has also been reported between major and minor constituents of essential oils with respect to feeding deterrence (Akhtar et al. 2012). To determine the potential contribution of individual constituents of cypress (*Cupressus sempervirens*, Cupressaceae) essential oil to the overall feeding deterrent effect of the oil, an artificial blend of cypress oil was created, including most of the major constituents of oil mimicking the natural oil, known as a "full mixture" (Fig. 11.1). The full mixture was compared with artificial blends, each lacking one constituent (Jiang et al. 2009).

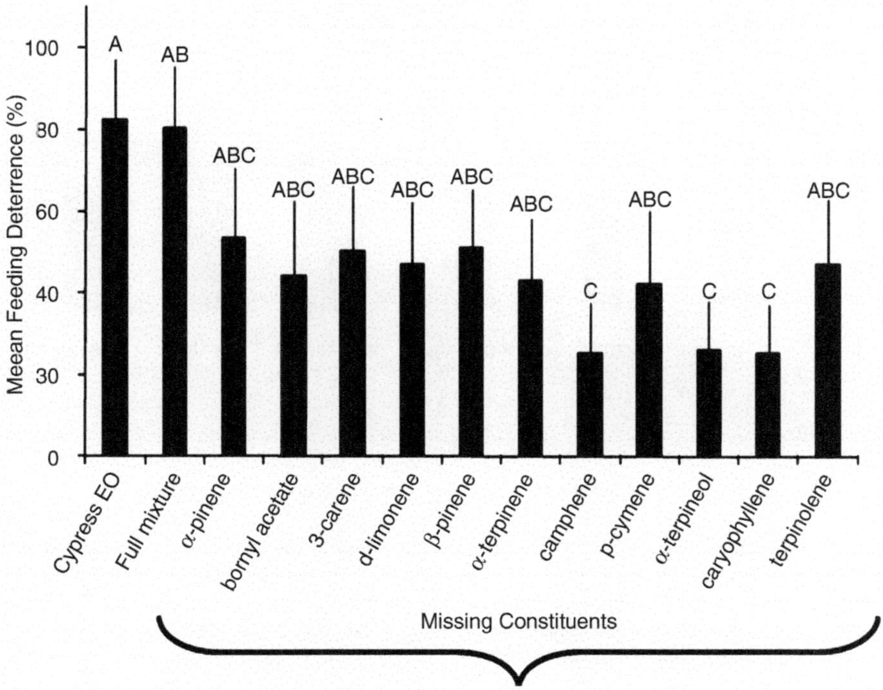

Fig. 11.1 Mean feeding deterrence caused by natural cypress essential oil, the full mixture, and selected blends of constituents of the oil to third-instars of *Trichoplusia ni* applied at levels equivalent to DC_{80} (concentration causing ~80% deterrence compared with the control) of the natural oil (DC_{80} = 170 μg/cm² for *T. ni*). *Error bars* represent the standard error of the mean of 24 larvae. Means corresponding to each treatment with *different letters* are significantly different from each other (LSD test, $p < 0.05$). The "*full mixture*" indicates a blend of 11 constituents, whereas *all others indicate* the full mixture minus the constituent noted

Blends were based on the natural composition of the essential oils and tested at a concentration at which the natural oil produced >80% feeding deterrence. Comparison of the deterrent activity of the full mixture with the artificial blends missing individual constituents demonstrated that minor constituents in a mixture can be as important as major constituents for the overall feeding deterrent effect. Bioassays with artificial mixtures showed that a blend containing all the known constituents of an essential oil ("full mixture") was the most active feeding deterrent and that the minor constituents may act as synergists (Fig. 11.1).

One of the limitations of using feeding deterrents as crop protectants is the potential for habituation in insects as a result of continuous exposure to the compounds. Several studies have indicated that habituation or the decrease in feeding deterrent response following prolonged exposure can be prevented by presentation of a complex mixture of substances (Jermy 1986). The role of mixtures in preventing habituation has been well demonstrated previously (Akhtar and Isman 2003). The effect of rearing larvae of *Trichoplusia ni* (neonates to third instars) on individual feeding

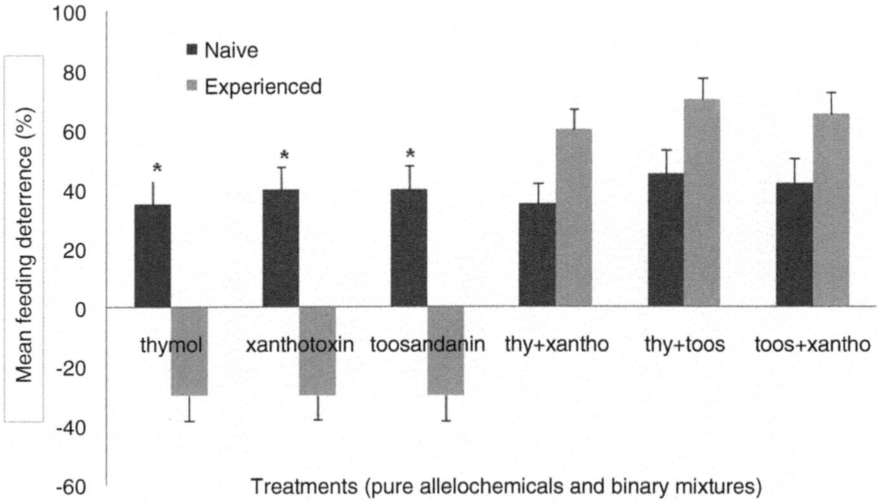

Fig. 11.2 Feeding responses of third instar *Trichoplusia ni* larvae to individual allelochemicals and binary mixtures thereof following previous dietary exposure to them individually or as binary mixtures (n = 52/treatment). Feeding deterrence means followed by *asterisks indicate* significant differences between experienced and naïve larvae (Tukey's test, $p < 0.05$); *thy* thymol, *toos* toosandanin, *xantho* xanthotoxin (Akhtar and Isman 2003)

deterrents or on binary mixtures of deterrents on their subsequent gustatory sensitivity was measured in paired choice leaf disc bioassays (Akhtar and Isman 2003). We have clearly demonstrated that third instar *T. ni* larvae showed a decrease in feeding deterrent response following prolonged exposure to individual allelochemicals (thymol, toosendanin or xanthotoxin) when presented singly but not as binary mixtures (Fig. 11.2, Akhtar and Isman 2003). We believe that this phenomenon likely occurs in natural insect-plant interactions as well. Such mixtures were synergistic in terms of their feeding deterrence to "experienced" larvae. Our experimental results support the hypothesis (Jermy 1986) that mixtures of deterrents can prevent decreased feeding deterrent response following prolonged exposure, and provides one explanation for the multiplicity of chemical defenses found in many plants.

These results corroborate earlier findings with neem (*Azadirachta indica*; Meliaceae) and its main constituent, azadirachtin (Bomford and Isman 1996). Larvae of *Spodoptera litura* showed a rapid decrease in feeding deterrent response to pure azadirachtin, but not to a refined neem seed extract containing a mixture of compounds including azadirachtin.

Plant defense chemicals (or combinations thereof) that exhibit more than one mode of action should be especially suitable for crop protection (Raffa 1987). Therefore, it seems logical to use mixtures of antifeedants for more durable crop protection rather than any single antifeedant.

3 Effects of Mixtures on Toxicity

Mixtures of compounds increase the insecticidal spectrum of action, because various species have variable responses to individual compounds (Singh et al. 2009). In this section, we discuss some examples showing the effect of mixtures on toxicity including mixtures of pure compounds, plant extracts or essential oils.

Synergy has been reported for binary mixtures of various constituents of plant essential oils. Bioassays using binary mixtures of essential oils revealed that combinations of thymol and linalool or 1,8-cineole; or terpineol and linalool or 1,8-cineole, were synergistic in terms of toxicity to third instar *Chilo partellus* larvae (Table 11.1). Linalool and 1,8-cineole exhibited an additive effect when combined with each other or with *trans*-anethole, respectively (Singh et al. 2009). Similarly, *trans*-anethole strongly synergized the toxicity of thymol, citronellal and α –terpineol against *S. litura* (Table 11.1, Hummelbrunner and Isman 2001). Binary mixtures of citronellal and α-terpineol also demonstrated synergy, whereas the remaining combinations of thymol with citronellal or α –terpineol were simply additive in effect (Table 11.1). Understanding the synergistic interactions between compounds could lead to the formation of new mixtures with enhanced efficacy. Since most of these compounds are widely distributed in essential oil-bearing plants, resource availability should not be an issue.

In another study, insect growth inhibition in the fall armyworm *Spodoptera frugiperda* and in the darkling beetle *Tenebrio molitor* resulted from the synergistic interaction of a binary mixture of the plant sterols peniocerol and macdougallin, obtained from the roots and aerial parts of *Myrtillocactus geometrizans* (Cactaceae) (Cespedes et al. 2005). Thyme oil, comprised largely of thymol and carvacrol, demonstrated greater sublethal effects on the growth of *S. litura* larvae than the individual compound (Hummelbrunner and Isman 2001).

Mixtures of plant extracts have also been very active in controlling insect pest populations. Binary mixtures of several plant extracts were investigated for efficacy in the management of two major post-flowering insect pests, *Maruca vitrata* (Maruca pod borer) and *Clavigralla tomentosicollis* (pod sucking bug) of cowpea (Oparaeke et al. 2005). There was a significant reduction in the number of *M. vitrata* and *C. tomentosicollis* and plant damage along with an increased yield in plots sprayed with binary mixtures of foliar extracts of neem (*Azadirachta indica*) and lemongrass (*Cymbopogon citratus*), neem and African basil (*Ocimum gratissimum*), neem and tomato (*Lycopersicum esculentum*), neem and bitter leaf (*Vernonia amygdalina*) and eucalyptus (*Eucalyptus citriodora*) and African bush tea (*Hyptis suaveolens*). In a related study (Sinzogan et al. 2006), mixtures of conventional insecticides at one half the recommended rate, and plant extracts of three local plants (*Azadirachta indica*, *Khaya senegalensis*, and *Hyptis suaveolens*) provided better protection of cotton against the bollworm, *Helicoverpa armigera*, than the conventional products or the plant extracts alone.

Table 11.1 Toxicity of binary mixtures of essential oil compounds to third instar *Chilo partellus* (Singh et al. 2009) and fourth instar *Spodoptera litura* (Hummelbrunner and Isman 2001) and measures of interactions

			Mortality (%)						
			Pure compounds		Binary mixtures				
Compound A	Compound B	Dose (μg/larva)	O A	OB	E	O	X^2	Effect	Reference
Thymol	Linalool	180+180	24.9	12.5	34.3	49.9	7.1	Synergy	Singh et al. (2009)
Thymol	1–8,Cineole	180+180	24.9	12.5	34.3	45.8	3.8	Synergy	
1–8,Cineole	Terpineol	402+402	29.2	24.9	46.9	62.5	5.2	Synergy	
Linalool	Terpineol	453+453	37.5	20.8	50.5	66.6	5.1	Synergy	
Thymol	*trans*-Anethole	35+35	37.5	12.5	45.3	100.0	66.0	Synergy	Hummelbrunner and Isman (2001)
Thymol	α-Terpineol	35+35	32.5	35.9	35.9	32.5	0.3	Additive	
Thymol	Citronellal	40+40	80.0	0	80.0	90.0	1.3	Additive	
Citronellal	α-Terpineol	110+110	10.0	15.0	23.5	65.0	73.3	Synergy	
Citronellal	*trans*-Anethole	70+70	15	60.0	66.0	100.0	17.5	Synergy	
α-Terpineol	*trans*-Anethole	60+60	32.5	37.5	57.8	95.0	23.9	Synergy	

OA = observed mortality of compound A, *OB* = observed mortality of compound B. Actual mortalities were compared to expected mortalities using the formula $E = Oa + Ob\,(1-Oa)$; The effects of mixtures were designated either additive, or synergistic by using the formula: $X^2 = (O-E)^2/E$ where O is observed mortality from the binary mixture and E is expected mortality; X^2 with df = 1 and $a = 0.05$ is 3.84. A pair with X^2 values >3.84 and having greater than expected mortality were considered to be synergistic, with X^2 values <3.84 representing additive effects

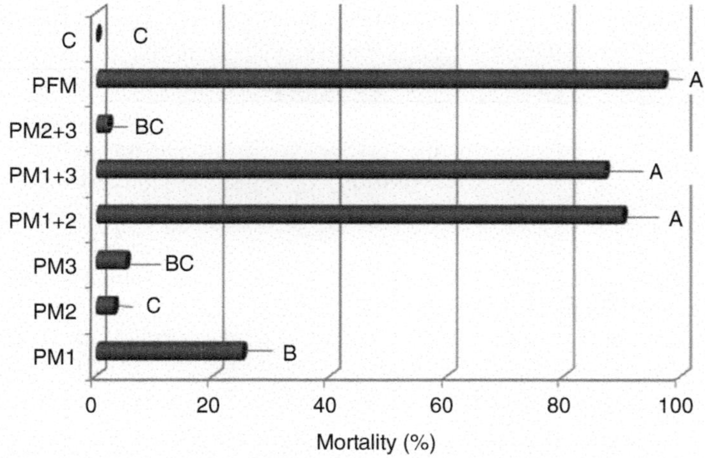

Fig. 11.3 Mortality caused by selected blends of active and inactive constituents of *L. pungens* oil to third-instar *Trichoplusia ni* larvae when applied at levels equivalent to those found in the 95% lethal concentration of the pure oil (LD_{95}=277.7 μg/larva for *T. ni*). *Error bars* represent the standard error of the mean of three replicates of ten larvae each. Means corresponding to each treatment with *different letters* are significantly different from each other (Tukey's test, $p < 0.05$). PM1 (very active constituents) = 1,8-cineole + carvone; PM2 (moderately active constituents); PM3 (inactive constituents); PM1 + 2 = PM1 + PM2; PM1 + 3 = PM1 + PM3; PM2 + 3 = PM2 + PM3; PFM = full mixture of all constituents. PM = *L. pungens* mixture (Jiang et al. 2009)

Toxicity of essential oils and blends of their major constituents have been studied against many insect pests. Although some of the constituents of the essential oils have strong biological activity, the presence of all constituents is necessary for the full activity of the oils. This was exemplified by a recent study involving blends of selected constituents of *Litsea pungens* and *L. cubeba* essential oils, demonstrating synergy among putatively active and inactive constituents, with the presence of all constituents necessary for full toxicity of the natural oils (Jiang et al. 2009) against third instars of cabbage looper (Fig. 11.3). Similar effects were observed earlier (Bekele and Hassanali 2001) with the essential oils of *Ocimium kilimandscharicum* and *O. kenyense* against *Sitophilus zeamais* and *Rhyzopertha dominica.*

4 Repellent Effects of Mixtures

Repellents are effective tools for protecting humans and domestic animals from bites by nuisance arthropods (Isman 2006). DEET (*N,N*-diethyl-3-methylbenzamide), is the most effective and widely used repellent. Due to problems associated with its use including an unpleasant odor, damage to plastics and synthetic rubber, as well as some potentially detrimental health effects (e.g., depression, urticaria, and contact dermatitis) (Katz et al. 2008), there has been a search for new insect repellents that

Table 11.2 Repellent effects of pure oils and their binary mixtures, with the addition of 10% vanillin, against *Aedes aegypti* females (Choochote et al. 2007)

Essential oil	Protection time (range, h)	
	Pure oil	Pure oil + 10% vanillin
C. zedoaria	0 (0)	0.75 (0–2)
K. galanga	0.25 (0–0.5)	1.25 (1–1.5)
Z. limonella	0 (0)	0.75 (0.5–2)
Z. piperitum	1 (0.5–1)	2.5 (1–2.5)
Essential oil mixtures (25:75)%	*Mixtures*	*Mixtures + 10% vanillin*
A. graveolens + K. galanga	0 (0)	0.5 (0.5)
A. graveolens + Z. piperitum	0.5 (0–0.5)	1.5 (1.5–2)
K. galanga + Z. piperitum	0 (0–0.5)	1 (1–2)

are safe, inexpensive, and odor-free or with a more pleasant aroma. Plant essential oils and their constituents have formed a basis for some alternative arthropod repellent products (Isman 2006) because of their status as minimum risk pesticides ([USEPA] U.S. Environmental Protection Agency 2010). The drawback of using plant-based repellents is that many of them are made up of relatively volatile constituents, limiting the duration of their protective effect.

Although there are many examples of the use of plant essential oils as repellents, here we will focus on mixtures of essential oils, their constituents or their mixtures with commercial products or other natural products for improved effects. A recent study (Hieu et al. 2010) compared the repellent effects of binary mixtures of seven essential oils and *Calophyllum inophyllum* (Clusiaceae) nut oil (tamanu oil) with DEET against female stable fly, *Stomoxys calcitrans* (Diptera: Muscidae). Exposed human hand bioassays showed that tamanu oil synergized the repellent effects of each essential oil tested. Protection time for a binary mixture of tamanu oil and lovage root essential oil (2.68 h) was significantly greater than for lovage root essential oil (1.13 h), tamanu oil (0.56 h), or DEET alone (2.20 h). The protection times (PT) of binary mixtures of each essential oil (clove bud, clove leaf, patchouli, savory, and white thyme) and tamanu oil (PT = 2.30–2.04 h) were almost identical to that for DEET (Hieu et al. 2010). The improved efficacy of the essential oils by the addition of tamanu oil might be attributed to the lower evaporation rate and the increased persistence of the mixture on the skin. *Lantana camara* flower extract in coconut oil provided 94.5% protection from *Aedes albopictus* and *Ae. Aegypti,* with no adverse effects on human volunteers for a 3-month period after application (Dua et al. 1996).

In many cases, the repellent activity of essential oils or their mixtures can be increased by the addition of vanillin (Tuetun et al. 2005). Essential oils from ten plant species were screened for repellence against *Age. aegypti* mosquitoes. Addition of 10% vanillin increased the protection time for *Zanthoxylum piperitum*, *Z. limonella*, *C. zedoaria* and *Kaempferia galanga*, against *Ae. aegypti* (Table 11.2). Binary mixtures of *Z. piperitum*, *A. graveolens* or *K. galanga*, also demonstrated increased protection time with the addition of 10% vanillin (Choochote et al. 2007). The repellent effect of *Apium graveolens* extract was also increased by the addition of 5% vanillin (Tuetun et al. 2005).

Oils from turmeric and hairy basil with addition of 5% vanillin repelled three species of mosquitoes under cage conditions for a period of 6–8 h depending on the mosquito species (Tawatsin et al. 2001). Increased protection time by the addition of vanillin might have resulted from a lower evaporation rate of repellent from the skin surface as described previously (Tawatsin et al. 2001).

In an effort to discover a new generation of compounds that overcome the limitations of repellents, two natural, host-derived compounds were combined together (Logan et al. 2010). A mixture of 6-methyl-5-hepten-2-one and geranylacetone provided better protection than that provided by the individual compounds alone or DEET. A mixture of the two natural products was significantly more repellent (87.1% repellency) than the compounds alone (6-methyl-5-hepten-2-one produced 34.1% repellency and geranylacetone produced 30.1% repellency at 0.1%) against *Anopheles gambiae* (Logan et al. 2010).

However, when applied under field conditions, *Zanthoxylum piperitum* oil + 5% vanillin was found to provide better protection against a wide range of natural mosquito populations (*Aedes gardnerii*, *Anopheles barbirostris*, *Armigeres subalbatus*, *Culex tritaeniorhynchus*, *Culex gelidus*, *Culex vishnui*, and *Mansonia uniformis*) than 25% DEET + 5% vanillin (Kamsuk et al. 2007).

One plant-based formulation incorporates two principal active ingredients: *p*-menthane-diol (PMD) derived from lemon eucalyptus (*Corymbia citriodora*) and lemongrass oil (LG), *Cymbopogon citratus* as a mosquito repellent. Both PMD and LG are effective repellents against *Anopheles darlingi.* To lower the cost of the repellent and maintain its efficacy, PMD and LG were combined with some low-cost ingredients (fixatives). The PMD/LG repellent significantly outperformed DEET, providing an average of 95% protection 6 h after application as opposed to 64% protection provided by DEET (Moore et al. 2007).

5 Mixtures of Plant Odors and Pheromones Attractants

Insects make use of pheromones and host plant odors, e.g., kairomones, to locate conspecifics for mating and host plants for feeding or oviposition respectively. Combinations of pheromones and host plant volatiles have been recommended for optimal trapping yield compared to the pheromones alone. Synergy between plant semiochemicals and pheromones can contribute to more successful mate finding and therefore it is likely to play an important role in reproductive isolation, for example in scolytid bark beetles (*Dendroctonus* spp.) and the sunflower moth (*Homoeosoma electellum*) (Landolt and Phillips 1997). In this section, we discuss the role of various mixtures of plant odors in host plant selection and mixtures of pheromones and plant odors as attractants for conspecifics.

Enhancement of attraction of male moths to the female sex pheromone by adding plant volatile compounds to the lure has been reported for several insects including *Plutella xylostella* (Reddy and Guerrero 2000), *Spodoptera exigua* (Deng et al. 2004), *Cydia pomonella* (Light et al. 1993; Yang et al. 2004), and *Helicoverpa zea*

(Light et al. 1993). A mixture of linalool and a green leaf volatile, (*Z*)-3-hexenol, increased the responses of pheromone olfactory receptor neurons (Ph-ORNs) of the male *H. zea* moth to (*Z*)-11-hexadecenal, the main pheromone component of the female sex pheromone (Ochieng et al. 2002). It is thought that synergy at the Ph-ORN level could have significantly contributed to the enhanced male behavioral response observed in several species (Landolt and Phillips 1997). An important basis for improving semiochemical-based trapping against pest insects is to understand the sensory mechanisms involved, especially when mixture interactions cause a dramatic change in the behavioral response (Said et al. 2011).

In the American palm weevil, *Rhynchophorus palmarum*, responses to aggregation pheromones were dramatically increased by the perception of host plant odors in the odor-baited traps in the field (Wertheim et al. 2005). Yang et al. (2004) reported that binary mixtures of (±) linalool, (*E*)-*ß*-farnesene or (*Z*)-3-hexenol with codlemone enhanced the attraction of male *C. pomonella* moth to codlemone in a wind tunnel. Addition of ethyl-acetate (EtOAc), one of the main constituents of the odor from raw plant baits, enhanced the responses of *Rhynchophorus palmarum* to the pheromone alone or when combined with plant volatiles (Rochat et al. 2000).

Aggregation of *Rhynchophorus palmarum* weevils on host plants is mediated by a male pheromone (rhynchophorol: R) and host-plant volatiles (PVs) acting in synergy. Synthetic PV blends synergizing pheromone contain acetoin (A) and ethyl acetate (EtAc). Traps with pheromone alone caught about one-tenth as many insects as combining pheromone and host-plant volatiles. Behavioral results support the role of acetoin as a pheromone synergist for *R. palmarum,* and electrophysiological data provide evidence of modulation of peripheral sensory responses to pheromone by acetoin (Saïd et al. 2005).

Monoterpenoids extracted from wood of Scots pine, *Pinus sylvestris* L, synergized the attraction of the old-house borer, *Hylotrupes bajulus* (L), to the male pheromone (3R)-3-hydroxy-2-hexanone ((3R)-ketol) + 1-butanol. Glasshouse experiments using ground traps baited with extracts derived from Scots pine wood or the monoterpenes (+)-α-pinene, (−)-verbenone, (−)-*trans*-pinocarveol and (+)-terpinen-4-ol attracted significantly more *H. bajulus* females, but caught fewer of them, than the synthetic pheromone mixture alone. However, a combination of (3R)-ketol + 1-butanol or (+/−)-3-ketol + 1-butanol with monoterpenes resulted in the capture of significantly more females than either the sex pheromone or the monoterpene mixture alone. Traps baited with a blend of the male sex pheromone or the monoterpenes attracted significantly more, but caught fewer, males than females (Reddy et al. 2005). Capture of three moth species including speckled cutworm (*Lacanobia subjuncta*), bertha armyworm *(Mamestra configurata*), and spotted cutworm, (*Xestia c-nigrum*) was significantly increased in traps baited with acetic acid, 3-methyl-1-butanol, and 3-methyl-1-pentanol. Binary mixture of acetic acid and 3-methyl-1-butanol was the most active mixture for the three moth species tested, whereas the binary mixture of acetic acid and 3-methyl-1-pentanol attracted significantly more *X. c-nigrum* moths than the individual constituents. Traps baited with binary mixtures of acetic acid and 3-methyl-1-butanol or acetic acid and 3-methyl-1-pentanol caught equal numbers of male and female moths of the three species (Landolt 2000).

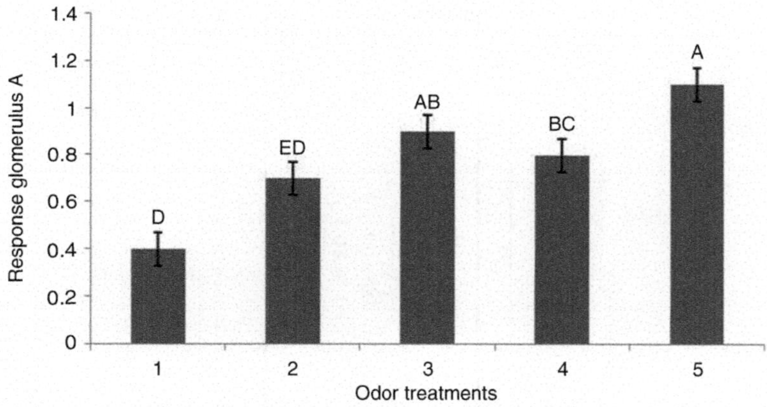

Fig. 11.4 Mean responses of female oriental fruit moth to synthetic mixtures of five odor treatments recorded in the mixture sensitive glomerulus A of the antennal lobe. *Error bars* representing *different letters* are not significantly different from each other (Tukey's HSD test; $p=0.05$) (Pinero et al. 2008)

At the behavioral level involving phytophagous insects, mixtures of host plant derived compounds play a major role in eliciting host specific response (Carlsson and Hansson 2003; Bruce et al. 2005). Synergistic interactions among the various constituents of an odor blend have been reported to be responsible for host plant selection in many insects. Synergistic interaction between the general green leaf volatiles and specific aromatic compounds have been documented in the attraction of *Cydia molesta* (Pinero and Dorn 2007; Pinero et al. 2008). A mixture of three green leaf volatiles, (Z)-3-hexen-1-ol, (E)-2-hexenal, (Z)-3-hexen-1-yl-acetate with benzonitrile and benzaldehyde was significantly more attractive to female oriental fruit moth, *C. molesta* than either of the compounds or the blend of three green leaf volatiles alone. Pinero et al. (2008) also demonstrated a strong positive correlation between the behavioral response and a physiological correlate of it in the form of increased glomerular activity in the antennal lobe of female *C. molesta* to the synergistic effect of 5-compound mixture (Fig. 11.4).

Synergy has also been reported among floral odorants in other insects. Phenylacetaldehyde (PAA) is a key floral odorant attractive to several moth species. Meagher and Landolt (2008) have demonstrated that traps containing binary mixtures of PAA and the floral odorants *cis*-jasmone, linalool, benzyl acetate, limonene, β-myrcene, methyl salicylate, and methyl 2-methoxybenzoate increased captures of several moth species (Table 11.3). Soybean looper moths, *Pseudoplusia includens*, most strongly responded to PAA + β-myrcene, Velvetbean caterpillar moths, *Anticarsia gemmatalis* responded most strongly to PAA+linalool. Positive responses to floral compound blends were also noted for several noctuids including golden looper (*Argyrogramma verruca*), yellow mocis moth (*Mocis disseverans*), soybean looper (*Pseudoplusia includens*), tobacco budworm (*Heliothis virescens*), southern armyworm (*Spodoptera eridania*) and a pyralid, the melonworm (*Diaphania hyalinata*).

Table 11.3 Mean (± SE) numbers of moths captured in traps (n=50) baited with floral compounds phenylacetaldehyde (PAA), *cis*-jasmone (CJ), linalool (LIN), methyl salicylate (MS), methyl 2-methoxybenzoate (M2MB), and β-myrcene (MYR) against soybean looper (*Pseudoplusia includes*), golden looper (*Argyrogramma verruca*), velvetbean caterpillar (*Mocis latipes*), yellow mocis moth (*Mocis disseverans*), tobacco budworm (*Heliothis virescens*), Southern armyworm (*Spodoptera eridania*) and the melonworm (*Diaphania hyalinata*)

Traps	*P. includes*	*A. verruca*	*M. latipes*	*M. disseverans*	*H. virescens*	*S. eridania*	*D. hyalinata*
Unbaited	0.0±0.0	0.06±0.03d	1.3±0.2c	0.02±0.02	0.0±0.0	0.02±0.0	0.0±0.0
PAA	5.1±1.3	3.5±0.5	3.6±0.9c	1.3±0.2	0. ±0.1	0.8±0.2	0.5±0.2
PAA+CJ	18.0*±4.4	5.7*±0.9	n.s	n.s	1.5*±0.4	1.4*±0.2	1.8*±0.6
PAA+LIN	n.s	n.s	n.s	n.s	n.s	0.6±0.1	1.3*±0.5
PAA+MS	n.s	n.s	n.s	n.s	n.s	1.5*±0.3	1.6*±0.5
PAA+M2MB	n.s	n.s	n.s	2.1*±0.4	1.4*±0.4	1.4*±0.3	1.4*±0.4
PAA+MYR	26.3*±4.9	n.s	11.8*±2.3	n.s	2.6*±0.6	1.5*±0.2	1.9*±0.4

(Meagher and Landolt 2008)

*number of moths captured in the traps baited with binary mixtures (PAA and other floral odorants) were significantly greater than traps baited with PAA alone or unbaited traps.

n.s = no significant difference in the number of moths trapped in the treatments and the control

Binary mixture of PAA + β-myrcene was the most active attractant for the moth species captured compared to PAA alone or other mixtures (Meagher and Landolt 2008). β-Myrcene, although weakly attractive or unattractive when presented alone, enhanced cabbage looper *(T. ni)* and alfalfa looper *(A. californica)* moth response to PAA (Landolt et al. 2001, 2006).

The attraction of female grapevine moth, *Lobesia botrana* was compared with specific and common (shared) odors from a wild host (*Daphne gnidium*) and a recently colonized host (*Vitis vinifera*). Attraction of females was elicited by a blend of compounds released from both host plants, and by two blends with the compounds released specifically from each host. However, more complete odor blends of the two plants elicited stronger attraction. The common compounds in combination with the specific compounds of *D. gnidium* were the most attractive (Tasin et al. 2010).

6 Conclusion

Mixtures of plant natural products can be effective, environmentally sound and cost effective insect control agents. Synergy has been repeatedly demonstrated between various constituents of mixtures including essential oils or other natural products. Such synergistic interactions among the various constituents of a mixture are considered to have a stronger and more durable effect (Chockalingam et al. 1990). Identifying synergistic compounds within mixtures may lead to the development of more effective insect control agents (acute toxicants, growth/feeding inhibitors, repellents and attractants) as well as the use of smaller amounts in the mixture to achieve satisfactory levels of efficacy. Combinations of compounds are more desirable due to increased benefits including broader insecticidal spectra, greater protection time and decreased residues, insect resistance or habituation and environmental effects.

In terms of insect management, more effective traps can be designed through combination of sex pheromones or aggregation pheromones with hostplant odors. Based on this strategy, the development of effective lures against a number of insects including several species of beetles is noteworthy (Reddy and Guerro 2004; Said et al. 2011). Moreover, it has been suggested that mating disruption dispensers could be developed for certain moth species by adding small amounts of expensive active pheromonal ingredients to the selected blend of inexpensive plant volatiles (Ochieng et al. 2002). In addition, the synergy between insect pheromones and plant odors can increase the attraction of natural enemies, offering new strategies for biological control (Reddy and Guerro 2004).

Plant essential oils and their constituents have formed a basis for alternative arthropod repellent products with effects lasting from several minutes to hours. Their active ingredients tend to be highly volatile, so although they are effective repellents initially, they rapidly evaporate leaving the user unprotected. However, this problem has been addressed by using fixatives or careful formulation to improve their longevity (Maia and Moore 2011). Addition of 5–10% vanillin to a mixture of oils increased protection time, possibly due to reduced evaporation of repellent from the skin surface.

Mixtures of compounds have also been able to diffuse the selection process mitigating resistance development compared with a single active ingredient. The green peach aphid, *Myzus persicae,* developed resistance to pure azadirachtin but not to a refined neem seed extract containing the same absolute amount of azadirachtin in the greenhouse (Feng and Isman 1995). This can be explained on the basis that different constituents in the mixture might have different modes-of-action or target sites in the insect or are capable of inhibiting the detoxification enzymes that normally degrade a single constituent.

In conclusion, mixtures of natural products (as found in plant essential oils or extracts) emulate natural insect-plant chemical interactions that have evolved for plant defense against herbivores or as mediators for host plant selection by insects. As such, a better understanding of these natural strategies should facilitate the development of more effective insect control agents (acute toxicants, growth/feeding inhibitors, repellents and attractants) based on mixtures of natural products.

Acknowledgement We thank Rod Bradbury (Ecosafe Natural Products Inc., Saanichton, BC, Canada) for chemical analysis of the essential oil of cypress. Supported by an NSERC Discovery grant (2729–06) to MBI.

References

Akhtar Y, Isman MB (2003) Binary mixtures of feeding deterrents mitigate the decrease in feeding deterrent response to antifeedants following prolonged exposure in the cabbage looper, *Trichoplusia ni* (Lepidoptera: Noctuidae). Chemoecology 13:177–182

Akhtar Y, Pages E, Stevens A, Bradbury R, de Camara C, Isman MB (2012) Effect of chemical complexity of essential oils on feeding deterrence in the cabbage looper, *Trichoplusia ni* larvae. Physiol Entomol 37:81–91

Bekele J, Hassanali J (2001) Blend effects in the toxicity of the essential oil constituents of *Ocimum kilimandscharicum* and *Ocimum kenyense* (Labiateae) on two post-harvest insect pests. Phytochemistry 57:385–391

Bernays EA, Chapman RF (2001) Taste cell responses in the polyphagous arctiid, *Grammia geneura*: towards a general pattern for caterpillars. J Insect Physiol 47:1029–1043

Bomford MK, Isman MB (1996) Desensitization of fifth instar *Spodoptera litura* to azadirachtin and neem. Entomol Exp Appl 81:307–313

Bruce TJA, Wadhams LJ, Woodcock CA (2005) Insect host location: a volatile situation. Trends Plant Sci 10:269–274

Carlsson MA, Hansson BS (2003) Plasticity and coding mechanisms in the insect antennal lobe. In: Blomquist GJ, Vogt RV (eds) Insect pheromone biochemistry and molecular biology. Elsevier Academic Press, San Diego, pp 699–728

Cespedes CL, Salazar JR, Martinez M, Aranda E (2005) Insect growth regulatory effects of some extracts and sterols from *Myrtillocactus geometrizans* (Cactaceae) against *Spodoptera frugiperda* and *Tenebrio molitor*. Phytochemistry 66:2481–2493

Chen WK, Isman MB, Chiu SF (1995) Antifeedant and growth inhibitory effects of the limonoid toosendanin and *Melia toosendan* extracts on the variegated cutworm, *Peridroma saucia* (Lep., Noctuidae). J Appl Entomol 119:367–370

Chockalingam S, Thenmozhi S, Nalina Sundari MS (1990) Larvicidal activity of different products against mosquito larvae. J Environ Biol 11:101–104
Choochote W, Chaithong U, Kamsuk K, Jitpakdi A, Tippawangkosol P, Tuetun B, Champakaew D, Pitasawat B (2007) Repellent activity of selected essential oils against *Aedes aegypti*. Fitoterapia 78:359–364
Deng J-Y, Wei H, Huang Y-P, Du J-W (2004) Enhancement of attraction to sex pheromones of *Spodoptera exigua* by volatile compounds produced by host plants. J Chem Ecol 30:2037–2045
Dua VK, Gupta NC, Pandey AC, Sharma VP (1996) Repellency of *Lantana camara* (Verbenaceae) flowers against Aedes mosquitoes. J Am Mosq Control Assoc 12:406–408
Feng R, Isman MB (1995) Selection for resistance to azadirachtin in the green peach aphid, *Myzus persicae*. Experientia 51:831–833
Hieu TT, Kim SI, Lee SG, Ahn YJ (2010) Repellency to *Stomoxys calcitrans* (Diptera: Muscidae) of plant essential oils alone or in combination with *Calophyllum inophyllum* nut oil. J Med Entomol 4:575–580
Hummelbrunner LA, Isman MB (2001) Acute, sublethal, antifeedant and synergistic effects of monoterpenoid essential oil compounds on the tobacco cutworm *Spodoptera litura* (Lep., Noctuidae). J Agric Food Chem 49:715–720
Isman MB (2004) Plant essential oils as green pesticides for pest and disease management. In: Nelson WM (ed) Agricultural applications in green chemistry, vol 887, ACS Symposium Series. American Chemical Society, Washington, DC, pp 41–51
Isman MB (2006) Botanical insecticides, deterrents, and repellents in modern agriculture and increasingly regulated world. Annu Rev Entomol 51:45–66
Isman MB, Matsuura H, MacKinnon S, Durst T, Towers GHN, Arnason JT (1996) Phytochemistry of the Meliaceae: so many terpenoids, so few insecticides. In: Romeo JT, Saunders JA, Barbosa P (eds) Phytochemical diversity and redundancy in ecological interactions. Plenum press, New York, pp 155–178
Jermy T (1986) The role of experience in the host selection of phytophagous insects. In: Chapman RF, Bernays EA, Stoffolano JG (eds) Perspectives in chemoreception and behaviour. Springer, New York, pp 143–157
Jiang ZL, Akhtar Y, Bradbury R, Zhang X, Isman MB (2009) Comparative toxicity of essential oils of *Litsea pungens* and *Litsea cubeba* and blends of their major constituents against the cabbage looper *Trichoplusia ni*. J Agric Food Chem 57:4833–4837
Kamsuk K, Choochote W, Chaithong U, Jitpakdi A, Tippawangkosol P, Riyong D, Pitasawat B (2007) Effectiveness of *Zanthoxylum piperitum*–derived essential oil as an alternative repellent under laboratory and field applications. Parasitol Res 100:339–345
Kang J, Caprio J (1991) Electro-olfactogram and multiunit olfactory receptor responses to complex mixtures of amino acids in the channel catfish, *Ictalurus punctatus*. J Gen Physiol 98:699–721
Katz TM, Miller JH, Hebert AA (2008) Insect repellents: historical perspectives and new developments. J Am Acad Dermatol 58:865–871
Koul O (2008) Phytochemicals and insect control: an antifeedant approach. Crit Rev Plant Sci 27:1–24
Landolt PJ (2000) New chemical attractants for trapping *Lacanobia subjuncta*, *Mamestra configurata*, and *Xestia c-nigrum* (Lepidoptera: Noctuidae). J Econ Entomol 93:101–106
Landolt PJ, Phillips WT (1997) Host plant influences on sex pheromone behaviour of phytophagous insects. Annu Rev Entomol 42:371–391
Landolt PJ, Adams T, Reed HC, Zack RS (2001) Trapping alfalfa looper moths (Lepidoptera: Noctuidae) with single and double component floral chemical lures. Environ Entomol 30:667–672
Landolt PJ, Adams T, Zack RS (2006) Field response of alfalfa looper and cabbage looper moths (Lepidoptera: Noctuidae, Plusiinae) to single and binary blends of floral odorants. Environ Entomol 35:276–281
Light DM, Flath RA, Buttery RG, Zalom FG, Rice RE, Dickens JC, Jang EB (1993) Host-plant green-leaf volatiles synergize the synthetic sex pheromones of the corn earworm and codling moth (Lepidoptera). Chemoecology 4:145–152

Logan JG, Stanczyk NM, Hassanali A, Kemei J, Santana AEG, Ribeiro KAL, Pickett JA, Mordue Luntz AJ (2010) Arm-in-cage testing of natural human-derived mosquito repellents. Malar J 9:239

Maia MF, Moore SJ (2011) Plant-based insect repellents: a review of their efficacy, development and testing. Malar J. doi:10.1186/1475-2875-10-S1-S11

Meagher RL, Landolt PJ (2008) Attractiveness of binary blends of floral odorant compounds to moths in Florida, USA. Entomol Exp Appl 128:323–329

Miresmailli S, Bradbury R, Isman MB (2006) Comparative toxicity of *Rosmarinus officinalis* L. essential oil and blends of its major constituents against *Tetranychus urticae* Koch (Acari: Tetranychidae) on two different host plants. Pest Manag Sci 62:366–371

Moore SJ, Darling ST, Sihuincha M, Padilla N, Devine GJ (2007) A low-cost repellent for malaria vectors in the Americas: results of two field trials in Guatemala and Peru. Malar J 6:101

Ochieng SA, Park KC, Baker TC (2002) Host plant volatiles synergise responses of sex pheromone-specific olfactory receptor neurons in male *Helicoverpa zea*. J Comp Physiol A 188:325–333

Oparaeke AM, Dike MC, Amatobi CI (2005) Botanical pesticide mixture for insect pest management on cowpea, *Vigna unguiculata* (L.) Walp plants – the pod borer *Maruca vitrata* FAB. (Lepidoptera: Pyralidae) and pod sucking bug, *Clavigralla tomentosicollis* STAL (Heteroptera: Coreidae). Agric Trop et Subtrop 38:33–38

Pinero JC, Dorn S (2007) Synergism between aromatic compounds and green leaf volatiles derived from the host plant underlies female attraction in the oriental fruit moth. Entomol Exp Appl 125:185–194

Pinero JC, Galizia CG, Dorn S (2008) Synergistic behavioural responses of female oriental fruit moths (Lepidoptera: Tortricidae) to synthetic host plant-derived mixtures are mirrored by odor-evoked calcium activity in their antennal lobes. J Insect Physiol 54:333–343

Raffa KF (1987) Influence of host plants on deterrence by azadirachtin of feeding by fall armyworm larvae. J Econ Entomol 80:384–387

Reddy GVP, Guerrero A (2000) Behavioral responses of the diamondback moth, *Plutella xylostella*, to green leaf volatiles of *Brassica oleracea* subsp capitata. J Agric Food Chem 48:6025–6029

Reddy GVP, Guerro A (2004) Interaction of insect pheromones and plant semiochemicals. Plant Sci 9:253–261

Reddy GVP, Fettköther R, Noldt U, Dettner K (2005) Enhancement of attraction and trap catches of the old-house borer, *Hylotrupes bajulus* (Coleoptera: Cerambycidae), by combination of male sex pheromone and monoterpenes. Pest Manag Sci 61:699–704

Rochat D, Nagnan-Le Meillour P, Esteban-Duran JR, Malosse C, Perthuis B, Morin JP, Descoins C (2000) Identification of pheromone synergists in American palm weevil, *Rhynchophorus palmarum*, and attraction of related *Dynamis borassi*. J Chem Ecol 26:155–187

Saïd I, Renou M, Morin JP, Ferreira JMS, Rochat D (2005) Interactions between acetoin, a plant volatile and pheromone in *Rhynchophorus palmarum*: behavioral and olfactory neuron responses. J Chem Ecol 31:1789–1805

Saïd I, Kaabiz B, Rochat D (2011) Evaluation and modeling of synergy to pheromone and plant kairomone in American palm weevil. Chem Cent J 5:14

Schoonhoven LM (1982) Biological aspect of antifeedants. Entomol Exp Appl 31:57–69

Schoonhoven LM, Blaney WM, Simmonds MSJ (1992) Sensory coding of feeding deterrents in phytophagous insects. In: Bernays E (ed) Insect-plant interactions, vol 4. CRC Press, Boca Raton, pp 59–79

Singh R, Koul O, Rup PJ, Jindal J (2009) Toxicity of some essential oil constituents and their binary mixtures against *Chilo partellus* (Lepidoptera: Pyralidae). Int J Trop Insect Sci 29:93–101

Sinzogan AAC, Kossou DK, Atachi P, van Huis A (2006) Participatory evaluation of synthetic and botanical pesticide mixtures for cotton bollworm control. Int J Trop Insect Sci 26:246–255

Tasin M, Bäckman AC, Anfora G, Carlin S, Ioriatti C, Witzgall P (2010) Attraction of female grapevine moth to common and specific olfactory cues from two host plants. Chem Senses 35:57–64

Tawatsin A, Wratten SD, Scott RR, Thavara U, Techadamrongsin Y (2001) Repellency of volatile oils from plants against three mosquito vectors. J Vector Ecol 26:76–82

Tuetun B, Choochote W, Kanjanapothi D, Rattanachanpichai E, Chaithong U, Chaiwong P, Jitpakdi A, Tippawangkosol P, Riyong D, Pitasawat B (2005) Repellent properties of celery, *Apium graveolens* L., compared with commercial repellents, against mosquitoes under laboratory and field conditions. Trop Med Int Health 10:1190–1198

[USEPA] U.S. Environmental Protection Agency (2010) Toxicity category. U.S. Environmental Protection Agency, 40 CFR 156.62, pp 75–580

Wertheim B, van Baalen EJA, Dicke M, Vet LE (2005) Pheromone-mediated aggregation in nonsocial arthropods: an evolutionary ecological perspective. Annu Rev Entomol 50:321–346

Yang ZH, Bengtsson M, Witzgall P (2004) Host plant volatiles synergize response to sex pheromone in codling moth, *Cydia pomonella*. J Chem Ecol 30:619–629

Chapter 12
Optical Manipulations: An Advance Approach for Reducing Sucking Insect Pests

David Ben-Yakir, Yehezkel Antignus, Yossi Offir, and Yosepha Shahak

1 Introduction

Insect pests are a major cause for reduction in the quantity and quality of crop plant products. Sucking insect pests that transmit viral diseases are an important cause of economic losses for growers of agricultural crops worldwide. Growers usually apply toxic insecticides to protect their crop plants from these pests. Frequent applications of insecticides create health hazards for workers, consumers, and the environment. Moreover, frequent applications of insecticides often induce resistance in the treated pest populations. Therefore, alternative methods for protecting crop plants from pests are constantly being sought. The use of mulches, traps and cladding materials that possess specific optical properties often reduced the infestation rates of pests and lowered the incidence of vector-borne viral diseases in crop plants. Recently, two comprehensive reviews were published on the effect of indirect and direct light on greenhouse pests by Vanninen et al. (2010) and Johansen et al. (2011), respectively. These reviews focus mainly on the potential for optical manipulation in high-technology year-round greenhouse production in northern Europe and Canada in which natural light is augmented with artificial light. In those greenhouses, pests may be manipulated by changing the light quality, quantity and photoperiod.

D. Ben-Yakir (✉) • Y. Antignus
Institute of Plant Protection, Agricultural Research Organization,
The Volcani Center, Bet Dagan 50250, Israel
e-mail: benyak@volcani.agri.gov.il; antignus@volcani.agri.gov.il

Y. Offir
Polysack Plastics Industries, Nir Yitzhak-Sufa, Israel
e-mail: yosi_o@polysack.com

Y. Shahak
Institute of Plant Sciences, Agricultural Research Organization,
The Volcani Center, Bet Dagan 50250, Israel
e-mail: shahaky@volcani.agri.gov.il

I. Ishaaya et al. (eds.), *Advanced Technologies for Managing Insect Pests*,
DOI 10.1007/978-94-007-4497-4_12, © Springer Science+Business Media Dordrecht 2012

In this chapter, we focus mainly on the manipulation of pests in open field and protected crops using the reflection of natural solar radiation. This approach is suitable for geographical regions that have a high intensity of direct sunlight during the seasons when sucking pests are active. In the Mediterranean region, sucking pests are mainly active from March to October and during this period most days are sunny and clear.

We propose that the optical cues of reflected light will be used to interfere with distant host finding by the pests. This is because it is expected that when pests are near or on the plants, other sensory cues, such as humidity gradient and plant odors, can substitute the optical cues. The optical modifications have to be compatible with optimal conditions for crop production. These conditions often include the undisturbed performance of beneficial insects such as biocontrol agents and pollinators.

Some of the experimental results and their interpretations presented in this chapter are based on our studies with the green peach aphid (*Myzus persicae* (Sulzer)), the cotton aphid (*Aphis gossypii* Glover), the sweet potato whitefly (*Bemisia tabaci* (Gennadius)), the onion thrips (*Thrips tabaci* Lindeman), the western flower thrips (WFT; *Frankliniella occidentalis* (Pergande)) and the chilli thrips (*Scirtothrips dorsalis* Hood). The crops were bell pepper (*Capsicum annuum* L.), tomato (*Lycopersicon esculentum* Miller), chives (*Allium schoenopransum* L.) and Lisianthus (*Eustoma russellianum* Salisb.).

2 Sucking Insect Pests

Sucking pests such as aphids and whiteflies feed by sucking fluids directly from the phloem vessels of plants. Thrips feed by breaking the epidermal cells of plants and sucking their contents. These pests cause injuries to plant tissue by the penetration of their mouthparts which can cause scars and often serve as ports of entry for bacterial and fungal pathogens. These pests also contaminate the surface of their host plants with their sticky sweet excrements (honeydew) that serve as a growing substrate for fungi. The majority of plant viruses are transmitted by sucking insect vectors such as aphids, whiteflies, and thrips (Hogenhout et al. 2008). Viruses transmitted by insect pests have various modes of transmission. Non-persistent viruses must be transmitted within minutes or a few hours after acquisition (Ng and Falk 2006). Insect-borne viral diseases often cause substantial economic damage to growers of crop plants (Raccah and Fereres 2009). The sweet potato whitefly, the onion thrips and the western flower thrips are important pests of many protected crops (Cohen and Berlinger 1986; Lewis 1997) and may be considered as quarantine pests.

At temperatures between 20°C and 25°C, these insects complete their life cycle in 1–4 weeks and can build up high populations on plants. Adult pests disperse to adjacent plants by walking or by short flights. Some of the adults migrate on long flight, aided by the wind, for colonizing new areas (e.g. Byrne 1999). In some species, extreme changes in weather conditions or massive drying of host plants induce swarming behavior (e.g. Matteson et al. 1992). Short flights usually occur just above the canopy of the host plants (e.g. Byrne 1999). In open and bare fields, most pests (about 80%) are trapped within 1.0 m above the soil level (e.g. BenYakir and Chen 2008). During long flight, these small pests are often are aided by the wind, at heights of 2–30 m above ground (e.g. Trevor 1997; Reynolds and Reynolds 2009 and references therein). While moving with the wind, they can be carried over greenhouses and enter them through the roof vents (Ben-Yakir et al. 2008b). Dispersal and migration flights are usually limited to a few hours every day (Ben-Yakir and Chen 2008; Ben-Yakir et al. 2008b). Take off and flight of small sucking pests, occur only when the wind velocity is low. The long flying female migrants are probably the main colonizers of new crops.

3 Natural Sunlight in the Agricultural Environment

The electromagnetic radiation emitted by the sun is filtered through the Earth's atmosphere before reaching the Earth's surface. If the sun's radiation is not obstructed, it reaches the surface as direct light (sunlight) but if it is refracted by clouds, dust etc., it reaches the surface as diffused light (skylight). Bright sunlight provides illuminance of approximately 100,000 lx (lumens per square meter), about 1,000 W/m^2 or photosynthetic photon flux density of near 2,000 μmol photons/m^2 s, at the Earth's surface (http://en.wikipedia.org/wiki/Sunlight). In Israel, maximum daily solar illuminance range between 70,000 and 100,000 lx from March to October (Manes et al. 1970). During this period about 80% of the solar illuminance is direct light and the rest is diffused. Sunlight reaching the land part of the Earth's surface is mostly absorbed by the soil and plants and only a small fraction of it is reflected. Light reflection is either mirror-like or diffused depending on the nature of the reflecting substrate (Björn 2008). Diffused light makes difficult to delineate the image of an object from its background. When sunlight is reflected off materials that are denser than the air (water surface, glass, metal), it undergoes a change in its polarity. Thus, the reflecting object determines the color (hue and saturation), intensity and polarity of the reflected light. Bare soil or soil covered with vegetation, reflects 20% or 15% of the sunlight, respectively. Light colored soils (sand, loess) reflect more sunlight than heavy dark soils. A smooth white surface can reflects up to 70% of the sunlight. The intensity of reflection is also affected by the daily and seasonal changes in the position of the sun relative to the reflecting object.

The sunlight reaching crop plants may be augmented by reflective soil covers or reduced by cladding materials. In the Mediterranean region, crops plants often need to be protected from excessive sunlight that causes sunburn and heat stress. Heat stress is especially severe in greenhouses and it is often alleviated by using shading nets.

4 Insect Vision

An updated review of insect vision by Johansen et al. (2011) was published recently. Aphids and whiteflies have light receptors in the ultraviolet (UV) region with peak sensitivity at 330–340 nm and in the green-yellow region with peak sensitivity at

520–530 nm (Doring and Chittka 2007; Coombe 1981, 1982; Mellor et al. 1997). Using the electroretinogram technique, Kirchner et al. (2005) noted that alate female summer-migrants of the aphid *M. persicae* have additional photoreceptor in the blue-green region (490 nm). Aphid color vision is achieved by possessing two to three classes of spectral receptors that either elicit direct response or are used in an opponent mechanism to 'compare' inputs from different spectral domains (Doring and Chittka 2007 and references therein). Thrips have light receptors in the yellow region (540–570 nm), the blue region (440–450 nm) and the UV region (350–360 nm) (Vernon and Gillespie 1990). Aphids and whiteflies do not possess receptors for red light (610–700 nm) and therefore their response to red is either neutral (Mellor et al. 1997) or inhibitory (Vaishampayan et al. 1975). However, alate green spruce aphids, *Elatobium abietinum* (Walker), were caught on red sticky traps more than on yellow or white traps (Straw et al. 2011), and females of the common blossom thrips, *Frankliniella schultzei*, are attracted to red flowers and to red traps (Yaku et al. 2007).

The response of insects to light is strongly affected by the intensity of radiation, the shape and contrast of the radiation source and the physiological state of the insect. Sucking pests usually require minimal light intensity for initiating a behavioral response. Lewis (1997) reported that thrips of the temperate climate require minimal light intensity of 1,000 lx for initiating flight. In contrast, high light intensity often inhibits the expected behavioral response to an attractive color. Aphids' preference for yellow over green may be explained by the higher reflectance of yellow in the green spectral domain (Prokopy et al. 1983). Indeed, when winged *Aphis fabae* were exposed to monochromatic lights of the same intensity, they preferred green (the peak receptor sensitivity) over yellow (Hardie 1989). Yellow usually has high reflectance in the long wavelengths (green to red spectrum) and low reflectance in the short wavelengths (UV to blue spectrum). Based on that, Doring and Chittka (2007) proposed that aphids employ an opponent mechanism to differentiate between yellow and other reflective colors like white or pink. In this mechanism, a positive input from the green receptor is coupled with a negative input from the UV or blue receptor resulting in the specific attraction to yellow.

Several studies have shown that in choice experiments, insects prefer to move to environments with a higher intensity of UV light (reviewed by Diaz and Fereres 2007). On the other hand, aphids and whiteflies seemed to be repelled by high intensity UV light (Summers et al. 2004). The attraction of thrips to yellow and blue traps was reduced by increasing the UV reflection (Vernon and Gillespie 1990). The attraction of WFT to colors was negatively affected when their UV reflectance was above 35% (Matteson et al. 1992). The attraction of the psyllid *Ctenarytaina thysanura* Ferris and Klyver to yellow cards was greatly reduced by diluting the yellow with white and lowering its hue (Mensah and Madden 1992). The reported attraction of sucking pests to white traps is very variable and it is probably affected by their reflection intensity and contrast. It appears that sucking pests are attracted to white traps over a dark background when the intensity of the solar radiation is low. In contrast, these pests are repelled by white color when the intensity of the reflected sunlight is high.

Circular and cylindrical traps were significantly more attractive for thrips than other shapes with the same color and size (Vernon and Gillespie 1995; Mainali and Lim 2010). Trapping efficiency was significantly higher for small sized traps (100 cm^2) that have high perimeter length to area ratio (Carrizo 2008). High contrast between the colored trap and its background (e.g. yellow over black) further enhanced the attraction of thrips. In a strawberry greenhouse, small circular yellow sticky traps (d = 5 cm) on a black background (12 cm × 12 cm) attracted 2.3–21.0 times more WFT than the commercial rectangular yellow sticky traps (5 cm wide × 8 cm length) (Mainali and Lim 2010). Similarly, yellow circles on a black background attracted about twofolds more *B. tabaci* per unit area than ordinary rectangular yellow sticky cards (Kim and Lim 2011). In pair wise choice tests, the WFT preferred yellow artificial flower shape to yellow geometrical patterns that had a similar size (Mainali and Lim 2011). Moreover, these thrips stayed on the artificial flower about four times longer than on the geometrical patterns. High contrast between the trap and its background enhance the attraction of aphids as well. In a Brussels sprouts field, more alate aphids were caught in yellow water-traps placed over bare soil than over weeds (Smith 1976). Aphids also landed more often on plants at low density because their contrast with the background soil was higher (A'Brook 1968; Bottenberg and Irwin 1992). On the other hand, a large area covered by a uniform material with an attractive color usually does not induce landing in pests (Ben-Yakir et al. 2012).

5 Light as a Modifier of Insect Behavior

Light is an important cue for insect orientation and for finding host plants. Radiation at the UV range stimulates flight activity in sucking pests (review by Kring 1972). During flight, these pests respond strongly to visual stimuli for orientation, navigation and host finding (Antignus and Ben-Yakir 2004). When aphids terminate their flight they lose their attraction to UV light and respond to yellow-green light for landing on potential host plants (Klingauf 1987). During the landing phase aphids are strongly attracted to intense (highly saturated) yellow light (Kennedy et al. 1961; Robert 1987; Fereres et al. 1999). Aphids locate host plants using the contrast between the soil background and the color reflected from the plant foliage (Kennedy et al. 1961; Doring et al. 2004). Whiteflies use similar optical cues during flight and host finding (Coombe 1982). Yellow and green reflected light are very attractive stimulus for orientation of *B. tabaci* during flight (Isaacs et al. 1999).

When aphids land on a yellow surface they are often induced to probe it, in an attempt to feed (Moericke 1950, as cited by Doring et al. 2004). As a result, most invading aphids that land on yellow objects are "arrested" on them (Bukovinszky et al. 2005). If the yellow object is not a plant, aphids usually fly away after a period of probing in vain (Kring 1972). Thrips are attracted to land on yellow, blue and white objects (Chu et al. 2006). WFT attraction to blue traps was enhanced by adding UV emitting diodes (LEDs) (Chu et al. 2005).

The attraction to color could be enhanced by plant odor over a short distance. For example, the carrot aphid, *Cavariella aegopodii* (Scopoli), was caught more often in water traps baited with carvone, a component of host odor, than in unbaited traps (Chapman et al. 1981). In greenhouse studies, yellow water traps with anisaldehyde caught 11–15 times more female WFT than yellow traps without anisaldehyde (Teulon et al. 1999).

The physiological state of the insects often affects their visual response. During dispersal, migration and swarming behavior insects are usually attracted to light in the UV range. Once aphids terminate their aerial transport they lose their attraction to UV light and respond to visual cues coming from potential host plants (Klingauf 1987). Among females of the carrot psyllid, *Trioza apicalis* Forster, gravids were more successful than virgins in visually selecting the carrot host plant (Nissinen et al. 2008). In wind tunnel experiments with WFT, older thrips (10–13 days post-adult emergence) landed twice as often on a yellow sticky trap compare with younger thrips (2–3 days post-adult emergence) (Davidson et al. 2006). In the same study, thrips that were starved for 4 h landed ten times more often on a yellow sticky trap compared with satiated thrips. On the other hand, no differences in color preferences were found between WFT males and females and between swarming and non-swarming thrips (Matteson et al. 1992).

6 Optical Manipulation of Pests

The optical manipulation proposed in this chapter is by using reflected sunlight to interfere with host finding by sucking pests. This can be achieved by repelling, attracting and camouflaging optical cues. Repelling cues include unattractive colors and high intensity reflection (glare). Attracting cues include attractive colors and shapes that divert the pests away from the hosts. Camouflaging cues reduce the contrast between the plants and their environment or block the visual cues from the plants before they reach the insect eye. Cues for optical manipulation are reflected from materials that are placed below or above the plants. Some of these cues can be reflected from the plant itself.

Experimental evidences for optical manipulations are difficult to compare and to interpret. This is because very diverse reflective materials were used and the actual sunlight reflections during these experiments were seldom reported.

6.1 Below the Plant

Covering the soil with colored polyethylene, straw or living plants have been used successfully to protect crops from sucking pests and the viral diseases they transmit (e.g. Hiljea and Stansly 2008). Highly reflective colored polyethylene mulches such as aluminum, silver, white and yellow have been used successfully to lower the

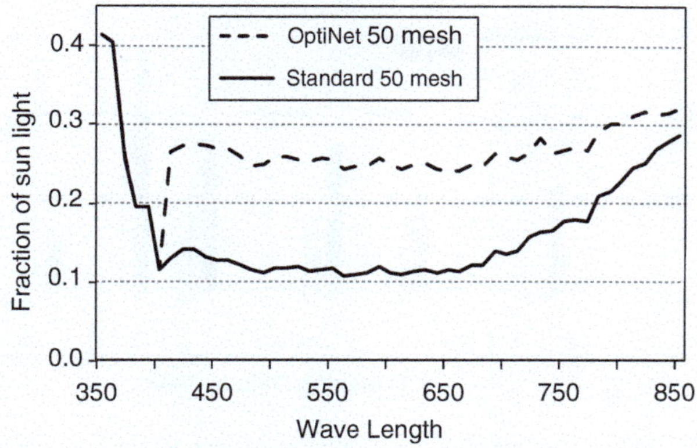

Fig. 12.1 Spectra of sunlight reflectance by the 50 mesh OptiNet® compared with a standard transparent 50 mesh net. The reflected spectra were divided by the sunlight spectrum that was measured at the same time (11:30 AM, August 5, 2005) in the Besor, Israel

infestation of sucking pests (e.g. Simmons et al. 2010). Metallic colored mulches are often referred to as "UV-reflecting" (e.g. Summers et al. 2004), however, they reflect a much wider range of the sunlight radiation. The actual color of the mulch appears to be less important than its brightness (Greer and Dole 2003). The overall effect of mulches on yield and fruit quality is dependent also on their effect on soil temperature, plant development and weed control (Csizinszky et al. 1995). When selecting a colored soil cover for reducing pests the other agronomic roles of this cover should be considered too.

As mentioned previously, sucking pests are repelled by high intensity UV and white light. The reflection level of sunlight from OptiNet® (50 mesh UV blocking net manufactured by Polysac Plastics Industries, Nir Yitzhak, Israel), at the range of 400–750 nm, is about 2.5 times greater than the reflection by standard 50 mesh net (Fig. 12.1). When we placed yellow sticky traps (10 × 10 cm) horizontally over OptiNet® used as a ground cover, they caught two- to threefolds fewer whiteflies (*B. tabaci*) compared with the same traps placed over a standard 50 mesh net (Fig. 12.2). When traps were placed over 50% OptiNet® (alternating 1 cm wide longitudinal bands with and without the UV blocking optical additives) the number of whiteflies caught was about half way between the number caught over OptiNet® and over standard net (Fig. 12.2). Thus, it appears that the intensity of light reflection by the screen, not only at the UV range, is negatively correlated with the likelihood that whitefly will land on an attractive target.

Covering the soil with yellow or green polyethylene sheets, straw or living plants probably camouflages the crop plants by reducing contrast. When using straw mulch or living plants as soil covers, it is likely that olfactory cues also play a role in modifying the insect behavior.

The visible area of the soil cover diminishes as the plants grow and their canopies cover the soil. Thus, the protective effect of reflective soil cover is limited to early growth stages or to widely spaced crop plants.

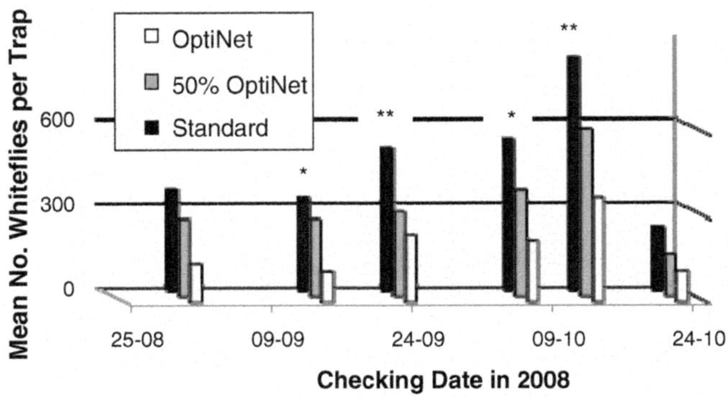

Fig. 12.2 The number of whiteflies caught on yellow sticky traps placed *horizontally* over various 50 mesh screens, Besor, 2008 (N = 3). *Bars* with * or ** *sign* over them are significantly different at the $P < 0.10$ or $P < 0.05$, respectively (ANOVA)

6.2 Above the Plant

Crop plants are often grown under protective cladding materials for improving production. In sub-tropical regions, protective plastic sheets and fine mesh nets are used mainly to physically exclude sucking pests (Berlinger et al. 2002). Covering crops with these cladding materials increases shading and reduces ventilation. The latter often results in heat stress for both crop plants and workers (Teitel 2007). Covering greenhouses with plastics or screens containing UV-blocking additives usually provides a greater protection against pests than standard cladding materials (reviewed by Antignus and Ben-Yakir 2004; Diaz and Fereres 2007; Johansen et al. 2011). The effects of other optical properties of the UV-blocking cladding materials, such as shading and reflection levels, have been ignored in most studies. Bionet® (Klayman Meteor, Petah Tikva, Israel) and OptiNet® are commercial nets containing UV-blocking additives. Bionet® provided a significant greater protection from whiteflies, *B. tabaci*, than standard net of the same density (Antignus et al. 1998). Kumar and Poehling (2006) reported that covering greenhouses with UV-blocking plastic and Bionet® significantly reduced both attraction and invasion of whiteflies (*B. tabaci*), aphids (*A. gossypii*) and thrips (*Ceratothripoides claratus*), compared to UV-transmitting materials. Growing lettuce under UV-blocking materials decreased aphid density and the spread of aphid-transmitted viruses (Legarrea et al. 2012). Ben-Yakir et al. (2008a) reported that covering walk-in tunnels with OptiNet® reduced (three- to ninefolds) thrips infestations (mainly *T. tabaci*) compare with standard net of the same density. Preliminary results indicate that covering growing tunnels with UV-blocking plastic and OptiNet® significantly reduced the invasion by the chilli thrips as well (Ben-Yakir et al. 2012). Growing tunnels covered with a 40 mesh OptiNet® screen had significantly fewer onion thrips (fivefolds) compared with tunnels covered with a standard 50 mesh screen (Fig. 12.3). Also, tunnels covered with a 30-mesh OptiNet® screen had significantly fewer whiteflies (two- to threefolds)

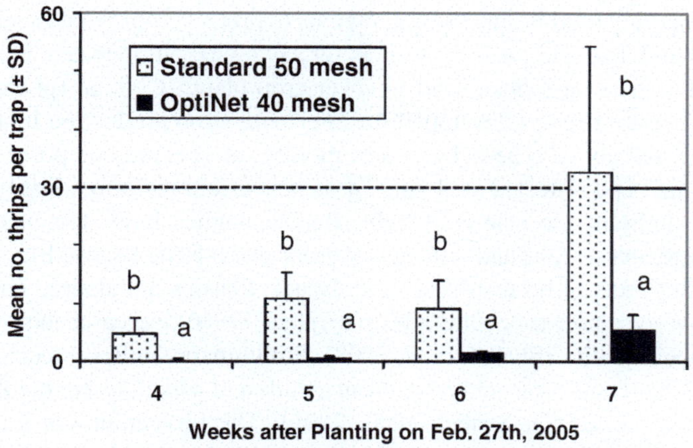

Fig. 12.3 The effect of photo-selective screen covering chives growing tunnels on the mean number of onion thrips (±SD) caught with blue sticky traps above plants. Besor. (N = 4)

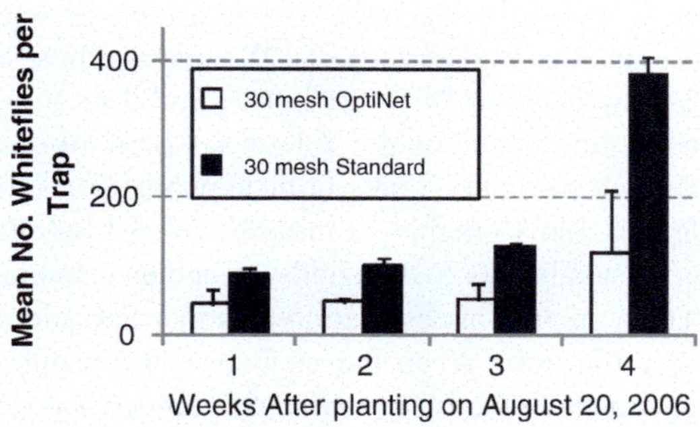

Fig. 12.4 The effect of screening with photo-selective additives on whiteflies infestation of tomato plants, Besor, 2006 (N = 4 tunnels/treatment; 2 traps/tunnel). On all weeks the bars are significantly different from each other at *P* <0.05 (*t*-test)

compared to tunnels covered with a standard 30-mesh screen (Fig. 12.4). Thus, it is possible to use these nets at a lower density than 50 meshes without increasing the risk of pests' invasion. The use of nets with larger holes is expected to improve ventilation and to reduce heat stress. Both Bionet® and OptiNet® screens, which absorbed and reflected high amount of UV radiation, provided protection against thrips, whiteflies and broad mites on pepper (Legarrea et al. 2010).

The mechanisms by which Bionet® and OptiNet® provide protection against sucking pests have not been elucidated. Many researchers attribute the protection to evidence that sucking pests prefer UV containing environment and that under low UV they disperse at a slower rate than under high UV (reviewed by Johansen et al. 2011). However, optical cues are expected to be less important to locate host plants from a short distance because pests can use other senses (olfactory, tactile) for that purpose. On the other hand, as shown in the previous section (Sect. 6.1), OptiNet® reflects high levels of incident sunlight (about 30%) which deter pests landing. This mechanism has been overlooked and its role in the protection that Bionet® and OptiNet® provide needs to be further investigated.

Coarse nets are used in sub-tropical regions to protect crops from excessive solar radiation, wind, hail and birds, as well as for saving irrigation water. Traditionally, black shading nets have been used to cover crop plants. Colored (photoselective) shading nets are currently developed for improving crop production in addition to their roles listed above. The colored nets modify the spectral composition of both the transmitted and reflected sunlight. These nets also transform a large portion of the direct sunlight into scattered light. Recent studies have demonstrated that growing vegetables, fruits and ornamental crops under Red, Yellow, Blue, Grey and Pearl shading nets (ChromatiNets™, Polysack Plastics Industries, Nir-Yitzhak, Israel, http://www.polysack.com/index.php?page_id=46) increases their yields and improves their quality (Shahak et al. 2008). Preliminary studies indicated that the Yellow and Pearl nets protected crops from aphids and whiteflies but not from thrips (Ben-Yakir et al. 2008a; Shahak et al. 2009). The protection from aphids and whiteflies and the viral diseases that they transmit to vegetable crops was studied from 2006 to 2010 (Ben-Yakir et al. 2012). These studies were conducted in the semi-arid, Besor region, in southern Israel. The plants were grown in 'walk-in' tunnels (6×6×2.5 m) that were covered by various colored nets with 35% shading capacity. These nets have large holes that permit free passage of sucking pests that are only 1–2 mm in length. The average hole size for the Black, Red, Pearl and Yellow nets are 7×9, 5×7, 4×7 and 4×6 mm, respectively. We also found that whiteflies landed on the Yellow net 20–40 times more often than on the other nets (Ben-Yakir et al. 2008a; Offir unpublished). Despite that, the infestation levels of aphids and whiteflies in tunnels covered by either the Yellow or Pearl nets were consistently two- to threefolds lower than in tunnels covered by the Black or Red nets. The reduction in pests led to a similar reduction in the incidences of viral diseases they transmit. When the incidence of cucumber mosaic virus (CMV) in pepper grown under the Black or Red nets ranged between 35% and 89%, they were two- to tenfolds lower under the Yellow or Pearl nets. Similarly, when the incidence of the necrotic strain of potato virus Y (PVY) in tomato grown under Black or Red nets ranged between 42% and 50%, they were two- to threefolds lower under the Yellow or Pearl nets (Fig. 12.5). Also, when the incidence of tomato yellow leaf curl virus (TYLCV) in tomato grown under the Black or Red nets ranged between 15% and 50%, they were two- to fourfolds lower under the Yellow or Pearl nets.

The mechanisms by which Yellow and Pearl nets provide protection against aphids and whiteflies are not known. We propose that the optical properties of these nets play a major role in this protection. The sunlight transmission and scattering characteristics of these nets were reported by Shahak et al. (2004) and Rajapakse and Shahak (2007). The sunlight reflections of these nets are described in Fig. 12.6. Covering crops with shading nets may interfere with the ability of flying pests to see the host plants under the nets, and to discern the plants from their background. Since the threads of the light colored nets are more translucent than the black threads, light colored nets have higher density of threads than Black nets of the same shading capacity. Therefore, the light color nets probably block the view and hide the plants to a greater extent than the Black net. However, the Red net did not provide any protection from pests although its threads density is about twice as high as the

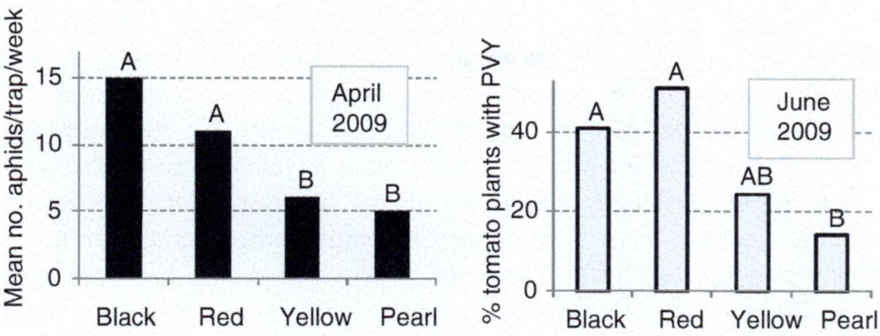

Fig. 12.5 The effect of colored shading nets on aphids' infestation and the rate of tomato plants with symptoms of PVY disease, Besor (N = 4 tunnels / net; 2 traps / tunnel; 70 plants / tunnel)

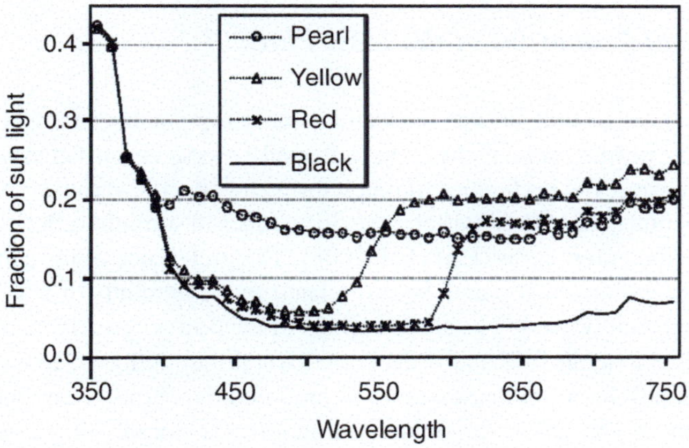

Fig. 12.6 Sunlight reflectance from colored 35% shading nets, Besor, 11:30 AM, December 10, 2007

Black nets. Therefore, hiding the crop plants does not seem to be a very important mechanism in the protection provided by the Yellow and Pearl shading nets. High reflection of sunlight deters the landing of both aphids and whiteflies (see Sect. 5). The reflection of sunlight in the range of 400–600 nm from Pearl nets is two- to fivefolds higher than that of the Black or Red nets (Fig. 12.6). Thus, the Pearl shading net can protect from pests by repelling them with its high glaring reflection. Yellow colored surfaces induce aphids and whiteflies to land, feed and settle (see Sect. 5). After the pests try to probe and feed in vain on the yellow plastic folia of the net, they usually fly away in what is termed a 'rejection flight' (Kring 1972). Thus, the Yellow shading net can protect from pests by attracting them away from the plants, delaying their entry to the growing area and, in turn, inducing them to fly away.

Similar protection from aphids and aphid-borne viral diseases was observed by Cohen (1981) in sweet pepper grown under coarse white, light grey or yellow nets in comparison to uncovered plants. The optical mechanisms that Cohen proposed included: (1) Interfering with the ability of pests to discern plants from their background, (2) Deterring landing by light colored nets, (3) Attraction of aphids away from host plants by the yellow net.

Floating crop covers are light weight synthetic fabrics that are placed over the plant beds after seeding for providing agronomic advantages to the plants grown under them. These covers can also protect plants growing under them from aphids and whiteflies and the viral diseases they transmit (Perring et al. 1989; Cradock et al. 2002; Qureshi et al. 2007). These covers provide physical barrier for pests but as they are usually colored white they probably also hide the plants and are highly reflective. Therefore, it is likely that the optical properties of the floating crop covers can contribute to their protection from sucking pests.

6.3 *Optical Properties of the Host Plant*

Characteristics of natural sunlight reflection from crop plants can affect the risk of infestation by pests. These include the reflected colors, visual patterns, contrasts and light intensity. The preference of onion thrips for specific varieties of white cabbage is determined, at least in part, by the differences in sunlight reflection between the head and the outer leaves (Fail et al. 2008). The preference of the cabbage seedpod weevil to various host plants is also related to the amounts of UV and yellow reflected from their flowers. The attractiveness of the flowers greatly increased when they reflected moderate UV and it decreased when they reflected low or high UV (Tansey et al. 2010). Visual assessment of onion cultivars indicated that those that were resistant to the onion thrips had yellow-green-colored foliage, whereas the susceptible cultivars had blue-green-colored foliage (Diaz-Montano et al. 2010). In curcubits, plants that have high pubescence that causes silvery reflection had a partial protection from aphids and aphid-transmitted viral diseases (Davis and Shifriss 1983). High reflection from crop plants for deterring pests can be produced artificially by spraying with highly reflective white kaolin-based particle film (e.g. Tsuchiya et al. 1995).

6.4 *Elsewhere in the Growing Environment*

Selectively modified light in the growing environment can disturb host finding by pests. An environment with low UV is not favored by sucking pests and it hinders their dispersal (see Sect. 5). Scattering and diffusion of light as well as enrichment of specific colors can reduce the contrast between host plants and their background. The Yellow and Pearl shading nets (described in Sect. 6.2) enrich the light passing through them with scattered and diffused light (Shahak et al. 2004). This may have also contributed to the protection that they provided against aphids and whiteflies.

Sticky boards and sheets with attractive colors are often placed near crop plants to divert pests away from the crop and to lower pest population by mass trapping. For example, in lettuce, mass trapping of the onion thrips and WFT with blue sticky cards provided significant protection from these pests (Natwick et al. 2007). Enhancing the reflection of colored traps with attractive light-emitting diodes (LED) has been demonstrated in several studies. Blue LEDs (peak emission at 465 nm) increased the trapping of WFT on blue sticky cards (Chen et al. 2004a). Yellow sticky card traps equipped with 530-nm lime green LED caught more whiteflies and leafhoppers (Chen et al. 2004b).

7 Future Research and Development

So far, optical manipulation of pests has been an unintentional byproduct of materials and methods that were developed for improving some aspects of plant production. We propose that optical manipulation of pests needs to be pursued as an independent topic for research and development.

Much information has been published about the visual response of aphids, whiteflies and thrips (see Sects. 4 and 5). However, more studies of pests' response to visual cues during migration, dispersal and host finding, in various agricultural environments, are required.

The optical manipulation proposed in this chapter is based on the use of reflected sunlight to interfere with host finding by sucking pests. This has been achieved already by using materials that are highly reflective, or materials that have attractive and camouflaging colors (see Sect. 6). However, currently the highly reflective materials used for covering crop plants are not selective enough and they block a significant amount of all the sunlight radiation. For example, Polyethylene sheets and nets containing the widely used UV blocking white pigment titanium dioxide are highly reflective (see OptiNet® reflection in Fig. 12.1). Therefore, currently used reflective covers increase shading and hinder plant development. Increased shading is particularly damaging for crops that are planted during the spring and fall. In those seasons, the intensity of sunlight is relatively low and the risk for infestation by sucking pests is very high. In the eastern Mediterranean, aphids and thrips are mostly abundant in the spring, and whiteflies are mostly abundant in the fall. Also, during the spring and fall plants are young and most susceptible to the viral diseases transmitted by sucking pests.

Plants mainly use the photosynthetically active radiation (PAR; ranging between 400 and 700 nm) of the sunlight. In general, sucking pests are most sensitive to radiation in the UV (330–350 nm) and in the green-yellow (520–550 nm) (see Sect. 4). Thus, covering material for optically manipulating pests should contain selective additives that let most of the PAR pass through and highly reflect the wavelengths that sucking pest can detect. The development of such selective additives will be a major advancement toward optically manipulating pests. Alternately, the highly reflective additives may not be distributed throughout the entire cladding materials but rather they will be limited to a few regions. For example, as a grid of reflective colored bands. Attractive colored materials may also be formed into attractive shapes over contrasting background and affixed on top of the cladding materials (e.g. yellow circles with a black ring around it). When the UV and the green-yellow portions of the sunlight are omitted from the growing environment it may negatively affect the performance of beneficial insects and mites that serve as natural enemies and pollinators. Here too, if the blocking additives or materials will be limited to a few regions of the covers it will alleviate the negative effect on the beneficials. To maximize the efficacy of optical cues, the reflective materials should face the sun at the peak time of pests' flight activity. Overall, covers designed for optical manipulation should be tailored to fit the specific crop, the major insect vector and the beneficial arthropods that are involved.

Inside protecting structures used for growing plants structural elements, boards or sheets, with attractive colors can be used for optical manipulation. Because these structures are densely packed with plants, non-visual senses can also be used by the pests to find their hosts. In protected crops, attractive optical cues may be enhanced by combining them with attractive odors, arresting glues or insecticides (attract and kill). Artificial lights may also be use to augment or to substitute the reflection of natural sunlight. Johansen et al. (2011) suggested using artificial light to attract and disrupt host-finding against whiteflies. In Japan they currently have a national research project entitled 'Elucidation of biological mechanisms of photo response and development of advanced technologies utilizing light'. Sucking pests like thrips and whitefly are being studied within the frame of this project. A team led by Dr. Masui, of the Shizuoka Research Institute of Agriculture and Forestry, is studying the effects of a single wavelength and mixed radiations on the behavior of *Thrips palmi* under laboratory and greenhouse conditions (Masui S personal communication).

Delay of sucking pests by arresting optical cues can be especially effective in protecting against stylet borne viruses such as CMV and PVY. These viruses must be transmitted within a short time (minutes to a few hours) after the aphids acquire them. Therefore, any delay of the infected aphids on arresting surfaces is expected to reduce the efficacy of viral transmission.

Some insects detect polarized light and are either attracted or deterred by certain types of polarization (Horváth and Varju 2004). The effects of polarized light on sucking pests need to be studied. Some plastics and glasses that are currently used for crop production change the polarity of sunlight and reflect polarized light to various degrees. This quality may also play a role in the optical manipulation of sucking pests.

8 Concluding Remarks

Manipulation of pests with optically modified cladding materials has been suggested by several authors (e.g. Doring and Chittka 2007; Antignus 2000). Pests may also be optically manipulated inside a greenhouse, using natural or artificial light, directly

(Reviewed by Johansen et al. 2011) or indirectly by affecting their host plants (Reviewed by Vanninen et al. 2010).

Reflecting natural sunlight to interfere with host finding by sucking pests has already been used in the form of reflective mulches. High reflection may be responsible in part for the protection from whiteflies and thrips by UV absorbing nets (see Sect. 6.1). Our recent studies show that pearl and yellow colored nets can also reduce infestations by aphids and whiteflies. Developing cladding materials that optically repel or arrest sucking pests is likely to be an effective strategy for plant protection. This technology could improve both crop production and pest management at the same time (Shahak et al. 2009).

The sunlight changes on seasonal and daily levels. Therefore, optical manipulation that is based on reflected sunlight may not very reliable method everywhere. In the Mediterranean region sucking pests are mainly active in open areas from March to October and during that period most days are sunny and clear. The reflective effect can be maximized if the reflective materials face the sun at the peak time of pests' flight activity.

It is unlikely that optical manipulation by itself will give sufficient protection for commercial crop production. Therefore, this technology should be integrated with other physical and chemical pest control methods. Optical manipulation can also be combined with varieties of crop plant that are less susceptible to viral diseases. The extra protection expected by the optical manipulation is likely to lower the infestation of sucking pests and reduce the viral diseases they transmit. This technology can help reducing the use of insecticides, which in turn, will slow down the development of insecticide resistance in whiteflies and thrips populations. Elucidating the mechanisms responsible for the plant protection by optical cues is likely to lead to the development of cladding materials that will provide a greater protection from sucking pests. The newly developed optically active materials must be compatible with optimal growing conditions. Materials and objects that optically repel or arrest pests can be important components of integrated pest management for both open field and protected crops.

References

A'Brook J (1968) The effect of plant spacing on the numbers of aphids trapped over the groundnut crop. Ann Appl Biol 61:289–294

Antignus Y (2000) Manipulation of wavelength-dependent behavior of insects: an IPM tool to impede insects and restrict epidemics of insect-borne viruses. Virus Res 71:213–220

Antignus Y, Ben-Yakir D (2004) Ultraviolet-absorbing barriers, an efficient integrated pest management tool to protect greenhouses from insects and virus diseases. In: Horowitz AR, Ishaaya I (eds) Insect pest management. Springer, Berlin, pp 319–335

Antignus Y, Lapidot M, Hadar D, Messika Y, Cohen S (1998) Ultraviolet-absorbing screens serve as optical barriers to protect crops from virus and insect pests. J Econ Entomol 91:1401–1405

Ben-Yakir D, Chen M (2008) Studies of thrips migratory flights in Israel. Acta Phytopathol Entomol Hung 43:243–248

Ben-Yakir D, Hadar MD, Offir Y, Chen M, Tregerman M (2008a) Protecting crops from pests using OptiNet (R) screens and ChromatiNet (R) shading nets. Acta Hortic 770:205–212

Ben-Yakir D, Teitel M, Tanny J, Chen M, Barak M (2008b) Optimizing ventilation of protected crops while minimizing invasion by whiteflies and thrips. Acta Hortic 797:217–222
Ben-Yakir D, Antignus Y, Offir Y, Shahak Y (2012) Colored shading nets impede insect invasion and decrease the incidences of insect born-transmitted viral diseases in vegetable crops. Entomol Exp Appl (in press)
Berlinger MJ, Taylor RAJ, Lebiush-Mordechi S, Shalhevet S, Spharim I (2002) Efficiency of insect exclusion screens for preventing whitefly transmission of tomato yellow leaf curl virus of tomatoes in Israel. Bull Entomol Res 92:367–373
Björn LO (2008) The nature of light and its interaction with matter. In: Björn LO (ed) Photobiology: the science of light and life. Springer, New York, pp 1–39
Bottenberg H, Irwin ME (1992) Canopy structure in soybean monocultures and soybean-sorghum mixtures: impact on aphid (Homoptera: Aphididae) landing rates. Environ Entomol 21:542–548
Bukovinszky T, Potting RPJ, Clough Y, Van Lenteren JC, Vet LEM (2005) The role of pre- and post- alighting detection mechanisms in the responses to patch size by specialist herbivores. Oikos 109:435–446
Byrne DN (1999) Migration and dispersal by the sweet potato whitefly, *Bemisia tabaci*. Agric For Meteorol 97:309–316
Carrizo P (2008) Effects of yellow trap size on sampling efficiency for western flower thrips (*Frankliniella occidentalis*) from pepper plants (*Capsicum annum*) grown in greenhouses. Cienc Investig Agrar 35:155–160
Chapman RF, Bernays EA, Simpson SJ (1981) Attraction and repulsion of the aphid, *Cavariella aegopodii*, by plant odors. J Chem Ecol 7:881–888
Chen TY, Chu CC, Fitzgerald G, Natwick ET, Henneberry TJ (2004a) Trap evaluations for thrips (Thysanoptera: Thripidae) and hoverflies (Diptera: Syrphidae). Environ Entomol 33:1416–1420
Chen TY, Chu CC, Henneberry TJ, Umeda K (2004b) Monitoring and trapping insects on poinsettia with yellow sticky card traps equipped with light-emitting diodes. Horttechnol 14:337–341
Chu CC, Chen TY, Natwick ET, Fitzgerald G, Tuck S, Alexander P, Henneberry TJ (2005) Light response by *Frankliniella occidentalis* to white fluorescent light filtered through color films and ultraviolet- and blue light-emitting diodes. SW Entomol 30:149–154
Chu CC, Ciomperlik MA, Chang NT, Richards M, Henneberry TJ (2006) Developing and evaluating traps for monitoring *Scirtothrips dorsalis* (Thysanoptera: Thripidae). Fla Entomol 89(1):47–55
Cohen S (1981) Reducing the spread of aphid-transmitted viruses in peppers by coarse-net cover. Phytoparasitica 9:69–76
Cohen S, Berlinger MJ (1986) Transmission and cultural control of whitefly-borne viruses. Agric Ecosyst Environ 17:89–97
Coombe PE (1981) Wavelength specific behaviour of the whitefly *Trialeurodes vaporariorum* (Homoptera: Aleyrodidae). J Comp Physiol 144:83–90
Coombe PE (1982) Visual behaviour of the greenhouse whitefly, *Trialeurodes vaporariorum*. Physiol Entomol 7:243–251
Cradock KR, da Graca JV, Laing MD (2002) Studies on the cultural control of virus diseases in zucchini crops. S Afr J Sci 98:225–227
Csizinszky AA, Schuster DJ, Kring JB (1995) Color mulches influence yield and insect pest populations in tomatoes. J Am Soc Hortic Sci 120:778–784
Davidson MM, Butler RC, Teulon DAJ (2006) Starvation period and age affect the response of female *Frankliniella occidentalis* (Pergande) (Thysanoptera: Thripidae) to odor and visual cues. J Insect Physiol 52:729–736
Davis RF, Shifriss O (1983) Natural virus infection in silvery and non-silvery lines of *Cucurbita pepo* L. Plant Dis 67:379–380
Diaz BM, Fereres A (2007) Ultraviolet-blocking materials as a physical barrier to control insect pests and plant pathogens in protected crops. Pest Technol 1:85–95
Diaz-Montano J, Fuchs M, Nault BA, Shelton AM (2010) Evaluation of onion cultivars for resistance to onion thrips (Thysanoptera: Thripidae) and iris yellow spot virus. J Econ Entomol 103:925–937

Doring TF, Chittka L (2007) Visual ecology of aphids-a critical review on the role of colours in host finding. Arthropod-Plant Interact 1:3–16

Doring TF, Kirchner SM, Kuhne S, Saucke H (2004) Response of alate aphids to green targets on coloured backgrounds. Entomol Exp Appl 113:53–61

Fail J, Zana J, Pénzes B (2008) The role of plant characteristics in the resistance of white cabbage to onion thrips: preliminary results. Acta Phytopathol Entomol Hung 43:267–275

Fereres A, Kampmeier GE, Irwin ME (1999) Aphid attraction and preference for soybean and pepper plants infected with potyviridae. Ann Entomol Soc Am 92:542–548

Greer L, Dole JM (2003) Aluminum foil, aluminium-painted plastic, and degradable mulches increase yields and decrease insect-vectored viral diseases of vegetables. Hort Technol 13:276–284

Hardie J (1989) Spectral specificity for targeted flight in the black bean aphid, *Aphis fabae*. J Insect Physiol 35:619–626

Hiljea L, Stansly PA (2008) Living ground covers for management of *Bemisia tabaci* (Gennadius) (Homoptera: Aleyrodidae) and tomato yellow mottle virus (ToYMoV) in Costa Rica. Crop Prot 27:10–16

Hogenhout SA, Ammar ED, Whitfield AE, Redinbaugh MG (2008) Insect vector interactions with persistently transmitted viruses. Annu Rev Phytopathol 46:327–359

Horváth G, Varju D (2004) Polarized light in animal vision: polarization patterns in nature. Springer, Berlin

Isaacs R, Willis MA, Byrne DN (1999) Modulation of whitefly take-off and flight orientation by wind speed and visual cues. Physiol Entomol 24:311–318

Johansen NS, Vanninen I, Pinto DM, Nissinen AI, Shipp L (2011) In the light of new greenhouse technologies: 2. Direct effects of artificial lighting on arthropods and integrated pest management in greenhouse crops. Ann Appl Biol 159:1–27

Kennedy JS, Booth CO, Kershaw WJS (1961) Host finding by aphids in the field. Ann Appl Biol 49:1–21

Kim S, Lim UT (2011) Evaluation of a modified sticky card to attract *Bemisia tabaci* (Hemiptera: Aleyrodidae) and a behavioural study on their visual response. Crop Prot 30:508–511

Kirchner SM, Doring TF, Saucke H (2005) Evidence for trichromacy in the green peach aphid, *Myzus persicae* (Sulz.) (Hemiptera: Aphididae). J Insect Physiol 51:1255–1260

Klingauf FA (1987) Host plant finding and acceptance. In: Minks AK, Harrewijn P (eds) Aphids: their biology, natural enemies and control, vol A. Elsevier, Amsterdam, pp 209–223

Kring JB (1972) Flight behavior of aphids. Annu Rev Entomol 17:461–492

Kumar P, Poehling HM (2006) UV-blocking plastic films and nets influence vectors and virus transmission on greenhouse tomatoes in the humid tropics. Environ Entomol 35:1069–1082

Legarrea S, Karnieli A, Fereres A, Weintraub PG (2010) Comparison of UV-absorbing nets in pepper crops: spectral properties, effects on plants and pest control. Photochem Photobiol 86:324–330

Legarrea S, Betancourt M, Plaza M, Fraile A, García-Arenal F, Fereres A (2012) Dynamics of nonpersistent aphid-borne viruses in lettuce crops covered with UV-absorbing nets. Virus Res 165:1–8

Lewis T (1997) Flight and dispersal. In: Lewis T (ed) Thrips as crop pests. CAB International, Wallingford, pp 175–196

Mainali BP, Lim UT (2010) Circular yellow sticky trap with black background enhances attraction of *Frankliniella occidentalis* (Pergande) (Thysanoptera: Thripidae). Appl Entomol Zoolog 45:207–213

Mainali BP, Lim UT (2011) Behavioral response of Western flower thrips to visual and olfactory cues. J Insect Behav 24(6):436–446. doi:10.1007/s10905-011-9267-7

Manes A, Tietelman A, Fruehling I (1970) Solar radiation and radiation balance. Ministry of Transport Publications, Bet Dagan

Matteson N, Terry I, Ascolichristensen A, Gilbert C (1992) Spectral efficiency of the western flower thrips, *Frankliniella occidentalis*. J Insect Physiol 38:453–459

Mellor HE, Bellingham J, Anderson M (1997) Spectral efficiency of the glasshouse whitefly *Trialeurodes vaporariorum* and *Encarsia formosa* its hymenopteran parasitoid. Entomol Exp Appl 83:11–20

Mensah RK, Madden JL (1992) Field studies on color preferences of *Ctenarytaina thysanura* in Tasmanian *Boronia* farms. Entomol Exp Appl 64:111–115

Natwick ET, Byers JA, Chu CC, Lopez M, Henneberry TJ (2007) Early detection and mass trapping of *Frankliniella occidentalis* and *thrips tabaci* in vegetable crops. SW Entomol 32:229–238

Ng JCK, Falk BW (2006) Virus-vector interactions mediating nonpersistent and semipersistent transmission of plant viruses. Annu Rev Phytopathol 44:183–212

Nissinen A, Kristoffersen L, Anderbrant O (2008) Physiological state of female and light intensity affect the host-plant selection of carrot psyllid, *Trioza apicalis* (Hemiptera: Triozidae). Eur J Entomol 105:227–232

Perring TM, Royalty RN, Farrar CA (1989) Floating row covers for the exclusion of virus vectors and the effect on disease incidence and yield of cantaloupe. J Econ Entomol 82:1709–1715

Prokopy RJ, Collier RH, Finch S (1983) Leaf color used by cabbage root flies to distinguish among host plants. Science 221:190–192

Qureshi MS, Midmore DJ, Syeda SS, Playford CL (2007) Floating row covers and pyriproxyfen help control silverleaf whitefly *Bemisia tabaci* (Gennadius) Biotype B (Homoptera: Aleyrodidae) in zucchini. Austr J Entomol 46:313–319

Raccah B, Fereres A (2009) Plant virus transmission by insects. In: Encyclopedia of life sciences. Wiley, Chichester

Rajapakse NC, Shahak Y (2007) Light quality manipulation by horticulture industry. In: Whitelam GC, Halliday KJ (eds) Light and plant development. Blackwell Publishing, Oxford, pp 290–312

Reynolds AM, Reynolds DR (2009) Aphid aerial density profiles are consistent with turbulent advection amplifying flight behaviours: abandoning the epithet 'passive'. Proc R Soc B-Biol Sci 276:137–143

Robert Y (1987) Aphids and their environment. In: Minks AK, Harrewijn P (eds) Aphids: their biology, natural enemies, and control, vol A. Elsevier, Amsterdam, pp 299–313

Shahak Y, Gussakovsky EE, Cohen Y, Lurie S, Stern R, Kfir S, Naor A, Atzmon I, Doron I (2004) ColorNets: a new approach for light manipulation in fruit trees. Acta Hortic 636:609–616

Shahak Y, Gal E, Offir Y, Ben-Yakir D (2008) Photoselective shade netting integrated with greenhouse technologies for improved performance of vegetable and ornamental crops. Acta Hortic 797:75–80

Shahak Y, Ratner K, Zur N, Offir Y, Matan E, Yehezkel H, Messika Y, Posalski I, Ben-Yakir D (2009) Photoselective netting: an emerging approach in protected agriculture. Acta Hortic 807:79–84

Simmons AM, Kousik CS, Levi A (2010) Combining reflective mulch and host plant resistance for sweet potato whitefly (Hemiptera: Aleyrodidae) management in watermelon. Crop Prot 29:898–902

Smith JG (1976) Influence of crop background on aphids and other phytophagous insects on Brussels sprouts. Ann Appl Biol 83:1–13

Straw NA, Williams DT, Green G (2011) Influence of sticky trap color and height above ground on capture of alate *Elatobium abietinum* (Hemiptera: Aphididae) in Sitka spruce plantations. Environ Entomol 40:120–125

Summers CG, Mitchell JP, Stapleton JJ (2004) Management of aphid-borne viruses and *Bemisia argentifolii* (Homoptera: Aleyrodidae) in zucchini squash by using UV reflective plastic and wheat straw mulches. Environ Entomol 33:1447–1457

Tansey JA, Dosdall LM, Keddie BA, Noble SD (2010) Contributions of visual cues to cabbage seedpod weevil, *Ceutorhynchus obstrictus* (Marsham) (Coleoptera: Curculionidae), resistance in novel host genotypes. Crop Prot 29:476–481

Teitel M (2007) The effect of screened openings on greenhouse microclimate. Agric For Meteorol 143:159–175

Teulon DAJ, Hollister B, Butler RC, Cameron EA (1999) Colour and odour responses of flying western flower thrips: wind tunnel and greenhouse experiments. Entomol Exp Appl 93:9–19

Trevor L (1997) Flight and dispersal. In: Lewis T (ed) Thrips as crop pests. CAB International, New York, pp 175–192

Tsuchiya M, Masui S, Kuboyama N (1995) Reduction of population-density of yellow tea thrips (*Scirtothrips dorsalis* Hood) on mandarin orange (*Citrus unshiu* Marc.) trees by application of white solution with or without reflective sheet mulching. Jpn J Appl Entomol Zool 39:305–312

Vaishampayan SM, Waldbauer GP, Kogan M (1975) Visual and olfactory responses in orientation to plants by the greenhouse whitefly, *Trialeurodes vaporariorum* (Homoptera: Aleyrodidae). Entomol Exp Appl 18:412–422

Vanninen I, Pinto DM, Nissinen AI, Johansen NS, Shipp L (2010) In the light of new greenhouse technologies: 1. Plant-mediated effects of artificial lighting on arthropods and tritrophic interactions. Ann Appl Biol 157:393–414

Vernon RS, Gillespie DR (1990) Spectral responsiveness of *Frankliniella occidentalis* (Thysanoptera, Thripidae) determined by trap catches in greenhouses. Environ Entomol 19:1229–1241

Vernon RS, Gillespie DR (1995) Influence of trap shape, size, and background color on captures of *Frankliniella occidentalis* (Thysanoptera, Thripidae) in a cucumber greenhouse. J Econ Entomol 88:288–293

Yaku A, Walter GH, Najar-Rodriguez AJ (2007) Thrips see red flower colour and the host relationships of a polyphagous anthophilic thrips. Ecol Entomol 32:527–535

Chapter 13
Recent Progress in Bed Bug Management

Kenneth F. Haynes and Michael F. Potter

1 The Status of the Bed Bug Resurgence

In North America, Europe, Australia and other parts of the developed world bed bugs, *Cimex lectularius* L. have reemerged as an urban insect pest after 50 years in the background (Potter et al. 2010b). Pest management professionals (PMPs) rarely encountered bed bugs just a decade ago, but now over 90% of PMPs in the U.S. have dealt with them, and consider bed bugs to be the most difficult urban pest to control. The near disappearance of bed bugs may be attributed primarily to the widespread use of DDT and other persistent insecticides in the period from the late 1940s to the 1970s (Potter 2011). There is no definitive explanation for their resurgence, but increasing world-wide travel, an increasingly mobile population within countries, changes in pest management procedures for other urban pests, lack of vigilance, and resistance to pyrethroid insecticides are possible explanations.

2 The Role of Insecticide Resistance in the Resurgence

In 2005 we collected *C. lectularius* from an apartment in Cincinnati, Ohio, USA (Romero et al. 2007). Some of these insects were resting on surfaces of furniture known to have been treated with a pyrethroid insecticide. After rearing these in the laboratory we conducted a dose–response residual assay with two pyrethroid insecticides (deltamethrin and λ-cyhalothrin). The assay results indicated that insects from this colony were thousands of fold resistant relative to a long-maintained

K.F. Haynes (✉) • M.F. Potter
Department of Entomology, College of Agriculture, University of Kentucky, Lexington, KY 40546, USA
e-mail: khaynes@uky.edu; mpotter@uky.edu

I. Ishaaya et al. (eds.), *Advanced Technologies for Managing Insect Pests*,
DOI 10.1007/978-94-007-4497-4_13,

laboratory colony (originating from Fort Dix, New Jersey and maintained by Harold Harlan for over 37 years) (Bartley and Harlan 1974). Other independently collected bed bugs originating in dwellings in California, Florida, Kentucky, Ohio, and Virginia were also resistant to deltamethrin. One colony from Los Angeles, California was susceptible as measured by a discriminating dose assay. Subsequently, we evaluated bed bugs sent to us from Massachusetts, Michigan, New Jersey, and New York. Only the group from New Jersey showed an intermediate level of resistance; the others were highly resistant. Pyrethroid resistance has been found throughout the world in both *C. lectularius* (Tawatsin et al. 2011; Boase et al. 2006; Kilpinen et al. 2008; Yoon et al. 2008; Zhu et al. 2010) and *Cimex hemipterus* (the tropical bed bug (Tawatsin et al. 2011; Karunaratne et al. 2007; Myamba et al. 2002)).

Yoon et al. (2008) established that two knockdown resistance (*kdr*) mutations were associated with deltamethrin resistance in a population of bed bugs from New York City. Zhu et al. (2010) found *kdr* mutations in 88% of populations sampled throughout the United States (over 100 populations). The original population that we collected from Cincinnati, Ohio did not have a *kdr* mutation; therefore, resistance mechanisms other than target site mutations were implicated. The toxicity of deltamethrin was enhanced by piperonyl butoxide (PBO) suggesting that P450-mediated detoxification was partially responsible for resistance. This hypothesis has gained further support by the reduction of deltamethrin toxicity by dsRNA prepared against the gene coding for *C. lectularius* NADPH-cytochrome P450 oxidoreductase. This enzyme is a conserved partner for diverse cytochrome P450s. Knockdown of this enzyme did not affect susceptibility in a susceptible strain. Because PBO synergism only accounted for a 40-fold increase in toxicity of deltamethrin, additional mechanisms of resistance are involved (in addition to *kdr* and P450).

The widespread distribution of pyrethroid resistance and the occurrence of at least two mechanisms of resistance is not a hopeful sign for reliance on pyrethroids alone. Already the industry is appropriately switching to different insecticide formulations with additional or new modes of action, and to non-insecticidal approaches.

3 Alternative Methods of Control

One effective and relatively safe, but expensive method of controlling bed bugs is to raise the temperature in infested areas. Bed bugs are well adapted to survive for long periods without access to a host, which is their source of water. It is postulated that their aggregation behavior is a mechanism that increases the humidity in the microhabitat. High temperatures are lethal to all stages of *C. lectularius* and *C. hemipterus*. Under laboratory conditions, How and Lee (2010) demonstrated that the survival time for egg, first instar and adult *C. hemipterus* was less than 1 day at 39°C, 40°C and 40°C respectively at 75% RH. Egg hatchability was 0% in *C. lectularius* at 37°C (Johnson 1941). Prior exposure of bed bugs to 30°C for 2 weeks or to 37°C for 1 h did not influence survival at high temperatures, suggesting that acclimation to high temperatures is not a concern. In dwellings the potential niches for bed bugs are varied so that raising temperature to kill bed bugs is much more complicated than simply raising average ambient temperature in a room, and therefore necessitates careful monitoring of various microhabitats where bed bugs are found, and maintaining lethal temperatures for long enough to ensure mortality. Pereira et al. (2009) found a linear relationship between temperature and Log time in minutes for 100% mortality (100 min for 100% mortality at 41°C; 10 min for 100% mortality at 45°C. etc.). Benoit et al. (2009a) demonstrated ca. 75–80% mortality at 1 h exposure to 46°C, with near 100% mortality at 48°C. There is a predictable relationship between temperature and duration of exposure for mortality. Under practical circumstances it would be necessary to maintain lethal temperature at a prescribed temperature in the most difficult to reach spots in a room. To achieve a temperature increase over normal room temperature requires the production of heat in proportion to the volume of the space. Typically this requires the use of specialized heating equipment in human dwellings. However, particular items of furniture could be enclosed in much smaller spaces and therefore require less expensive heating equipment, time, and energy consumption. Precautions must be taken during the heating process to ensure that household items are not damaged by excessive temperatures, and that bed bugs do not seek refuge in adjacent areas that are unheated. Heating a dwelling or even a single room to lethal temperature is energy intensive and therefore expensive (with costs proportional to the volume of the space to be heated). More modest heating to 36°C leads to a reduced reproductive ability and a loss of symbiotic bacteria (Chang 1974), but this exposure is sublethal and would need to be prolonged.

Clothing and bed clothes can be rendered bed bug free in a clothes dryer (high temperature setting, for 30 min) (Naylor and Boase 2010). Potter et al. (2007) have reported that less time is required at a high setting, but they also achieved much higher temperatures at this setting (82°C vs ca. 44°C). Because dryers vary in their temperature setting and some have automatic shutoff features, care must be taken when using this approach. A common recommendation is to leave potentially infested items in a closed vehicle (no air conditioning or air circulation) exposed to direct sunlight. Temperature in a vehicle is influenced by the external ambient temperature and the duration of exposure to solar radiation (McLaren et al. 2005). Under a range of ambient temperatures, the internal vehicle temperature was shown to increase by ca. 22°C in 1 h, at which time the rate of increase in temperature slows. Ambient temperatures below 22°C will not reach a lethal temperature in 1 h, and insulated materials may require more time to reach lethal temperatures. In temperate climates this approach may only be effective in warmer seasons.

Freezing is an alternative to heating for de-infestation of objects. Benoit et al. (2007) found no survival of adult female bed bugs at −18°C for 1 h. However, household freezers do not achieve this level of cooling, with a more typical −10°C resulting in very low mortality (<10%) at 1 h. Prolonged cooling at −10°C will result in high mortality after days. Direct exposure of bed bugs to solid CO_2 (dry ice −78.5°C) is nearly instantaneously lethal, which has led to the use of dry ice delivery systems that can apply CO_2 crystals to mattress seams and other harborage locations (www.cryonite.net).

4 Insecticides with Alternative Modes of Action

In parts of the world where carbamate and organophosphate insecticides are still options for control of bed bugs, the consensus view is that bed bugs are a manageable pest (Potter et al. 2010b). These two insecticides target synaptic acetyl cholinesterase, rather than axonal sodium ion channels, which are a target of pyrethroids (Yu 2008). In the United States the volatile insecticide 2,2-dichlorovinyl dimethyl phosphate (DDVP, dichlorvos) has a role in management of bed bugs, and it is the only organophosphate that is labeled for use. Its utility comes from its volatility. Unlike residuals and direct sprays, there is no need for physical contact between the source of the insecticide and the insect. It can be used in a manner parallel to gaseous fumigants or heat to kill bed bugs on or within potentially infested items. Both eggs and all mobile stages are susceptible to DDVP, but at room temperature lethal exposure may take time (up to 2 weeks or more) and if the insecticide vapors do not penetrate to where the bed bugs are, it will be ineffective. Because its use label prohibits long term human exposure to the insecticide, it is used in restricted volume spaces without occupants. We successfully killed individuals from a pyrethroid resistant strain inside several items, such as a clock radio, under a suitcase liner, or under frames of artwork (Potter et al. 2010a). Mortality at 2 weeks was lower inside a computer keyboard, in the toe of a shoe, or in a book-binding, suggesting that longer exposures, adequate air circulation or adequate temperatures (volatility is reduced at lower temperatures) may be required for DDVP to be effective. Because bed bugs may seek harborage in cracks, crevices and other hidden locations where air circulation is minimal, DDVP may not always be the appropriate choice.

Two insecticide formulations that include both a pyrethroid and neonicotinoid insecticides are labeled in the U.S. for bed bug treatments. Neonicotinoids target the postsynaptic niciotinergic receptor sites for acetyl choline (Tomizawa 2004). Temprid® contains imidacloprid and β-cyfluthrin and Transport® contains acetamiprid and bifenthrin. In laboratory tests we have seen improved residual efficacy for both products against strains of bed bugs known to be resistant to deltamethrin. In the absence of pyrethroid resistance, the long-term efficacy of these products might be further enhanced because they combine insecticides with two modes of action. In theory it would be more difficult for two adaptations to arise simultaneously. However the widespread distribution of pyrethroid resistance might diminish this advantage for resistance management because evolution of resistance to the neonicotinoid would translate into resistance to the product, unless there is an unexpected tradeoff in the factors that contribute to resistance to these two types of insecticides.

Chlorfenapyr is a pyrrole insecticide that affects production of ATP in mitochondria, and thus impacts cellular energy resources (Yu 2008). Chlorfenapyr must be activated by P450 enzymes to yield the toxic form, and thus has been called a proinsecticide. Phantom® (the formulated commercial form) is slower acting than the combination products mentioned above. However, pyrethroid resistant populations succumb to this insecticide over days of constant exposure (Romero et al. 2010b). This insecticide is not repellent and does not act as a locomotor excitant or initiator (sensu Miller et al. (2009)). Bed bugs given a choice between a Phantom treated refuge and an untreated refuge resided on these with equal probability (Romero et al. 2009b). This contrasts with Suspend® (containing the pyrethroid deltamethrin) that led to reduced numbers of residents on the treated refuge, because they were more likely to leave once they encountered the residues (note this impact of deltamethrin does not deter bed bugs from entering treated aggregation sites or cross an insecticide treated barrier to get to a food source). A deposit of Phantom remains toxic for months (Romero et al. 2010b). Because bed bugs must be exposed for extended periods, a key issue is whether such long term exposure is likely to occur in the field. Certainly good coverage of the areas where bed bugs are likely to reside is extremely important with Phantom or any other insecticide that requires long exposure.

Juvenile Hormone Analogues (JHAs) mimic hormones that are involved in embryonic development, metamorphosis, and reproduction in insects (Minakuchi and Riddiford 2006). Because juvenile hormones are a unique characteristic of insects and other arthropods, analogs should show low non-target effects, particularly indoors. Gentrol® and Precor® are two JHAs that are registered for use on bed bugs. Todd (2006) showed that Gentrol had little or no impact on development of nymphs, but did lead to considerable mortality after they molted to adults. These lethal effects may have been due to developmental abnormalities in the final molt. Another study (Naylor et al. 2008) detailed experiments in which bed bug nymphs and adults were exposed to dry residuals of (S)-methoprene, the active ingredient in Precor. They demonstrated high efficacy against both laboratory-maintained bugs and insecticide-resistant field strains, using higher than label rates, which is not permissible in commercial practice

Dusts without neurotoxic insecticides, such as silica gel or diatomaceous earth can kill insects by their actions on the outermost layers of the exoskeleton of insects. They can be adsorbents of the epicuticular wax or abrasive. Both actions lead to increased desiccation (Ebeling 1971). For terrestrial arthropods with a large surface area to volume ratio (a ratio that is greater in the smallest insects) water loss through the cuticle is critical, especially when access to ambient water is limited. Bed bugs are very effective in conserving internal water, which explains their ability to survive extended periods without access to a host (Usinger 1966). Romero et al. (2009a) found that desiccant dusts and pyrethroid-laced dusts were effective in killing bed bugs even in strains known to be pyrethroid resistant. Benoit et al. (2009b) found that silica gel (Dri-die), which is thought to adsorb cuticular lipids, resulted in more rapid loss of internal water than did diatomaceous earth (a cuticle abrasive). The action of these dusts was synergized with a blend of two bed bug produced compounds, (*E*)-2-hexenal and (*E*)-2-ocental, that have been referred to as alarm pheromones (Benoit et al. 2009b; Levinson et al. 1974). This synergism is likely the result of increased locomotor activity that enhances the impact of the desiccant. At least in the short-term it is unlikely that this pheromone synergism will be used in the field, but desiccant dusts have an ongoing role.

5 Detection

Since bed bugs are both cryptic and nocturnal, visual inspections often fail to reveal their presence. Consequently, developing other approaches to detect their presence is extremely critical. Unlike recurring agricultural pests where monitoring of population density are the hallmark of decision-making, the assumption of pest management professionals and the public at large is that there is zero tolerance for an infestation. Detection of infestations is critical at two decision points. Because infestations tend to spread out from an initial focal point around the bed as populations grow, early detection facilitates control efforts, especially those that rely on labor intensive approaches (e.g., steam, Cryonite®, direct spray, vacuuming). Confirmation or refutation of a suspicion of an infestation requires information that the pest manager and client can trust. If the presence of bed bugs is confirmed then action can be taken using the available technologies. The second decision point is when continuing action is no longer necessary. Like the original decision that course of action is critical. A decision to suspend treatment before the last mated female is killed, or the last eggs or nymphs are eliminated leaves the risk of reestablishment of the infestation. The more sensitive the detector is the better the decision that will follow.

The olfactory sensitivity of dogs is extraordinary (Krestel et al. 1984). As a result they are used increasingly to detect explosives, drugs, cancers and now bed bugs. Bed bugs release a characteristic scent when they are disturbed that can be detected by the human nose if the population density is high. This odor includes (*E*)-2-hexenal and (*E*)-octenal (Levinson et al. 1974) and may include other compounds (Siljander et al. 2008) mediating aggregation. However in modern well-ventilated spaces odor is seldom the first detected sign of an infestation. In laboratory training exercises, well-trained dogs can discriminate between bed bugs (and eggs) and other insects, with a high rate of positive identifications and a low rate of false positives (Pfiester et al. 2008). Using dogs in bed bug detection is increasingly common, but is complicated by the requirement of ongoing training, need for experienced dog handlers, and expenses associated with housing and transporting the animals. Another challenging factor is the need to maintain a constant supply of live bed bugs to reinforce training of the canine. Under real field conditions, the effectiveness of dogs has recently been brought into question (Wang and Cooper (2011)).

Bed bugs use a variety of cues to find their host, including heat and CO_2. They also engage in circadian mediated movement that occurs at night even in the absence of host stimuli (Romero et al. 2010a). Pitfall traps, such as the ClimbUp® Insect Interceptor, which allow bed bugs to crawl into the trap, but not escape because of the slippery inclined sides of the trap can trap many bed bugs in heavily infested apartments (Wang et al. 2011). A pitfall trap with dry ice to allow CO_2 to slowly sublimate overnight led to capture of more bed bugs over one night than more sophisticated traps that use a combination of heat and CO_2. These stimuli as well as others found to be effective for other blood-feeding insects have been incorporated into traps of various designs (Anderson et al. 2009; Wang et al. 2011, 2009).

Bed bugs use semiochemicals to reaggregate in refuges around the bed (Siljander et al. 2007, 2008; Weeks et al. 2011a, b; Levinson and Barilan 1971). Both volatile and non-volatile compounds play a role in stimulating or maintaining the aggregation. The hope is that a trap baited with aggregation pheromones would be effective in detecting low level populations of bed bugs. In theory such traps could be simpler and easier to maintain than traps dependent on release of CO_2.

6 Mating Behavior

Bed bugs have an unusual form of mating behavior that has been called traumatic or hypodermic insemination. The male punctures the exoskeleton of the female with a sickle-shaped paramere. The puncture occurs at a V-shaped ectospermalege on the female's abdomen. Mating is both essential and dangerous to the female. Cuticular punctures outside of the ectospermalege reduce life-time egg production (Morrow and Arnqvist 2003). Ongoing mating results in decreased female survival (Stutt and Siva-Jothy 2001) and female dispersal from optimal positions near the host (Pfiester et al. 2009). Mating has the potential to introduce environmental pathogens or insect specific pathogens directly into the female body cavity (Reinhardt et al. 2005). In dwellings insect pathogens may be uncommon, therefore their introduction could reduce population numbers. Using mating behavior as a vehicle for sexual transmission of insect pathogens may be an attractive concept, but its application would be complicated by issues surrounding their introduction into human dwellings, and the zero tolerance for even very low bed bug numbers.

Because mating is traumatic and males will attempt to mate with larger nymphs and other males, both nymphs and males have evolved a signal that discourages ineffective copulations (Ryne 2009; Harraca et al. 2010). Avoiding male-male and male-nymph inseminations is beneficial to both the signaler and the receiver. Seminal fluid is a limited resource (Reinhardt et al. 2011) therefore avoiding ineffective copulations could be critical. Nymphs and adult males produce distinct pheromone signals that discourage copulation (Feldlaufer et al. 2010; Ryne 2009; Harraca et al. 2010). These results suggest that mating disruption using anti-mating pheromones may be possible either to discourage male–female matings or to encourage male-male and male-nymph matings. However, because males will mate many times (Reinhardt et al. 2011), mating disruption would need to be very effective to impact the population.

7 Summary

Because pyrethroid resistance is common in bed bug populations, bed bug control is facilitated by utilizing approaches that target different vulnerabilities. These include the use of heat and cold, insecticides that target different neuronal targets and the hormonal system and water balance. Early detection of infestations and monitoring of elimination of populations will be cornerstones of any control program. Understanding bed bug behavioral responses is likely to be a key to bed bug detection. Manipulation of mating behavior seems possible, but it is too early to determine if it could be effective enough to meet consumer expectations. The resurgence of bed bugs was likely due to many interacting factors. Similarly the future of bed bug control will likely be dependent on multiple tactics based on scientific information. The risk is high that ineffective false remedies will proliferate because consumer concern is so high.

References

Anderson JF, Ferrandino FJ, McKnight S, Nolen J, Miller J (2009) A carbon dioxide, heat and chemical lure trap for the bedbug, *Cimex lectularius*. Med Vet Entomol 23(2):99–105. doi:10.1111/j.1365-2915.2008.00790.x

Bartley J, Harlan H (1974) Bed bug infestation: its control and management. Mil Med 139:884–886

Benoit JB, Del Grosso NA, Yoder JA, Denlinger DL (2007) Resistance to dehydration between bouts of blood feeding in the bed bug, *Cimex lectularius*, is enhanced by water conservation, aggregation, and quiescence. Am J Trop Med Hyg 76(5):987–993

Benoit JB, Lopez-Martinez G, Teets NM, Phillips SA, Denlinger DL (2009a) Responses of the bed bug, *Cimex lectularius*, to temperature extremes and dehydration: levels of tolerance, rapid cold hardening and expression of heat shock proteins. Med Vet Entomol 23(4):418–425

Benoit JB, Phillips SA, Croxall TJ, Christensen BS, Yoder JA, Denlinger DL (2009b) Addition of alarm pheromone components improves the effectiveness of desiccant dusts against *Cimex lectularius*. J Med Entomol 46(3):572–579

Boase CJ, Small G, Naylor R (2006) Interim report on insecticide susceptibility status of UK bed bugs. Prof Pest Controll 2006(Summer):6–7

Chang KP (1974) Effects of elevated-temperature on mycetome and symbiotes of bed bug *Cimex lectularius* (Heteroptera). J Invertebr Pathol 23(3):333–340

Ebeling W (1971) Sorptive dusts for pest control. Annu Rev Entomol 16:123–158. doi:10.1146/annurev.en.16.010171.001011

Feldlaufer MF, Domingue MJ, Chauhan KR, Aldrich JR (2010) 4-Oxo-aldehydes from the dorsal abdominal glands of the bed bug (Hemiptera: Cimicidae). J Med Entomol 47(2):140–143. doi:10.1603/me09210

Harraca V, Ryne C, Ignell R (2010) Nymphs of the common bed bug (*Cimex lectularius*) produce anti-aphrodisiac defence against conspecific males. BMC Biol 8:121. doi:10.1186/1741-7007-8-121

How YF, Lee CY (2010) Effects of temperature and humidity on the survival and water loss of *Cimex hemipterus* (Hemiptera: Cimicidae). J Med Entomol 47(6):987–995. doi:10.1603/me10018

Johnson CG (1941) The ecology of the bed-bug, *Cimex lectularius* L, in Britain – report on research, 1935–40. J Hyg 41(4):345–461

Karunaratne S, Damayanthi BT, Fareena MHJ, Imbuldeniya V, Hemingway J (2007) Insecticide resistance in the tropical bedbug *Cimex hemipterus*. Pestic Biochem Physiol 88(1):102–107

Kilpinen O, Jensen K-MV, Kristensen M (2008) Bed bug problems in Denmark, with a European perspective. In: Robinson WH, Bajomi D (eds) Proceedings of the sixth international conference on urban pests. OOK-Press, Papa, pp 395–399

Krestel D, Passe D, Smith JC, Jonsson L (1984) Behavioral determination of olfactory thresholds to amyl acetate in dogs. Neurosci Biobehav Rev 8(2):169–174. doi:10.1016/0149-7634(84)90037-x
Levinson HZ, Barilan AR (1971) Assembling and alerting scents produced by bedbug *Cimex lectularius*. Experientia 27(1):102
Levinson HZ, Levinson AR, Maschwitz U (1974) Action and composition of alarm pheromone of bedbug *Cimex lectularius* L. Naturwissenschaften 61(12):684–685
McLaren C, Null J, Quinn J (2005) Heat stress from enclosed vehicles: moderate ambient temperatures cause significant temperature rise in enclosed vehicles. Pediatrics 116(1):E109–E112. doi:10.1542/peds.2004-2368
Miller JR, Siegert PY, Amimo FA, Walker ED (2009) Designation of chemicals in terms of the locomotor responses they elicit from insects: an update of Dethier et al. (1960). J Econ Entomol 102(6):2056–2060
Minakuchi C, Riddiford LM (2006) Insect juvenile hormone action as a potential target of pest management. J Pestic Sci 31(2):77–84. doi:10.1584/jpestics.31.77
Morrow EH, Arnqvist G (2003) Costly traumatic insemination and a female counter-adaptation in bed bugs. Proc R Soc Lond B Biol Sci 270(1531):2377–2381
Myamba J, Maxwell CA, Asidi A, Curtis CF (2002) Pyrethroid resistance in tropical bedbugs, *Cimex hemipterus*, associated with use of treated bednets. Med Vet Entomol 16(4):448–451
Naylor R, Bajomi D, Boase C (2008) Efficacy of (S)-methoprene against *Cimex lectularius* (Hemiptera: Cimicidae). In: Robinson WH, Bajimi D (eds) Proceedings of the sixth international conference on urban pests. OOK-Press Kft, Papa, pp 115–121
Naylor RA, Boase CJ (2010) Practical solutions for treating laundry infested with *Cimex lectularius* (Hemiptera: Cimicidae). J Econ Entomol 103(1):136–139. doi:10.1603/ec09288
Pereira RM, Koehler PG, Pfiester M, Walker W (2009) Lethal effects of heat and use of localized heat treatment for control of bed bug infestations. J Econ Entomol 102(3): 1182–1188
Pfiester M, Koehler PG, Pereira RM (2008) Ability of bed bug-detecting canines to locate live bed bugs and viable bed bug eggs. J Econ Entomol 101(4):1389–1396
Pfiester M, Koehler PG, Pereira RM (2009) Sexual conflict in the extreme: traumatic insemination in bed bugs. Am Entomol 55(4):244–249
Potter MF (2011) The history of bed bug management with lessons from the past. Am Entomol 57(1):14–25
Potter MF, Romero A, Haynes K, Hardebeck E (2007) Killing them softly: battling bed bugs in sensitive places. Pest Control Technol 35(1):24–32
Potter MF, Haynes KF, Goodman MH, Stamper S, Sams S (2010a) Bed bugs: a blast from the past. Pest Manag Prof 78(3):46–52
Potter MF, Rosenberg B, Henriksen M (2010b) Bugs without borders: defining the global bed bug resurgence. Pestworld September/October:8–20
Reinhardt K, Naylor RA, Siva-Jothy MT (2005) Potential sexual transmission of environmental microbes in a traumatically inseminating insect. Ecol Entomol 30(5):607–611
Reinhardt K, Naylor R, Siva-Jothy MT (2011) Male mating rate is constrained by seminal fluid availability in bedbugs, *Cimex lectularius*. PLoS One 6(7). doi:10.1371/journal.pone.0022082
Romero A, Potter MF, Potter DA, Haynes KF (2007) Insecticide resistance in the bed bug: a factor in the pest's sudden resurgence? J Med Entomol 44(2):175–178
Romero A, Potter MF, Haynes KF (2009a) Are dusts the magic bullet? Pest Manag Prof 77(5):22, 23, 26, 28, 30
Romero A, Potter MF, Haynes KF (2009b) Behavioral responses of the bed bug to insecticide residues. J Med Entomol 46(1):51–57
Romero A, Potter MF, Haynes KF (2010a) Circadian rhythm of spontaneous locomotor activity in the bed bug, *Cimex lectularius* L. J Insect Physiol 56(11):1516–1522. doi:10.1016/j.jinsphys.2010.04.025
Romero A, Potter MF, Haynes KF (2010b) Evaluation of chlorfenapyr for control of the bed bug, *Cimex lectularius* L. Pest Manag Sci 66(11):1243–1248. doi:10.1002/ps.2002

Ryne C (2009) Homosexual interactions in bed bugs: alarm pheromones as male recognition signals. Anim Behav 78(6):1471–1475. doi:10.1016/j.anbehav.2009.09.033

Siljander E, Penman D, Harlan H, Gries G (2007) Evidence for male- and juvenile-specific contact pheromones of the common bed bug *Cimex lectularius*. Entomol Exp Appl 125(2):215–219

Siljander E, Gries R, Khaskin G, Gries G (2008) Identification of the airborne aggregation pheromone of the common bed bug, *Cimex lectularius*. J Chem Ecol 34(6):708–718. doi:10.1007/s10886-008-9446-y

Stutt AD, Siva-Jothy MT (2001) Traumatic insemination and sexual conflict in the bed bug *Cimex lectularius*. Proc Natl Acad Sci USA 98(10):5683–5687

Tawatsin A, Thavara U, Chompoosri J, Phusup Y, Jonjang N, Khumsawads C, Bhakdeenuan P, Sawanpanyalert P, Asavadachanukorn P, Mulla MS, Siriyasatien P, Debboun M (2011) Insecticide resistance in bedbugs in Thailand and laboratory evaluation of insecticides for the control of *Cimex hemipterus* and *Cimex lectularius* (Hemiptera: Cimicidae). J Med Entomol 48(5):1023–1030

Todd RG (2006) Efficacy of bed bug control products in lab bioassays: do they make it past the starting gate? Am Entomol 52(2):113–116

Tomizawa M (2004) Neonicotinoids and derivatives: effects in mammalian cells and mice. J Pestic Sci 29(3):177–183. doi:10.1584/jpestics.29.177

Usinger RL (1966) Monograph of Cimicidae, vol 7, The Thomas Say Foundation. Entomological Society of America, College Park

Wang CL, Cooper R (2011) Detection tools and techniques. Pest Control Technol 39(8):72,74, 76,78–79

Wang CL, Gibb T, Bennett GW, McKnight S (2009) Bed bug (Heteroptera: Cimicidae) attraction to pitfall traps baited with carbon dioxide, heat, and chemical lure. J Econ Entomol 102(4): 1580–1585

Wang CL, Tsai WT, Cooper R, White J (2011) Effectiveness of bed bug monitors for detecting and trapping bed bugs in apartments. J Econ Entomol 104(1):274–278. doi:10.1603/ec10141

Weeks ENI, Birkett MA, Cameron MM, Pickett JA, Logan JG (2011a) Semiochemicals of the common bed bug, *Cimex lectularius* L. (Hemiptera: Cimicidae), and their potential for use in monitoring and control. Pest Manag Sci 67(1):10–20. doi:10.1002/ps.2024

Weeks ENI, Logan JG, Gezan SA, Woodcock CM, Birkett MA, Pickett JA, Cameron MM (2011b) A bioassay for studying behavioural responses of the common bed bug, *Cimex lectularius* (Hemiptera: Cimicidae) to bed bug-derived volatiles. Bull Entomol Res 101(1):1–8. doi:10.1017/s0007485309990599

Yoon KS, Kwon DH, Strycharz JP, Hollingsworth CS, Lee SH, Clark JM (2008) Biochemical and molecular analysis of deltamethrin resistance in the common bed bug (Hemiptera: Cimicidae). J Med Entomol 45(6):1092–1101

Yu SJ (2008) The toxicology and biochemistry of insecticides. CRC Press, Boca Raton

Zhu F, Potter MF, Palli R, Palli SR, Haynes KF, Romero A, Wigginton J, Ferguson K, Moore A (2010) Widespread distribution of knockdown resistance mutations in the bed bug, *Cimex lectularius* (Hemiptera: Cimicidae), populations in the United States. Arch Insect Biochem Physiol 73(4):245–257. doi:dx.doi.org/10.1002/arch.20355

Chapter 14
Advanced Methods for Controlling Insect Pests in Dry Food

Moshe Kostyukovsky and Eli Shaaya

1 Introduction

Insect damage in stored grain and other durable commodities may account to 10–40% in developing countries, where modern storage technologies have not been introduced (Raja et al. 2001). Fumigation is still one of the most effective methods for the protection of stored grain and dry food products from insect infestation. Phosphine is mainly in use today after the phase out of methyl bromide in developed countries due to its ozone depletion effects (WMO 1995; Shaaya and Kostyukovsky 2006). However, some limitations such as low temperature and relatively long exposure time limit the use of Phosphine. Therefore, there is an urgent need for new strategies to improve the Phosphine application and to focus on the search for alternatives for the control of stored product insects.

Some techniques for direct applications of gaseous Phosphine from cylinders (ECO_2Fume, $VAPORPH_3OS^{TM}$) and on-site generator sources (Horn generator, Chinese generator, QuickPHlo-R Phosphine generator, QuickPHlo-C technology and the Degesch Phosphine Generator) and others, have been developed to solve these problems (Williams et al. 2000; Mathews and Luzaich 2003; Waterford and Asher 2003; Waterford 2004; Horn and Horn 2006; Steuerwald et al. 2006; Rajendran and Sriranjini 2007; Ryan et al. 2010). Each of them aimed for different purposes and has advantages and disadvantages. In order to improve Phosphine application, especially for low temperatures and for shorter treatment time, a special device, so-called "Speedbox" has been developed by Detia Degesch GmbH Germany (Jakob et al. 2006). The Speedbox is a waterproof aluminum box containing a heater and a ventilator. It has been designed to be used exclusively with Degesch Phostoxin Plates®. For studying the effectiveness of Phosphine fumigation using Speedbox,

M. Kostyukovsky (✉) • E. Shaaya
Department of Food Quality and Safety, Agricultural Research Organization,
The Volcani Center, Bet Dagan 50250, Israel
e-mail: inspect@volcani.agri.gov.il; vtshaaya@volcani.agri.gov.il

I. Ishaaya et al. (eds.), *Advanced Technologies for Managing Insect Pests*, 279
DOI 10.1007/978-94-007-4497-4_14,

we have conducted two kinds of experiments: one in a fumigation room (pilot) and other in commercial warehouses.

The use of contact insecticides as grain protectants against stored product insect pests is a common and effective treatment worldwide. However, the demands for residue-free food and environmental safety, as well as the development of insect resistance to residual insecticides, have led to attempts to search for alternative protectants non-toxic to human and environmentally friendly. Diatomaceous earth (DE) is known as one of the most promising alternatives to the traditional residual insecticides (Athanassiou et al. 2003, 2004, 2007, 2008; Athanassiou and Korunic 2007; Vayias and Stephou 2009). DE is a non-toxic, safe, natural origin material with a unique, non-chemical mode of action against insects, which die through desiccation (Korunic 1998; Subramanyam and Roesli 2000).

Today, DE is widely used for various products and processes, from toothpaste to cigars, plastics to paprika, filter media in swimming pools to home fish tanks, as well as insects and parasites control in animals and grains. The efficacy of commercial formulations of DE has been proven also against a number of stored product insect pests. However, the DE efficacy often varies with the formulation, the treated commodity and other factors (Desmarchelier and Dines 1987; Subramanyam et al. 1994; Subramanyam and Roesli 2000; Athanassiou et al. 2003, 2004, 2007, 2008; Vayias and Athanassiou 2004; Athanassiou and Kavallieratos 2005; Kavallieratos et al. 2005).

The bioactivity of essential oils, the major volatiles in aromatic plants and their constituents, have been well studied against a large number of stored product insects (Regnault-Roger and Hamraoui 1995; Raja et al. 2001; Ogendo et al. 2008). In our laboratory, by screening a large number of essential oils from aromatic plants, it was possible to isolate two very active fumigants from Labiatae plants. The main component of one of the oils is Pulegone, the other is not yet identified we call it SEM-76 (Shaaya et al. 1991, 1994, 1997; Shaaya and Kostyukovsky 2006). In this chapter, we report on the toxicity of these two oils against external and internal stored product insects and the potential use of the oils as fumigants in grain bins for insect control.

2 Studies with Speedbox

Comparative studies using a concentration of 2 g/m^3 and 24 h exposure time showed the fumigation with Speedbox is much more efficient than with tablets. In the case of Speedbox, maximum concentration of 600 ppm arrived after 10 h, compared with 200 ppm using tablets and 445 ppm after 25 h (Fig. 14.1).

Two types of experiments were performed to evaluate the potential of using the Speedbox. The first experiment using fumigation room, volume 15 m^3, 30% of the area filled with wheat grain bags. The fumigation was performed by connecting the Speedbox, which contains 1–3 Degesch plates to the fumigation room. Each plate contains 56% magnesium phosphide, weights 117 g and evolved 33 g of

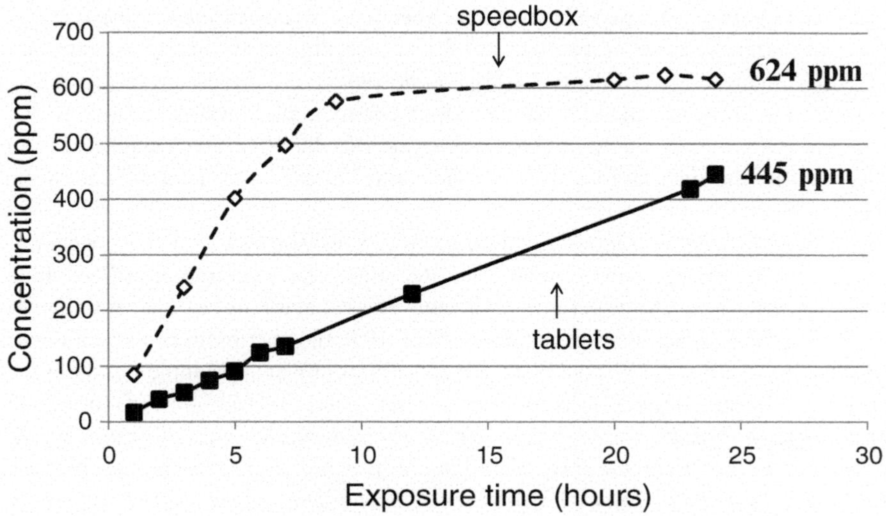

Fig. 14.1 Comparison the Phosphine concentration between Speedbox and tablets measured in the air space. 2 g/m^3 × 24 h

Phosphine gas (about 2 g of Phosphine gas per m^3). The target concentration was 2–6 g of Phosphine gas per m^3. The plates were heated to 36°C. The produced hydrogen phosphide was first blown into the fumigation room and then pumped back to the Speedbox for recirculation. The exposure time was 1–4 days. The Phosphine concentration was monitored during fumigation by Bedfont device model 415. Six Phosphine gas-sampling points were located at the top, middle and bottom of the fumigation room, at the point of entry of Phosphine gas into the room and at two places between the bags containing wheat. The temperatures of inter-granular air and of the room space were also recorded. Adults, pupae and late larvae of common insects were used. The test insects were placed between the bags in three replicates. The control insects were kept outside of the fumigation room under the same temperature conditions. The mortality of external stages was recorded 1, 7 and 30 days following treatment and the mortality of internal stages was counted 7, 14 and 30 days after treatment. The number of the hatched eggs was recorded 3 and 7 days after treatment. The progeny of the tested insects at all stages were counted 6 weeks after treatment. All the treated insects were reared and maintained under laboratory conditions at 28 ± 0.5°C and R.H. of 65 ± 5%.

In the second kind of experiments, the fumigation was carried out in commercial warehouses in the southern and central Israel with hot and mild climate conditions, respectively. Bags of wheat, rice, sunflower, beans, peanuts, pistachio and nuts, each weighing 40–50 kg, were sealed using plastic sheets. The range of stack volume was 15–60 m^3. The outdoor temperatures were between 15°C and 23°C at the beginning of the experiments and 8–25°C during the experiments. The temperature of the above-mentioned treated products was 6–17°C at the beginning of the experiments and 6–19°C during the experiments. The Speedbox was connected to the stack of the bags and the number of the plates used was according to the target concentration of Phosphine gas. The produced hydrogen phosphide was blown into the stack and pumped out from the stack to the Speedbox for recirculation. The dosage of the Phosphine gas was 2–4 g/m^3, and exposure time was 48–96 h. The test insects were inserted into the stack in three places. The control insects were kept outside of the stack in the same warehouse.

Studies performed in the fumigation room (15 m^3) showed, that at a concentration of 4 g/m^3 and exposure time of 48 h, the effective concentration of Phosphine was reached in a short period. Half-hour after beginning of the fumigation 95–115 ppm of Phosphine was recorded. After 10 h, the maximum concentration of the gas has reached 1,252 ppm. The gas concentrations in six sample points in the space and in the commodity were very similar. Total (100%) mortality of adults and larvae of *Sitophilus oryzae* and *Rhyzopertha dominica,* and all tested stages of *Oryzaephilus surinamensis*, *Tribolium castaneum*, *Trogoderma granarium*, *Callosobruchus maculatus*, *Plodia interpunctella* and *Ephestia cautella* was recorded. In the case of pupae of *S. oryzae* and *R. dominica* 88 and 97% kill was recorded respectively (Tables 14.1 and 14.2).

In the field experiment in commercial warehouse with 60-m^3 stack of wheat grain bags, using a concentration of 4 g/m^3 and exposure time of 2 days, the concentration of Phosphine has reached 414 and 1,480 ppm after 2 and 24 h respectively. The commodity temperature was 17–20°C. Total (100%) kill was recorded of adults, larvae and pupae of *S. oryzae*, *R. dominica, O. surinamensis*, *T. castaneum, C. maculatus* and *T. granarium*, as well as larvae and pupae of *P. interpunctella* and *E. cautella*. It should be mentioned that though the fumigation was performed at the commodity temperature of as low as 6–8°C, 100% mortality of all tested insect species at adult, larvae and pupae stages was obtained (no table of the results is given).

The current results show that the use of "Speedbox" significantly decreased the period of Phosphine release from the Degesch plates. Maximum Phosphine concentration in the treated area was achieved just after 12–17 h from the beginning of the fumigation. In contrast, without the Speedbox the degassing rate of the Degesch plates was only 60% after 24 h (Jakob et al. 2006). By the use of Speedbox the effective concentration of Phosphine arrive in a shorter time than by the tablets which made it possible to reduce the exposure time. The Speedbox also allows the recirculation of the gas and its distribution evenly in the treated stack. All developmental stages of the tested insect, except pupae, were totally killed at a target concentration of 4 g/m^3 for 48 h both in pilot and field fumigations. Our data is consistent with the results obtained by the others (Mills et al. 2003). It is important to underline that in our field fumigations the temperatures of the outdoor and treated products were as low as 8–16°C and 6–8°C, respectively. Despite the strong gas sorption by the treated commodities, especially with high lipids content, at these temperatures, the concentration of Phosphine during the treatment was high enough to achieve total mortality of tested insects. The current results are consistent with the findings, that with the Speedbox, only negligible degassing differences were observed between 10°C and 20°C (Jakob et al. 2006).

Table 14.1 Mortality (%) of major internal stored product insects at maximum concentrations of Phosphine (ppm) obtained 10–20 h fumigation by Speedbox depending on the various dosages and exposure time

		Dosages and exposure time ($g/m^3 \times h$)						
Insect	Stage	2 $g/m^3 \times 24$ h 624 ppm 10 h	4 $g/m^3 \times 48$ h 1,252 ppm 10 h	4 $g/m^3 \times 72$ h 1,335 ppm 20 h	4 $g/m^3 \times 96$ h 1,290 ppm 20 h	6 $g/m^3 \times 24$ h 1,742 ppm 15 h	6 $g/m^3 \times 36$ h 1,711 ppm 10 h	6 $g/m^3 \times 72$ h 1,680 ppm 10 h
Sitophilus oryzae	Adult	59	100	100	100	99	100	100
	Larvae	77	100	100	100	100	100	100
	Pupae	50	88	84	93	62	68	91
Rhyzopertha dominica	Adult	95	100	100	100	100	100	100
	Larvae	76	100	100	100	80	100	100
	Pupae	82	97	100	100	75	88	100
Callosobruchus maculatus	Adult	100	100	100	100	100	100	100
	Larvae	95	97	100	100	100	100	100
	Pupae	–	99.8	100	100	–	95	100
	Egg	100	100	100	100	100	100	100

15 m^3 -fumigation room was used

Table 14.2 Mortality (%) of major external stored product insects at maximum concentrations of Phosphine (ppm) obtained 10–20 h fumigation by Speedbox depending on the various dosages and exposure time

		Dosages and exposure time ($g/m^3 \times h$)						
		$2\ g/m^3 \times 24$ h	$4\ g/m^3 \times 48$ h	$4\ g/m^3 \times 72$ h	$4\ g/m^3 \times 96$ h	$6\ g/m^3 \times 24$ h	$6\ g/m^3 \times 36$ h	$6\ g/m^3 \times 72$ h
		624 ppm	1,252 ppm	1,335 ppm	1,290 ppm	1,742 ppm	1,711 ppm	1,680 ppm
Insect	Stage	10 h	10 h	20 h	20 h	15 h	10 h	10 h
Oryzaephilus surinamensis	Adult	100	100	100	100	100	100	100
	Larvae	100	100	100	100	100	100	100
	Pupae	100	100	100	100	100	100	100
Tribolium castaneum	Adult	100	100	100	100	100	100	100
	Larvae	100	100	100	100	100	100	100
	Pupae	100	100	100	100	100	100	100
	Egg	100	100	100	100	100	100	100
Trogoderma granarium	Adult	100	100	100	100	100	100	100
	Larvae	90	100	100	100	100	100	100
	Pupae	79	100	100	100	100	100	100
Plodia interpunctella	Larvae	100	100	100	100	100	100	100
	Pupae	100	100	100	100	100	100	100
	Egg	83	100	100	100	80	98	100
Ephestia cautella	Larvae	100	100	100	100	100	100	100
	Pupae	100	100	100	100	100	100	100
	Egg	71	98	100	100	21	87	100

15 m^3 -fumigation room was used

3 Studies with Diatomaceous Earth

In the current study, the commercial formulation of Diatomaceous Earth (DE) Detia Degesch Diatomaceous Earth – DDDE- Inerto (Detia Degesch GmbH) was used. The infested untreated and treated wheat of 12% moisture content were kept at 28±0.5°C and R.H. of 65±5% for 2, 3- and 4-weeks exposure time. The number of adult progeny was counted 9 weeks after treatment.

At a concentration of 0.5 g/kg and exposure time of 30 days, 94% and 88% mortality of *S. oryzae* and *O. surinamensis* were recorded respectively. In contrast, only 47% and 5% kill were recorded for *R. dominica* and *T. castaneum* respectively. A concentration of 1 g/kg and exposure time of 30 days caused 100% kill of *S. oryzae* and *O. surinamensis* and 90% and 82% kill of *R. dominica* and *T. castaneum* respectively. Using higher concentration of 2 and 4 g/kg resulted in minor increase in mortality. At all concentrations tested, lower mortality was recorded at exposure time of 14 and 21 days compared to 30 days (Table 14.3).

Progeny production of *S. oryzae* and *T. castaneum* at a concentration of 2 g/kg was highly suppressed and only few adults were recorded compared to the control (Table 14.4). Similarly, the progeny production of *O. surinamensis* was also notably reduced. In contrast, the progeny of *R. dominica* was decreased less than 50% of the control. In the case of larvae of *T. castaneum* at a concentration of 2 and 4 g/kg the number of larvae developed to adults was 11 and 4% compared to control, respectively (Table 14.4).

Studies with DE indicate that concentration, insect species (external or internal feeder), developmental stage and exposure time to the treated commodity influenced the efficacy of the DE. Among adults, *S. oryzae* and *O. surinamensis* were found to be the most susceptible to DE, even at the lowest concentration tested of 0.5 g/kg. In contrast, *T. castaneum* and *R. dominica* were much more tolerant, for the control of adults higher concentration of 4 g/kg is needed. These findings are consistent with the results obtained by other researches (Korunic 1998; Fields and Korunic 2000; Arthur 2001, 2002; Vayias and Athanassiou 2004; Arnaud et al. 2005; Athanassiou et al. 2007). In contrast to adults, the larvae of *T. castaneum,* were very susceptible to DE. It should be mentioned that, even in the cases where larvae survived and reached the adult stage, no progeny was produced. Therefore, it is expected, that despite the tolerance of adults to DEs, susceptibility of larvae may slowly control *T. castaneum* populations. Also for the confused flour beetle, *Tribolium confusum*, DE was much more effective against larvae than against adults (Vayias and Athanassiou 2004). From the internal feeders, *S. oryzae* was by far more susceptible than *R. dominica*. This could be attributed to the fact that *R. dominica* adults are less mobile, which may reduce the overall contact with the DE particles (Fields and Korunic 2000). It is known, that in general, mobile species, such as the rusty grain beetle, *Cryptolestes ferrugineus*, are more susceptible to DE than less mobile species (Rigaux et al. 2001; Vardeman et al. 2007). The current results confirm the findings from previous studies, about the rank of stored-product insect species according to their susceptibility to DEs (Korunic 1998;

Table 14.3 The efficacy of the DDDE – Inerto against adults of major stored product insects

Concentration (g/kg)	Exposure time (d)	Adult mortality (%)			
		S. oryzae	*O. surinamensis*	*R. dominica*	*T. castaneum*
0.5	14	82	67	23	2
	21	92	86	37	3
	30	94	88	47	5
1	14	96	92	61	13
	21	100	97	77	59
	30	100	100	90	82
2	14	93	96	67	11
	21	100	100	84	72
	30	100	100	86	96
4	14	100	100	75	52
	21	100	100	90	96
	30	100	100	96	98
Control	14	2	3	0	0
	21	2	8	11	2
	30	8	19	13	2

Twenty individuals of each tested species were separately inserted into each glass jar of 1 l capacity, filled with 500 g of treated or not treated wheat grain. The data is average from three replicates

Table 14.4 Effect of Inerto on the progeny of adults and larvae. Exposure time 2 months

Concentration (g/kg)	No. of adults (F_1)				No. of larvae (F_1)
	T. castaneum	*S. oryzae*	*O. surinamensis*	*R. dominica*	*T. castaneum*
0.5	–	–	–	171	45
1	20	10	0	179	25
2	5	1	10	102	8
4	0	0	0	85	3
Control	90	330	120	173	70

20 insects were introduced to 500 g of treated or not treated wheat grain. The data is average from three replicates

Fields and Korunic 2000; Subramanyam and Roesli 2000; Athanassiou et al. 2004; Vayias and Athanassiou 2004; Athanassiou and Kavallieratos 2005; Kavallieratos et al. 2005).

4 Studies with Essential Oils

Screening a large number of essential oils and their monoterpenes, showed that the monoterpene Pulegone and SEM 76 the major component, 80–90% in the oil, were found most active from all the oils and monoterpenes tested against a large number of stored product insects. The bioassay to evaluate the activity of the essential oils was space fumigation in glass chambers of 3.4 L capacity (for details see Shaaya

Table 14.5 Toxicity of a number of active essential oils against adults of major stored products insects. Space fumigation. Exposure time 24 h

Essential oil	Concentration (μl/l)	% mortality 7 days following treatment *S. oryzae*	*R. dominica*	*T. castaneum*	*O. surinamensis*
SEM-76	0.5	100	100	100	97
Pulegone	0.5	97 (93–100)	97 (93–100)	100	100
Limonene	0.5	27 (14–40)	24 (20–27)	27 (24–30)	0
Control	–	0	0	0	0

The data is average of five replicates after Abbot's correction. The numbers in brackets are the mortality range

Table 14.6 Toxicity of essential oils SEM-76 and Pulegone against larvae of stored products insects. Space fumigation. Exposure time 24 h

Essential oils	Concentration (μl/l)	Mortality (%) days following treatment											
		T. castaneum			*T. granarium*			*P. interpunctella*			*E. cautella*		
		0	2	7	0	2	7	0	2	7	0	2	7
SEM-76	0.5	26	40	60	13	15	36	7	18	30	3	10	23
	1.5	76	79	90	74	87	90	56	57	90	59	55	60
Pulegone	0.5	40	50	90	20	47	60	26	40	60	20	47	90
	1.5	99	100	100	89	99	100	85	92	100	97	100	100
Control	–	0	0	0	0	0	0	0	0	5	0	0	5

The data is average from five replicates after Abbot's correction

et al. 1991). The active components were tested in pilot experiment using 0.25-t bins filled by 70% with wheat using air circulation, with and without the addition of carbon dioxide (CO_2). For studies with essential oils on various developmental stages of *S. oryzae*, a large amount of wheat infested for 24 h was used. Samples each of 10 g were drawn at intervals according to the age of the insect. Fumigation was performed in fumigation chambers of 300 ml filled by 70% with wheat and the infested sample. In the case of *C. maculatus* we followed the development from eggs 4 h old to adult. For each test 18–21 eggs were used, in space fumigation.

In space fumigation a concentration of 0.5 μl/l air (equivalent to 0.5 g/m^3) was enough to cause 97–100% kill of adults of *T. castaneum*, *O. surinamensis*, *S. oryzae* and *R. dominica*, both with Pulegone and SEM-76. For comparison, we also tested Limonene, which showed much lower toxicity, around 25% to all the tested insects (Table 14.5). It should be mentioned that Limonene is regarded an active compound and it is used widely for insect control.

In the case of larvae, a higher concentration of 1.5 μl/l air is needed to get 90–100% kill of *T. castaneum*, *T. granarium*, *P. interpunctella* and *E. cautella*. The mortality of the larvae was low immediately after fumigation and increased 2 days and then 7 days following treatment (Table 14.6).

Using space fumigation, a study on Pulegone toxicity was also conducted on various developmental stages of *C. maculatus* such as eggs on the seeds, and

Table 14.7 Toxicity of Pulegone on various developmental stages (inside the seeds) of *Callosobruchus maculatus*. Space fumigation. Exposure time 24 h

Developmental stage	Concentration (μl/l)	No. off eggs used (average)	Adult F_1 No.	Adult F_1 % of control
Eggs 20–26 h old	0.5	21	0	0
Larvae prior to penetrate	0.5	19	2	13
inside the seed	1	19	0	0
Larvae 1 day old	0.5	19	0.5	3
inside the seed	1	20	0	0
Larvae 3 day old	1.5	18	6	40
inside the seed	3	20	4	27
Larvae 7 day old	1.5	19	13	87
inside the seed	3	19	13	87
Larvae 11 day old	1.5	20	11	73
inside the seed	3	19	10	67
Pupae 3 day old	1.5	20	14	93
	3	20	16	106
Control	–	20	15	75

The data is average of three replicates, each 18–21 eggs

Table 14.8 Toxicity of Pulegone on various developmental stages (inside the seeds) of *Sitophilus oryzae*. 300 ml fumigation chambers. Exposure time 24 h

Treatment after infestation (days)	Stage	Concentration (μl/l)	No. of adult F_1	% of control
2	Eggs	100	94	10
10	Young larva	100	27	3
15	Old larvae	100	5	1
20	Pupae	100	0	0
Control	Eggs	0	1,035	100

The data is average of three replicates

larvae and pupae inside the seeds. The eggs and young larvae were most sensitive, a concentration of 0.5 μl/l air was enough to cause 100% kill. A concentration of one μl/l air was enough to cause 100% kill of larvae shortly before penetrating the seed. Older larvae and pupae were more tolerant to the compound using 3 μl/l air and exposure time of 24 h, the mortality of larvae 7-days-old and older and pupae was only slightly higher than in the control (Table 14.7).

Fumigation with Pulegone in 300 ml fumigation chambers filled with 70% wheat contain a sample of infested wheat was found very toxic against various developmental stages (inside the seeds) of *S. oryzae*. A concentration of 100 μl/l air and exposure time of 24 h caused 90% reduction of F_1 of treated eggs, in the case of larvae and pupae a minor number of adults appear in F_1 compared with the control. Eggs and young larvae were more tolerant compared with old larvae and pupae (Table 14.8).

Pilot experiment in 0.25-t bins filled 70% with wheat, using SEM-76 at 100 g/m^3 and exposure time of 7 days with and without the addition of CO_2 (dry ice) and air circulation, was done. The result showed that there is a need of the addition of 200 g/m^3 CO_2 and circulation to get 100% kill of adult insects of *S. oryzae*, *R. dominica* and *O. surinamensis*. In the case of *T. castaneum* only 70–90% kills was achieved (Table 14.9). In the case of larvae of *T. castaneum* and *T. granarium*, also 100% kill was obtained (Table 14.10).

EOs and their constituents are known to possess insecticidal (Wilson and Shaaya 1999; Shaaya et al. 1997; Ogendo et al. 2008) and insect repellent activity (Jilani et al. 1988; Ogendo et al. 2008), and cause reduction in progeny (Regnault-Roger and Hamraoui 1995). For example, the fumigant toxic activity, antifeedant and reproduction inhibition induced by a number of EOs and their monoterpenoids were evaluated against the bean weevil *Acanthoscelides obtectus* and *C. maculatus* (Klingauf et al. 1983; Regnault-Roger and Hamraoui 1995; Raja et al. 2001).

In our laboratory, in order to isolate active EOs, we screened a large number of EOs extracted from aromatic plants and isolated their main constituents. We have isolated many such compounds from the EOs of a large number of aromatic plants (Shaaya et al. 1991, 1994, 1997; Shaaya and Kostyukovsky 2006). Using space fumigation (see Shaaya et al. 1997), two EOs obtained from Labiatae plants, were found as the most potent fumigants from all oils tested. The main component of one of the oils was Pulegone, the other is not yet identified and it is called SEM-76. Pulegone and SEM-76 were also found most potent fumigants compared with all monoterpenes tested. Moreover, they had a high bioactivity against *T. castaneum* and *S. oryzae*, which were found the most tolerant of all insects tested.

5 Conclusion

The Speedbox allows: to optimize the Phosphine fumigation of stacked bags in controlling all developmental stages of the major stored product insects; to decrease significantly the period of Phosphine release from the Degesch Phostoxin Plates and hence the exposure time; to enable effective Phosphine fumigation at commodity temperatures as low as 6–8°C; to achieve even distribution of the gas in the treated space. The use of the Speedbox opens novel possibilities for Phosphine fumigation also as quarantine treatment.

DE is effective against stored-grain pests, at minimum dose rate of 1 g/kg. A longer exposure time may alleviate the need for increased doses in order to control species that are tolerant to DEs. The progeny production of a number of insects was highly depressed by DE. It is well established that DE has low reactivity with the environment, which makes DEs ideal candidates for long-term protection (Korunic 1998; Subramanyam and Roesli 2000; Vayias et al. 2006).

EOs are highly selective to insects, since they are probably targeted to the insect-selective octopaminergic receptor, a non-mammalian target (Kostyukovsky et al. 2002). In this chapter, we report that Pulegone and SEM-76 have high bioactivity against internal stages of *S. oryzae* and eggs and young larvae of *C. maculatus*. In field experiments, using 0.25-t-bins filled by 70% with wheat grain, SEM-76 at a concentration of 100 g/m^3 with supplementation of CO_2 and air circulation

Table 14.9 Fumigant toxicity of SEM-76 in pilot experiment using 0.25 t bin filled 70% with wheat against four stored product adult insects with and without the addition of CO_2 (200 g/m^3) and air circulation

	Adult mortality (%) 7 days following treatment											
	S. oryzae			*R. dominica*			*T. castaneum*			*O. surinamensis*		
Insect placed in:	Circulation	+ CO_2	Circulation + CO_2	Circulation	+ CO_2	Circulation + CO_2	Circulation	+ CO_2	Circulation + CO_2	Circulation	+ CO_2	Circulation + CO_2
10 cm from the surface of grain	100	100	100	75	82	100	30	44	90	40	60	100
50 cm from the surface of grain	62	86	100	47	77	100	10	37	86	50	66	100
In the bottom of the bin	65	82	100	45	58	100	5	22	70	15	44	100

Concentration 100 g/m^3, exposure time 7 days

Cages with 20 insects and food in each one were used. The results are average of three replicates. CO_2 was added as dry ice

Table 14.10 Fumigant toxicity of SEM-76 in pilot experiment using 0.25 t bin filled 70% with wheat against larvae of *T. granarium* and *T. castaneum* with and without the addition of CO_2 and air circulation

	Larval mortality (%) 7 days following treatment					
	T. castaneum			*T. granarium*		
Insect placed in:	Circulation	+CO_2	Circulation +CO_2	Circulation	+CO_2	Circulation +CO_2
10 cm from the surface of grain	75	88	100	63	75	100
50 cm from the surface of grain	88	95	100	68	82	100
In the bottom of the bin	57	75	100	48	57	100

Cages with 20 insects and food in each one were used. The results are average of three replicates. CO_2 was added as dry ice
Concentration 100 g/m^3, exposure time 7 days

caused 100% kill of a number of major stored product insects. It should be mentioned that in high bins addition of CO_2 is also essential in Methyl bromide or Phosphine fumigation for the gas penetration.

The worldwide availability of plant EOs and their terpenoids, their use in cosmetics and as flavoring agents in food and beverages, is a good indication of their relative safety to warm-blooded animals and humans. The ultimate goal is the introduction of these phytochemicals with low toxicity, which comply with health and environmental standards, as alternatives to methyl bromide and Phosphine for the preservation of grain and dry food. We should keep in mind that it is very difficult to introduce the broad-spectrum fumigants like Methyl bromide or Phosphine. In this context, alternative fumigants could be developed against particular species of insects or to be used for specific food product commodity.

References

Arnaud L, Lang HTT, Brostaux Y, Haubruge E (2005) Efficacy of diatomaceous earth formulations admixed with grain against populations of *Tribolium castaneum*. J Stored Prod Res 41:121–130

Arthur FH (2001) Immediate and delayed mortality of *Oryzaephilus surinamensis* (L.) exposed on wheat treated with diatomaceous earth: effects of temperature, relative humidity and exposure interval. J Stored Prod Res 37:13–21

Arthur FH (2002) Survival of *Sitophilus oryzae* (L.) on wheat treated with diatomaceous earth: impact of biological and environmental parameters on product efficacy. J Stored Prod Res 38:305–313

Athanassiou CG, Kavallieratos NG (2005) Insecticidal effect and adherence of PyriSecs in different grain commodities. Crop Prot 24:703–710

Athanassiou CG, Korunic Z (2007) Evaluation of two new diatomaceous earth formulations, enhanced with abamectin and bitterbarkomycin, against four stored-grain beetle species. J Stored Prod Res 43:468–473

Athanassiou CG, Kavallieratos NG, Tsaganou FC, Vayias BJ, Dimizas CB, Buchelos CT (2003) Effect of grain type on the insecticidal efficacy of SilicoSec against *Sitophilus oryzae* (L) (Coleoptera: Curculionidae). Crop Prot 22:1141–1147

Athanassiou CG, Kavallieratos NG, Andris NS (2004) Insecticidal effect of three diatomaceous earth formulations against adults of *Sitophilus oryzae* (Coleoptera: Curculionidae) and *Tribolium confusum* (Coleoptera: Tenebrionidae) on oat, rye and triticale. J Econ Entomol 97:2160–2167

Athanassiou CG, Kavallieratos NG, Meletsis CM (2007) Insecticidal effect of three diatomaceous earth formulations, applied alone or in combination, against three stored-product beetle species on wheat and maize. J Stored Prod Res 43:303–334

Athanassiou CG, Kavallieratos NG, Vayias BJ, Panoussakis EC (2008) Influence of grain type on the susceptibility of different *Sitophilus oryzae* (L.) populations, obtained from different rearing media, to three diatomaceous earth formulations. J Stored Prod Res 44:279–284

Desmarchelier JM, Dines JC (1987) Dryacide treatment of stored wheat: its efficacy against insects, and after processing. Aust J Exp Agric 27:309–312

Fields P, Korunic Z (2000) The effect of grain moisture content and temperature on the efficacy of diatomaceous earths from different geographical locations against stored-product beetles. J Stored Prod Res 36:1–13

Horn P, Horn F (2006) Large scale grain fumigations using pure cylinderized Phosphine together with the HORN DILUPHOS SYSTEM. In: Lorini I, Bacaltchuk B, Beckel H, Deckers D, Sundfeld E, dos Santos JP, Lorini I, Bacaltchuk B, Beckel H, Deckers D, Sundfeld E, Biagi JD, Celaro JC, Faroni LRD'A, Bartolini LdeOF, Sartori MR, Elias MC, Guedes RNC, De-Fonseca RG, Scussel VM (eds) Proceedings of the ninth international working conference on stored product protection, 15–18 October 2006, Sao Paulo, Brazil. Brazilian Post-harvest Association, Campinas

Jakob G, Dierks-Lange H, Heck FW, Schmitt S (2006) The speedbox – an innovative application device for the Degesch plates. In: Lorini I, Bacaltchuk B, Beckel H, Deckers D, Sundfeld E, dos Santos JP, Biagi JD, Celaro JC, Faroni LRD'A, Bartolini LdeOF, Sartori MR, Elias MC, Guedes RNC, De-Fonseca RG, Scussel VM (eds) Proceedings of the ninth international working conference on stored product protection, 15–18 October 2006, Sao Paulo, Brazil. Brazilian Post-harvest Association, Campinas

Jilani G, Saxena RC, Rueda BP (1988) Repellent and growth inhibiting effects of turmeric oil, sweetflag oil, neem oil and Margosan-O on red flour beetle (Coleoptera: Tenebrionidae). J Econ Entomol 81:1226–1230

Kavallieratos NG, Athanassiou CG, Paschalidou FG, Andris NS, Tomanović Ž (2005) Influence of grain type on the insecticidal efficacy of two diatomaceous earth formulations against *Rhyzopertha dominica* (F) (Coleoptera: Bostrychidae). Pest Manag Sci 61:660–666

Klingauf F, Bestmann HJ, Vostrowsky O, Michaelis K (1983) Wirkung von altherishcen Olen auf Schadinsekten. Mitt Dtsch Gesselschaft fuer Allg Angew Entomol 4:123–126

Korunic Z (1998) Diatomaceous earths, a group of natural insecticides. J Stored Prod Res 34:87–97

Kostyukovsky M, Rafaeli A, Gileadi C, Demchenko N, Shaaya E (2002) Activation of octopaminergic receptors by essential oil constituents isolated from aromatic plants: possible mode of activity against insect pests. Pest Manag Sci 58:1–6

Mathews M, Luzaich G (2003) Treatment of an empty fumigation chamber using the Degesch Phosphine generator. In: Credland PF, Armitage DM, Bell CH, Cogan PM, Highley E (eds) Advances in stored product protection. Proceedings of the eighth international working conference on stored-product protection, 22–26 July 2002, York, UK. CAB International, Wallingford

Mills KA, Wontner-Smith TJ, Cardwell SC, Bell CH (2003) The use of Phosphine as an alternative to Methyl bromide for the disinfestation of palm dates. Advances in stored product protection. In: Credland PF, Armitage DM, Bell CH, Cogan PM, Highley E (eds) Advances in stored product protection. Proceedings of the eighth international working conference on stored-product protection, 22–26 July 2002, York, UK. CAB International, Wallingford

Ogendo O, Kostyukovsky M, Ravid U, Matasyoh JC, Deng AL, Omolo EO, Kariuki ST, Shaaya E (2008) Bioactivity of African basil (*Ocimum gratissimum* L.) oil and two constituents against fife insect pests of stored food products. J Stored Prod Res 44:328–334

Raja N, Albert S, Ignacimuthu S, Dorn S (2001) Effect of plant volatile oils in protecting stored cowpea *Vigna unguiculata* (L.) Walpers against *Callosobruchus maculatus* (F.) (Coleoptera: Bruchidae) infestation. J Stored Prod Res 37:127–132

Rajendran S, Sriranjini VR (2007) Use of fumigation for managing grain quality. Stewart Postharvest Rev 3:6, Article 9

Regnault-Roger C, Hamraoui A (1995) Fumigant toxic activity and reproductive inhibition induced by monoterpenes upon *Acanthoscelides obtectus* Say (Coleoptera), bruchid of kidney bean (*Phaseolus vulgaris* L). J Stored Prod Res 31:291–299

Rigaux M, Haubruge E, Fields PG (2001) Mechanisms for tolerance to diatomaceous earth between strains of *Tribolium castaneum*. Entomol Exp Appl 101:33–39

Ryan RF, Shore W, Newman C (2010) Phosphine generator trial using external air dilution. In: Carvalho OM, Fields PG, Adler CS, Arthur FH, Athanassiou CG, Campbell JF, Fleurat-Lessard F, Flinn PW, Hodges RJ, Isikber AA, Navarro S, Noyes RT, Riudavets J, Sinha KK, Thorpe GR, Timlick BH, Trematerra P, White NDG (eds) Proceedings of the tenth international working conference of stored product protection, 27 June-2 July 2010, Estoril, Portugal. Julius Kühn-Institut, Berlin

Shaaya E, Kostyukovsky M (2006) Essential oils: potency against stored product insects and mode of action. Stewart Postharvest Rev 2(4):5:1–6

Shaaya E, Paster N, Juven B, Zisman U, Pisarev V (1991) Fumigant toxicity of essential oils against four major stored-product insects. J Chem Ecol 17:499–504

Shaaya E, Kostjukovsky M, Ravid U (1994) Essential oils and their constituents as effective fumigants against stored-product insects. Isr Agrisearch 7:133–139

Shaaya E, Kostjukovsky M, Eilberg J, Sukprakan C (1997) Plant oils as fumigants and contact insecticides for the control of stored-product insects. J Stored Prod Res 33:7–15

Steuerwald R, Dierks-Lange H, Schmitt S (2006) The degesch Phosphine generator – a fast Phosphine application. In: Lorini I, Bacaltchuk B, Beckel H, Deckers D, Sundfeld E, dos Santos JP, Biagi JD, Celaro JC, Faroni LRD'A, Bartolini LdeOF, Sartori MR, Elias MC, Guedes RNC, De-Fonseca RG, Scussel VM (eds) Proceedings of the ninth international working conference on stored product protection, 15–18 October 2006, Sao Paulo, Brazil. Brazilian Post-harvest Association, Campinas

Subramanyam B, Roesli R (2000) Inert dusts. In: Subramanyam Bh, Hagstrum DW (eds) Alternatives to pesticides in stored-product IPM. Kluwer Academic, Dordrecht

Subramanyam B, Swanson CL, Madamanchi N, Norwood S (1994) Effectiveness of Insecto, a new diatomaceous earth formulation in suppressing several stored-grain insect species. In: Highley E, Wright EJ, Banks HJ, Champ BR (eds) Proceedings of the sixth international working conference on stored-product protection, 17–23 April 1994, Canberra, Australia. CAB International, Wallingford

Vardeman EA, Campbell JF, Arthur FH, Nechols JR (2007) Behavior of female *Rhyzopertha dominica* (Coleoptera: Bostrichidae) in a mono-layer of wheat treated with diatomaceous earth. J Stored Prod Res 43:297–301

Vayias BJ, Athanassiou CG (2004) Factors affecting efficacy of the diatomaceous earth formulation SilicoSec against adults and larvae of the confused beetle *T. confusum* du Val (Coleoptera: Tenebrionidae). Crop Prot 23:565–573

Vayias BJ, Stephou VK (2009) Factors affecting the insecticidal efficacy of an enhanced diatomaceous earth formulation against three stored-product insect species. J Stored Prod Res 45:226–231

Vayias BJ, Athanassiou CG, Kavallieratos NG, Tsesmeli CD, Buchelos CT (2006) Persistence and efficacy of two diatomaceous earth formulations, and a mixture of diatomaceous earth with natural pyrethrum against *Tribolium confusum* Jacquelin du Val (Coleoptera: Tenebrionidae) on wheat and maize. Pest Manag Sci 62:456–464

Waterford C (2004) QuickPHlo-C™ technology delivers Phosphine. Outlooks Pest Manag 15:120–121

Waterford CJ, Asher PP (2003) Slow generation of Phosphine using QuickPHlo-C™ technology. In: Credland PF, Armitage DM, Bell CH, Cogan PM, Highley E (eds) Advances in stored product protection. Proceedings of the eighth international working conference on stored-product protection, 22–26 July 2002, York, UK. CAB International, Wallingford

Williams P, Hepworth G, Goubran F, Muhunthan M, Dunn K (2000) Phosphine as a replacement for methyl bromide for postharvest disinfestation of citrus. Postharvest Biol Technol 19:193–199

Wilson L, Shaaya E (1999) Natural plant extracts might sub for methyl bromide. Agric res 47:14–15

WMO (1995) Scientific assessment of ozone depletion: 1994. World Meteorological Organization global ozone research and monitoring projects. Report no. 37, WMO, Geneva, Switzerland

Chapter 15
Nanotechnology: An Advanced Approach to the Development of Potent Insecticides

Katrin Margulis-Goshen and Shlomo Magdassi

1 Introduction

Nanoparticles can be generally defined as structures with dimensions of less than 100 nm, although some sources refer to a wider size range, below 1 μm. Due to the enormous increase in surface area following nanosizing, nanoparticles exhibit unique behavior. Even the very fundamental physico-chemical properties of the substance composing the nanoparticles are often altered (Fendler and Meldrum 1995; Muller and Keck 2004). Optical, magnetic, electrical, mechanical, thermal and chemical properties of nanoparticles may significantly differ from those of the bulk material. Nanoparticles have a much higher chemical activity and their solubility is significantly larger than that of the bulk material. As a result, the biological activity, mobility and bioavailability are tremendously increased (Muller and Keck 2004; Sasson et al. 2007; Horn and Rieger 2001). These unique phenomena motivate the massive research, which is conducted on nanotechnology, to engage in many application fields. In the field of pharmaceutics, several drugs comprised of nanoparticles are already in the market, and many others are in their stages of development (Jinno et al. 2008). The main drive for the implementation of nanoparticles in drug delivery is the significant increase in dissolution rate and bioavailability of poorly water-soluble drugs, which can be achieved by this process. More than 40% of pharmaceutically active substances, identified through combinatorial screening programs, are poorly soluble in water, and thus scantly available (Merisko-Liversidge and Liversidge 2008). Very often, they do not reach the drug commercialization stage due to their poor solubility. In many cases, it is much more cost-effective to formulate the poorly-soluble drug into a better-dissolving nanometric formulation than to

K. Margulis-Goshen • S. Magdassi (✉)
Casali Institute of Applied Chemistry, Institute of Chemistry, Center for Nanoscience and Nanotechnology, The Hebrew University of Jerusalem, Jerusalem, Israel
e-mail: magdassi@mai.huji.ac.il

I. Ishaaya et al. (eds.), *Advanced Technologies for Managing Insect Pests*,
DOI 10.1007/978-94-007-4497-4_15,

Table 15.1 Comparison of dissolution profiles of the poorly-soluble drug celecoxib from nanoparticles, bulk material, and bulk material in presence of surface active agents based on the results presented by Margulis-Goshen et al. 2010

Time	1 min (%)	2 min (%)	5 min (%)	10 min (%)	15 min (%)	30 min (%)
% Released drug						
Celecoxib in nanoparticles	60	71	89	92	94	96
Bulk celecoxib	0	0	0.2	0.3	0.4	0.4
Bulk celecoxib in the presence of surfactants	1	4	7	8	9	9

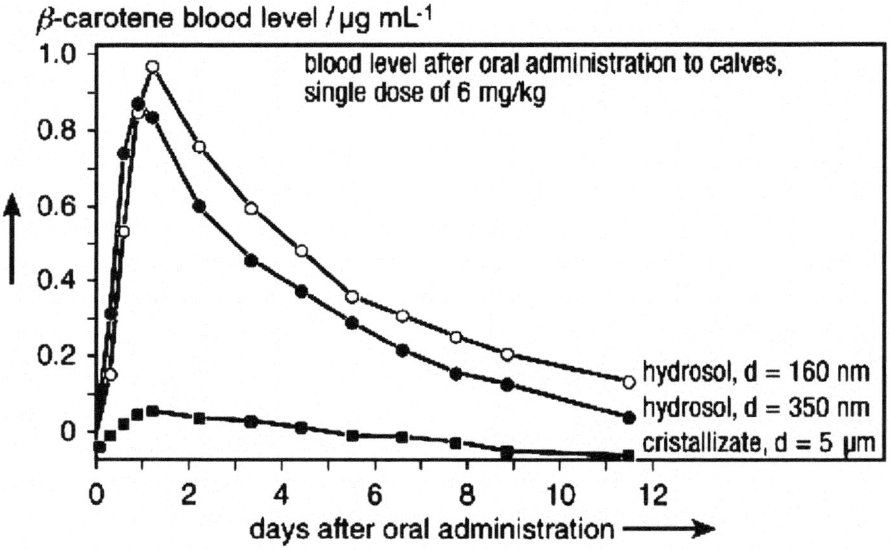

Fig. 15.1 Bioavailability of ß-Carotene as a function of particle size (Reproduced with permission from Horn and Rieger 2001)

search for a new chemical entity (Sasson et al. 2007; Merisko-Liversidge and Liversidge 2008).

For example, the increase in dissolution rate in water of a poorly water-soluble drug, celecoxib formulated in nanoparticles, can be seen in Table 15.1 (Margulis-Goshen et al. 2010).

The theoretical basis behind the increase in solubility and in dissolution rate of nanometric particles will be explained below.

Another example is the increase in the bioavailability of a practically insoluble vitamin, betacarotene (vitamin A), as a function of particle size reduction (Fig. 15.1 Horn and Rieger 2001).

It can be seen that decreasing particle sizes to 160 nm yields a tremendous increase in vitamin concentration in plasma, after oral administration.

Abraxane® is an example for a commercial nanometric drug preparation. This is an intravenous suspension of 100 nm solid particles of the antitumor drug Paclitaxel that are stabilized by the amphiphilic protein albumin at their surface (Desai et al. 2006).

Nanoparticles are also used for specific targeting to various tissues. Thus, specific targeting delivery of drugs to tumors or inflamed tissues can be achieved by formulating the nanoparticles within a size range that is compatible with the size of the large gaps that exist in the vascular wall of blood vessels supplying blood to tumors and inflamed tissues (leaky vasculature). This, however, does not apply to normal blood vessels. The particles will penetrate specifically into the target organ and will not infiltrate to the normal tissues (Nagayasu et al. 1999).

Throughout this chapter we will describe and discuss insecticide particles in the size range below 100 nm which meet the usual size-based definition of nanoparticles, as well as particles or aggregates below 500 nm, which should actually be defined as submicron particles. Nanoporous structures filled with insecticidal substance will also be regarded as nanoparticles.

2 Nanotechnology in Insecticidal Formulations

While the concept of nanotechnology and dispersion systems based on nanotechnology is already considerably applied in the field of pharmaceutics, it is noticeably less implemented in the field of agrochemistry.

Just under half of the market value of agrochemicals is generated by compounds which are practically insoluble in water (Whitehouse and Rannard 2010). Many of the insecticides known today are organic compounds poorly soluble in water (having aqueous saturation solubility of less than 1% w/v). This includes many of the pyrethroids, benzoyl urea derivatives, polychlorinated hydrocarbons, carbomates, diamide insecticides, organophosphates, antraniliproles, formamidine insecticides, macrocyclic lactones, oxidiazines, phthalimide insecticides, neonicotinoids, pyrazoles, pyrimidinamines, juvenile hormone analogues, pyrroles, tetronic acid derivatives, thiazoles and plant derived insecticides. Large amounts of organic solvents are usually employed to dissolve these materials in order to achieve effective and uniform application in the field. However, this leads to environmental contamination with the volatile solvents (Stackelberg et al. 2001) and increases the personal exposure of agricultural workers to various organic compounds. Organic diluents are also much more expensive than water. Moreover, many of the resultant organic solutions are incompatible with aqueous environment, such as aquatic pools, where control over mosquito larvae growth is desired. Another disadvantage of using organic solutions is localized precipitation of the hydrophobic pesticide in presence of aqueous microenvironment in the field. This can lead to lack of the application uniformity and cause insufficient effectiveness (Timothy et al. 1999). In such a case, a pesticide is used in amounts greater than its minimal effective dose and it results in excessive pollution of groundwater, soil, air and waste, in addition to the contamination caused by organic solvents used for pesticide dissolution. Furthermore, the overexposure to insecticide leads to greater insect resistivity towards this chemical.

Nanotechnology presents an appealing way to overcome these problems. While no chemical alteration to insecticide molecule is actually made, formulating the material as nanoparticles allows a significant increase in water solubility, dissolution rate and dispersion uniformity upon application. Diminishing particle size to the nanoscale boosts up saturation solubility of the material, according to the Oswald-Freundlich equation (Eq. 15.1), which is based on the vapor condensation theory proposed by Kelvin (Muller and Keck 2004; Knapp 1921; Simonelli et al. 1970). Though several corrections to this equation have been proposed (Knapp 1921; Mihranyan and Stromme 2007; Wu and Nancollas 1998), it is still regarded as the main tool by which one can predict an increase in saturation solubility with particle size reduction.

$$\frac{S(d)}{S(0)} = exp\frac{\gamma V_m}{RTD} \tag{15.1}$$

where S (d) is the solubility of particles with the diameter D at temperature T, molar volume Vm, and surface energy γ, R is the gas constant, and S (0) is the solubility of the bulk material. It can be seen that the solubility increases with smaller particle size.

According to the Noyes-Whitney equation (Noyes and Whitney 1897) (Eq. 15.2), the dissolution rate of a solid particle in aqueous media is inversely proportional to the particle radius. Thus, the dissolution rate is also increased by reducing the particle size.

$$\frac{dW}{dt} = \frac{DA(Cs - C)}{L} \tag{15.2}$$

where $\frac{dW}{dt}$ is the rate of dissolution, D is the diffusion coefficient of the solid, L is the effective diffusion layer thickness, A is the surface area of the solid, C_S is concentration of the solid in the diffusion layer surrounding the solid, and C is the concentration of the solid in dissolution medium. Upon a significant reduction of the particle radii achieved by nanosizing, A, the surface area, tremendously increases, causing a much faster dissolution.

Additional enhancement of solubility by nanosizing is achieved when utilizing nanoparticle preparation methods which lead to a partially or fully amorphous product. It is well known that amorphous materials are markedly more soluble than their crystalline counterparts (Hancock and Parks 2000; Yu 2001). Provided that sufficient stability of the preparation is obtained, additional solubility advantages of nanoparticles are achieved when compared with a crystalline bulk material.

When compared with the bulk, nanoparticles are also much more mobile, enabling better penetration into insect tissues and raising the induction of insecticide activity. This can be achieved either by faster penetration by direct contact through the insect's cuticle, or by ingestion and penetration through the digestive tract (Sasson et al. 2007). In addition, it was proven that nanoparticles can penetrate better into plants, and consequently into insects which are harmful to these plants (Boehm et al. 2003). This can potentially convert the insecticide delivery process into a much more efficient one. Moreover, when the nanoparticles are deposited on the surface of the leaf following the evaporation of the solvents, the uniform and continuous coverage of the leaf surface is usually achieved. Nanoparticles also exhibit improved diffusion properties to soil due to their enhanced bonding and mobility characteristics (Cameron and Mitchell 2007). At the same time colloidal properties of nanoparticles and their large surface area compared with bulk material can potentially cause higher adhesiveness of nanoparticles to soil and impede their mobility.

Water-soluble pesticides may also be good candidates for encapsulation in nanoparticulate systems, which will provide their multi-staged, controlled release pattern. One example for such protective-encapsulation nanoparticles are mesoporous silica nanoparticles (Liu et al. 2006). Insecticides sensitive to UV and other environmental degradation factors may also benefit from encapsulation into such porous silica nanoparticles (Li et al. 2006), or monolith silicon dioxide (Chen et al. 2011). Many essential oils, which on one hand, have insecticidal activity, but, on the other hand, are extremely volatile and sensitive to degradation, are suitable for such nano-encapsulation. Loading them into protective nanoparticles will provide controlled release and retard their fast evaporation and degradation (Lai et al. 2006; Yang et al. 2009). Extended release over time may be achieved and the frequency of the required pesticide applications may be reduced.

An additional benefit of using nanoparticulate insecticides is the option of combining several agents in the same formulation. It is well known that many agents have synergistic effect, and quite often, control over more than one type of pest is desired. Applying two different pesticide formulations separately may lead to their instability in each other's presence in terms of chemical incompatibility with each other and with the diluting solvents. However, several agents can possibly be loaded simultaneously into nanoparticles and be protected by encapsulation, if required (Storm et al. 2001).

It is worth mentioning that the development of nanoparticles for agricultural use is much cheaper and much easier compared with the development of a nanoparticulate drug product. One of the reasons is that a wide variety of surface-active agents, crystallization inhibitors, particle stabilizers, wetting and dispersing agents and other auxiliary chemicals are approved for use in agriculture, but are not approved in pharmaceutics. This expanded choice of chemicals, which are at the formulator's disposal, guarantees a lower price for nanometric agrochemicals as opposed to nanometric pharmaceuticals. For a reference of all the accepted crop-protective inert ingredients in the USA, one may consult the Electronic Code of Federal regulations database, Title 40 (Protection of Environment),

Chapter 1 (Environmental Protection Agency), Subchapter E (Pesticide Programs), Part 180 (Tolerances and Exemptions for Pesticide Chemical Residues in Food) at the website: http://ecfr.gpoaccess.gov.

3 Preparation Methods

3.1 Mechanical Reduction of Particle Size

The most straightforward methods for obtaining submicron particles and nanoparticles are the top-down disintegration methods, which depend on the input of mechanical energy and high shear forces required to break the particle to the nanometric size particles. Examples of these methods are wet/dry milling, high pressure homogenization and sonication.

The main disadvantages of these methods are: the necessity to invest high energy and shear forces to reach the sub-micron dimensions; the in-process heat generation due to friction, which could be harmful to the milled material and should be controlled by a constant cooling; long procedure time (usually 30–60 min), since the shorter milling time can lead to a higher variance in particle sizes; and the need for special, high-cost equipment. In addition, it should be noted that most of these methods lead to particles above 100 nm, depending on the method and the active ingredient properties.

On the other hand, these methods are simple; in certain systems the particle size polydispersity is very low; high load of the pesticide in the nanoparticles can be achieved; no auxiliary organic solvents are required; the methods are applicable to materials that are poorly soluble in both water and organic media; and the scale-up process is simple with a minimal batch-to-batch variations (Sasson et al. 2007).

Chin et al. (2011) prepared nanoparticles of a poorly water-soluble carbamate insecticide, carbofuran, by homogenization in the first step and sonication in the second step, followed by several stages of milling at 3,000 rpm for 120 min by a high-energy-intensive ball mill with milling beads of zirconium oxide (0.6–0.8 mm). The obtained average particle size was 30–60 nm.

Storm and et al. (2001) from Dow AgroScience used milling technologies in the presence of surface-active agents to obtain stable suspensions of various insecticides with the mean volume particle size below 500 nm. Elevated efficacy of spinosad insecticide against spider mite was demonstrated for nanometric particles. It was found that less amount of the insecticide was required to kill 50% of the spider mites at smaller particle sizes.

Lai et al. (2006) produced solid lipid nanoparticles (SLN) of Artemisia arborescens L essential oil which is used as a safe, natural insecticide. SLN were produced by high-pressure homogenization of melted lipid with the dissolved essential oil into the aqueous solution of surfactants. Particles around 200 nm were obtained and they showed controlled release of the oil and good protection against oil evaporation.

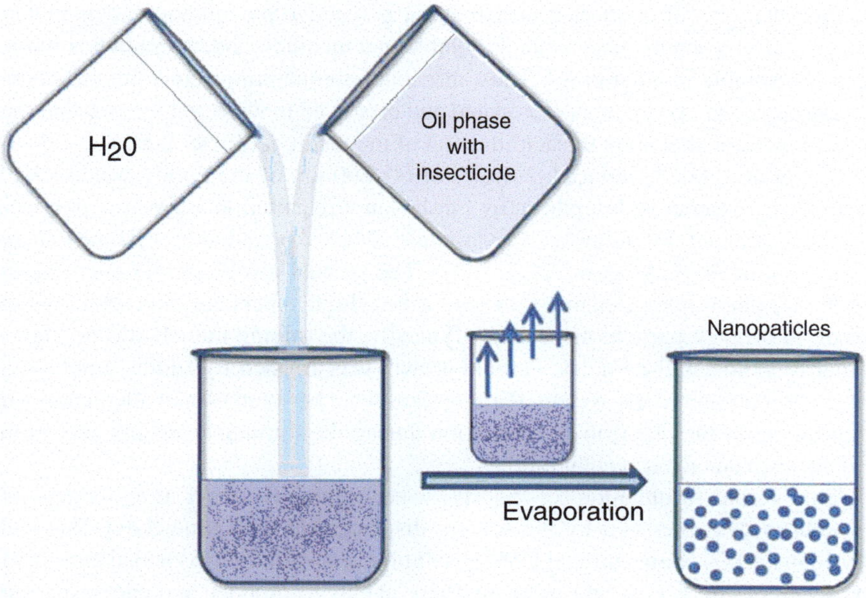

Fig. 15.2 Obtaining nanoparticles by a controlled precipitation process

3.2 Precipitation Methods

Controlled precipitation (anti-solvent precipitation or solvent displacement) is a process that can be applied to prepare nanoparticles of insecticides which are insoluble in water, but are soluble in organic solvents miscible with water. Hydrophobic insecticide is first dissolved in a suitable, water-miscible organic solvent. During encountering of this organic solution with water, by a drop- wise addition or by turbulent mixing, the organic solvent diffuses into the aqueous solution, the active material becomes insoluble, phase separation occurs and results in the production of fine particles. The organic solvent is further evaporated. The size of the final particles is determined by the characteristics of the stabilizers which are added either to water or to the organic phase and by the conditions of the process (Horn and Rieger 2001). The process is schematically shown in Fig. 15.2.

The described method is simple, with no need for a high energy investment or expensive equipment. Usually a high load of the active substance in the particles can be achieved and the method is easily scalable. On the other hand, auxiliary organic solvent has to be used for dissolving the substance and the process is time consuming, because the solvent has to be removed (either by evaporation or by multiple washing) after the mixing. Particle sizes are usually polydisperse and strongly dependant on the process conditions. Also the method is suitable only for the compounds that have a sufficient solubility in organic solvents miscible with water.

Another type of controlled precipitation is reactive precipitation. This method may be used when the insecticide is soluble in a medium, under certain conditions, and is insoluble in other cases. Thus, insecticides containing acidic moieties (such as carboxylic acid) can be deprotonated and dissolved in water at high pH, and then rapidly precipitated upon the acidification of the medium (Texter 2001).

Liu et al. (2008) produced particles of 60–200 nm of the poorly water-soluble pyrethroid insecticide, bifenthrin, by Flash NanoPrecipitation, a precipitation technique developed by Johnson, Prudhomme, Macosko and others (Johnson and Prud'homme 2004; Macosko et al. 2007). The process involves rapid micromixing in a controlled mixing chamber to gain a fast, high supersaturation which would result in a quick particle nucleation. Typically, the mixing rate should be high in order to yield small particles. Particle growth was arrested by adding amphiphilic block co-polymer to the system. Particle sizes were between 60 and 200 nm, mostly depending on the bifentrin concentration during the formation process and on the bifentrin-to-stabilizer weight ratio.

Submicron formulation of poorly water-soluble benzoyl urea compound, novaluron, was prepared by the solvent displacement technique (Levy-Ruso and Toledano 2007). Concentrated DMSO solution of novaluron was added to water in presence of emulsifiers and other auxiliary agents to produce a stable suspension with the mean particle size of 350 nm. This concentrated suspension could be directly diluted with water and applied onto selected targets. It was found that this nanoparticular dispersion was more potent against *Spodoptera littoralis* than a commercial suspension.

3.3 Supercritical Fluids Based Methods

The main techniques for nanoparticle formation based on supercritical fluids are:

- Rapid expansion of supercritical solutions (RESS) – this process consists of solvating the product in the supercritical fluid, such as liquid carbon dioxide, followed by a rapid release of the pressure resulting in a fast particle nucleation.
- Gas antisolvent (GAS)- this method is based on decreasing the solvent power of a polar liquid solvent in which the substrate is dissolved, by saturating it with carbon dioxide in supercritical conditions, causing substrate precipitation or re-crystallization.
- Particles from Gas-Saturated Solutions (or Suspensions) (PGSS) – this process is based on either dissolving a supercritical fluid in a liquid substrate, a solution of the substrate in a solvent, or a suspension of the substrate in a solvent, followed by a rapid depressurization of this mixture through a nozzle, which causes the formation of solid particles or liquid droplets, according to the system (Jung and Perrut 2001).

By using this method it is possible to dissolve the hydrophobic active substance and subsequently form nanoparticles without using any liquid organic solvent.

By this way the need for solvent evaporation or washing is eliminated and the method is much more environmentally friendly. On the other hand, high energy and expensive equipment have to be used to liquefy the gas. This aspect makes the process somewhat challenging for a high-scale fabrication.

An example of a modified process for insecticidal nanoparticles preparation partially based on the rapid expansion of supercritical solution can be seen in the work of Li et al. (2006). Preformed nanometric particles of porous hollow silica were loaded with avermectin, a macrocyclic lactone derivative with insecticidal properties, by first solvating avermectin in supercritical carbon dioxide with acetone as cosolvent, then mixing this solution with the pre-made silica particles in a high-pressure adsorption apparatus, and finally obtaining trapped nanoparticles upon the pressure relief. Such 140–180 nm porous hollow silica particles, filled with avermectin, showed an increased protection against UV-degradation of avermectin as well as controlled release properties, depending on the shell thickness.

3.4 *Spraying Techniques*

Spray drying is another technique that can be utilized to gain submicron pesticide particles. The active material can be dissolved in a suitable volatile solvent and the solution can be then carried by a pressurized heated inert gas through an atomizer to evaporate the solvent. The rapid evaporation of the solvent leads to the supersaturation of the active material in each droplet, followed by the formation of solid particles, which are then separated from the gas. The size of the particles obtained is highly dependent on the type of atomizer, the temperature, the concentration and the speed of the fed solution. For example, an atomizer such as piezoelectric driven droplet actuator with a thin vibrating mesh can be used to produce fine, precisely-sized droplets as small as 8 μm, leading to the formation of the ultrafine solid particles (Li et al. 2010a, b) during evaporation of the liquid from the droplets. Alternatively, a colloidal system which carries insecticide in its inner phase can be subjected to the spray drying process (Elek et al. 2010; Magdassi et al. 2008). In such a system, the final particle size will be dictated by the size of the dispersed phase rather than by the atomizer size. Example for such system will be given below (Sect. 3.5, formation of novaluron nanoparticles from volatile oil-in-water microemulsion by spray-drying). The UK company IOTA NanoSolutions produced nanoparticles of Lambda-cyhalothrin by spray drying from a solution of Lambda –cyhalothrin, polyethylene glycol block co-polymer and polyvinylpyrrolidone in dichloromethane. Polymers were used here to create hydrophilic marix that would facilitate the dispersion of nanometric clusters of the insecticide created during the spray process. The resultant powder could be dispersed in water at 0.1%w/v to produce nanoparticles with z-average size of 94 nm (Angus et al. 2007).

The main advantage of the spray drying is a very rapid transformation of dissolved material into nanoparticles. The process is quick, simple and easily scalable. The disadvantages are the need for the spray drying equipment and elevated

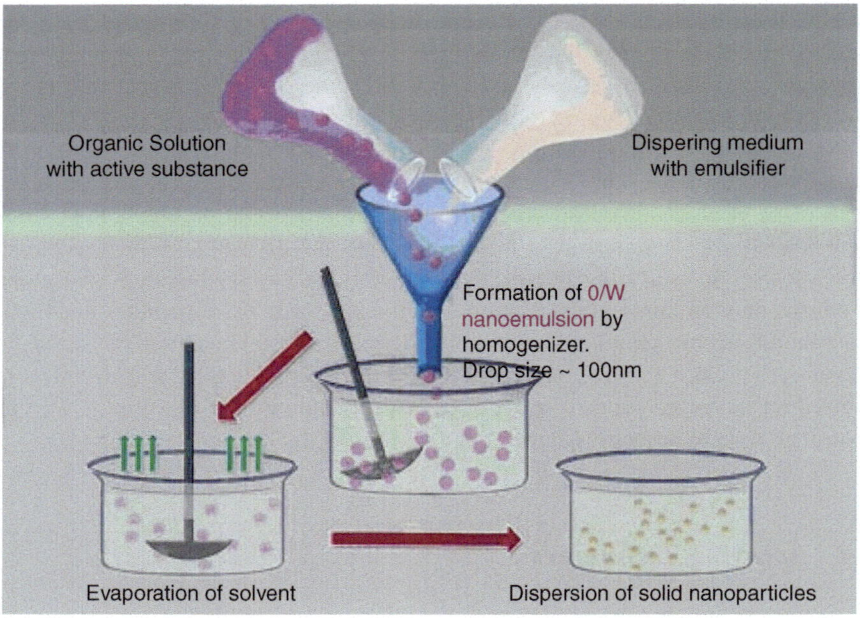

Fig. 15.3 Obtaining nanoparticles by solvent evaporation from emulsions. In order to form nanoparticles, the size of the emulsion droplets has to be reduced by of high-shear forces (*e.g.* by using a high pressure homogenizer)

temperatures required for the drying process. The technique is applicable only to solvents with a high evaporation rate.

Another possible modification to this method is spray freezing of the solution into a cryogenic liquid or spray freeze-drying (Rogers et al. 2003).

3.5 Confined Structures Based Methods

By these methods many dispersion systems can be employed, in which the inner phase serves as the template for production of nanoparticles. Examples of such systems are emulsions, nanoemulsions, microemulsions and fine colloidal dispersions, such as micellar solutions, dispersions of nanometric hollow porous silica particles, polymeric capsules and carbon or amphiphilic nanotubes. Some of those templates are spontaneously formed (e.g. microemulsions) while others require an investment of high-shear forces and mechanical energy for their formation (e.g. emulsions, Fig. 15.3).

The clear advantage of these methods is the relative ease with which those templates can be formed. Either produced spontaneously (e.g. microemulsions) or with the aid of the high shear forces, their formation requires much less energy compared with milling down solids into nanoparticles. The disadvantage is the utilization of the auxiliary solvents and various additives as the building blocks of these confined structures.

One example is the preparation of nanometric avermectin in hollow, porous silica particles, which was previously discussed (Li et al. 2006). Another example is the formation of nanoparticles of water-insoluble pyrethroid insecticide, permethrin, by solvent evaporation from a spontaneously formed oil-in-water microemulsion (Anjali et al. 2010). Microemulsion was formed by a simple mixing of a volatile organic solvent, permethrin, natural surfactants and water. Permethrin was pre-dissolved and therefore trapped within the dispersed nanometric droplets composed of the organic solvent. Both the volatile solvent and the water were removed by freeze-dying to obtain a dry powder composed of nanometric particles. The resultant particles were freely dispersible in water, and had an average size of 150 nm. It was proven that the nano-permethrin powder was at least six times more potent against *A. aegypti* mosquito larvae compared with bulk permethrin. Another example is the formation of nanoparticles of novaluron, from volatile oil-in-water microemulsion by spray-drying (Elek et al. 2010). Following the spray-drying process, a fine powder was obtained, which was freely dispersible in water to aggregates having an average size of 250 nm, composed of smaller, 20–80 nm particles. Biological tests performed on *S littoralis* 1st instars showed that the efficiency of the solvent-free formulation with the nanoparticles was similar to that of a commercial organic solvent-based formulation, thus presenting a prospect for solvent-free pest control formulation which is more environmentally friendly.

Another example relates to the use of submicron emulsion or nanoemulsion, which was prepared by high pressure homogenization, as the template for nanoparticle preparation. Lambda-cyhalothrin particles of 220 nm were prepared by solvent evaporation under vacuum from the nanometric size emulsion (Magdassi et al. 2008). The resultant product was an aqueous dispersion of the nanoparticles that can be further converted into a powder, if required, by lyophilization or spray drying.

Another method utilizing emulsions to prepare nanoparticles of various insecticides was described by BASF Corporation researchers (Martin et al. 2007). This method results in polymeric core-shell nanometric or micronic structures, with a mean diameter of 50–2,000 nm, with amorphous insecticide trapped in the core together with one or more polymers. The method involves: (a) dissolving a pesticide in a water-immiscible organic solvent; (b) dissolving a core polymer in the water-immiscible organic solvent; (c) emulsifying the mixture resulting from (a) and (b) with aqueous solution containing the components of shell matrix; (d) evaporation of the organic solvent after the emulsification.

Also going through emulsion system, Shenzhen Noposion Agrochemicals Manufacturing Company patented the production of solid lipid particles of emamectin benzoate by first dissolving it with fatty acid in organic solvent, followed by emulsification of the organic phase with aqueous solution containing polyvinyl alcohol, and eventually drying both, water and organic solvent (Yuan et al. 2009).

ViveNano Vive Crop Protection is a Canadian company specializing in nanoencapsulation of crop- protective materials. In 2010 it patented polymeric polyelectrolyte nanoparticles of size less than 100 nm, which entrap agricultural active compound (Li et al. 2010a, b). It offers to produce nanoparticles by a variety of methods, mainly based on collapse of a dissolved polymer around the active material, thus forming a confined structure.

There are many commercial insecticidal microemulsion formulations currently on the market. They have a significant insecticidal activity with the reduced impact on environment (Zhao and Yaqian 2011). They contain active substance inside the oil nanodroplets dispersed in the continuous water medium or as the microemulsion concentrate, which has to be diluted with water to produce homogeneous, transparent liquid. Most of these formulations contain pyrethroid insecticides.

3.6 Other Methods

The melt-dispersion method was utilized to prepare nanoparticles of garlic essential oil, a volatile insecticide used against store-product pests (Yang et al. 2009). Garlic oil was simply mixed with molten polyethylene glycol, cooled naturally, ground manually and sieved. It was reported that roundly-shaped particles with an average size below 240 nm, were obtained. It was also proven that over time, the insecticidal effect of nanoparticles against *T. castaneum* was superior to that of the free oil, probably due to decreased volatility. IOTA NanoSolutions produced nanoparticles of Lambda-cyhalothrin by freeze drying from a solution of Lambda –cyhalothrin, polyethylene glycol and polyethylene glycol block co-polymer in chloroform. Reported particle sizes were 44–168 nm for 10 wt.% loading of the active substance in the hydrophilic matrix, 91–128 nm for 20 wt.% loading and 81–241 nm for 30 wt.% loading (Dunculf et al. 2008). Inorganic nanopesticides can also be prepared by methods applicable to the preparation of inorganic nanoparticles. For instance, silicon dioxide particles were prepared by a sol–gel process and showed insecticidal activity (Goswami et al. 2010), and PMA-capped silver nanoparticles were synthesized by photoreduction and exhibited insecticidal activity against *A. aegypti* mosquito larvae (Sap-Iam et al. 2010).

It should be emphasized that it is beyond the scope of this chapter to provide a full, comprehensive review of all possible methods for preparation of nano and submicron particles. Here, we have surveyed only the methods that were recently employed to form nanoparticles of insecticides. It is important to note that numerous other techniques can be utilized to form organic nanoparticles, and we suggest that prior to initiation of a nanopesticide project, to evaluate the methods which are employed for nanosized drug delivery systems. The actual preparation process should take into consideration the physico-chemical properties of the insecticidal material, its mode of action, as well as the expected formulation budget and scalability.

4 Current Legal Status

Despite all these benefits, at present only very few commercial agrichemical formulations are formally associated with nanotechnology. The reason for that is the lack of proper and clear legislation and product evaluation standards, in addition to insufficient information about the environmental and health impacts of nanotechnology in agriculture. All this leads to potential public concern and reluctance towards manufactured goods containing traces of nanoparticles. Nanoparticles of silver or silver composite structures (jointly referred to as "nanosilver") are well-known for their antimicrobial action, as well as for their insecticidal activity (Sap-Iam et al. 2010). They are the most widely discussed "nanopesticide" in the US market. About a year ago, the US Environmental Protection Agency (EPA) officially registered a product containing nanosilver, granting it the permission to be sold in the US market for a period of 4 years. It is essential to note that among other products containing silver, which are currently sold in the USA, several may already contain nanoparticles (Federal Register Vol. 76, No. 117 2011).

In reality, many of the commercialized and registered insecticide products are exploiting the advantages of increasing the surface area and the aqueous solubility of hydrophobic chemicals, by reducing their particle sizes. Dispersion concentrates emulsifiable concentrates, suspoemulsions, wettable powders and other formulations employing surface active agents may potentially contain submicron particles. For many years, the absence of sufficient regulatory status and ambiguous public opinion on nanoscale materials prevented the declaration of this information, although the presence of such particles may bring significant advantages, as previously described.

Only recently, in June 2011, public concern in the USA has led to an official EPA policy proposal regulating pesticide products that are either manufactured by using nanotechnology, or contain ingredients that are themselves the result of nanotechnology (Federal Register Vol. 76, No. 117 2011). This includes the proposed regulatory requirements and applicable processes under the Federal Insecticide, Fungicide, and Rodenticide Act (FIFRA) for registration of new pesticide products, as well as those for already approved products, which contain nanometric ingredients.

The goal of this policy is to determine whether the registration of pesticide formulations may cause unreasonable adverse effect on the environment or human health. Whenever any active or inert ingredient of pesticide formulation or any component parts thereof are intentionally produced to have at least one dimension that measures between 1 and 100 nm, regardless of the aggregation or agglomeration state of the final material, it will be subjected to this regulation (Federal Register Vol. 76, No. 117 2011). In determining whether an ingredient meets this description, EPA may review particle size data. It may also examine the manufacturing process in order to determine whether its aim is to create or enhance the proportion of the nanoscale material in the product, as compared with other processes used to produce similar products. At this stage, the Agency does not intend to include in this regulation biological nanoscale materials (*e.g.*, DNA, RNA, proteins) or materials which are naturally nanometric. It seems that EPA puts efforts in listing the potential benefits of nanosized pesticides and tries to encourage manufacturers to exploit these benefits. Also, it seems to be very aware of the fact that probable risks and adverse effects of one nanoscale pesticide are not necessarily applicable to another, and it could be that some nanomaterials are identical or even less risky than their larger sized counterparts. The effects of nanoscale are also likely to be specific to the properties of every material type under each specific exposure scenario (Federal Register Vol. 76, No. 117 2011). At the same time, the EPA proposes to apply an initial presumption that active and inert ingredients, which are nanoscale versions of non-nanoscale materials already present in registered pesticide products, are potentially different from their conventionally-sized counterparts. Furthermore, since the size, the shape, the surface coating and other characteristics of nanometric ingredients prepared by diverse methods are likely to vary, each new ingredient is presumed to differ from other registered nanoscale versions of the same material. In such cases the applicants for EPA approval will not be able to enjoy the benefits and alleviations of a so called "me-too" application, an application to register a pesticide formulation that claims to have the same composition and uses as an already registered product. In these circumstances, the Agency would most likely require the applicant to provide the types of data typically required for an assessment of the potential hazards and exposure to the new active or inert ingredient. This initial presumption, however, could be rebutted on a case-by-case basis. For this purpose, the applicant may, among other possibilities, submit bridging data or other information demonstrating that the nanoscale material's properties, which are relevant in assessing the potential risks to human health and environment, are substantially similar to an already registered form of that material, or different only in ways that do not significantly increase the risk of unreasonable adverse effects on the environment and human health (Federal Register Vol. 76, No. 117 2011). If this is demonstrated to EPA's satisfaction, the application will be processed as a "me-too" application. It is anticipated that the new regulations will apply to all new pesticide product applications and to all registered products, which are known to contain nanoscale materials. It is also expected that these regulations will be applied to already registered products whenever a change from non-nanometric to nanometric particles is considered. Since only a proposed policy statement is currently being published, many important definitions and issues remain unclear. Currently, the Office of Pesticide Programs has described a nanoscale material as an active or inert ingredient of a pesticide and any component parts thereof, intentionally produced to have at least one dimension that measures between approximately 1 and 100 nm. This means that submicron particles with dimensions greater than 100 nm are not regarded as nanopesticides, although their characteristics may significantly differ from the bulk material, as shown above. However, EPA considers expanding this definition to any material of submicron range intentionally manufactured to have characteristics dissimilar to bulk. It may also specify the minimal concentration of the nanoscale particles required to define a product as nanometric. Moreover, it may require additional data to evaluate the products safety under FIFRA, as well as various characteristics of the product, such as morphology, shape, crystal structure, surface chemistry and reactivity, solubility, and so on. It should be emphasized that in any case, the official updated EPA guidelines should be consulted regarding the registration regulations of the nanoscale pesticides in the United States and the relevant countries.

In European Union, REACH (**R**egistration, **E**valuation, **A**uthorization and **R**estriction of **Ch**emical substances) regulation entered in force on June 2007. It is extensively dealing with Nanotechnology in chemical substances. REACH provides legislation applicable to the manufacture, marketing and use of the chemical substances. Nanomaterials are covered by the definition of a "substance" in REACH. For substances produced or imported in quantities of 1 t or more per year, the manufacturers and importers must elaborate and submit a technical dossier and, for substances in quantities of 10 t or more, a chemical safety report, based on a chemical safety assessment. Where substances that are already on the market as bulk substance, are produced or imported at the nano scale without modifications, they will, for registration purposes, not be considered as different from the bulk material. Manufacturers and importers would therefore have to cover the nano form in the same registration as the bulk substance. The following information about the nano form would be required in cases where properties or uses differ between the nano and the bulk: (1) the information about the properties and uses, (2) safety assessment for the nano form, (3) any different or additional classification with regard to hazardous properties of the nano form, and (4) any risk management measure and operational conditions required. In order to address the specific hazards associated with the nano form, additional testing or information may be required.

Directive 98/8/EC on Biocidal Products provides requirements for the biocidal products on the European market (Directive 98/8/EC of the European Parliament and of the Council 1998). The definition of a biocidal product covers also nanomaterials, so the later fall under the scope of the Directive if used for insecticidal purposes. Comprehensive data requirements are detailed in the Directive. These requirements are applicable to nanomaterials as well, since no other reference is made to any specific test guidelines in this case. Nanomaterials can also be used as co-formulants in insecticidal products. Full composition of the insecticide needs to be presented when applying for its authorization. When necessary, competent authorities can request additional data on 'substances of concern'.

Based on the above, the Biocides Directive in its current form does not provide specific mandatory characteristics of nanopesticides. There is an ongoing assessment done by Commission working groups of the necessity of regulatory change on this subject. EU legislation (REACH) concerning nanotechnology will be subjected to the extensive review in 2012. By November 2012 final project report is expected. Similarly to the EPA guidelines in the US, the official updated REACH guidelines should be consulted regarding the registration regulations of the nanoscale pesticides in Europe. (http://ec.europa.eu/enterprise/sectors/chemicals/reach/nanomaterials/index_en.htm).

It is anticipated that a scientific, comprehensive approach will be established for nanotechnology in pest control, and will build up public trust and enthusiasm towards the nanoscale agrochemicals.

5 Safety and Environmental Impact

Numerous scientific reports have been published regarding human health and environmental effects of nanoparticles. The majority of the studies focus on the impacts produced by a particular type of nanometric particle on a specific health or environmental aspect. Occupational Safety organizations are usually concerned by the fact that the airborne nanoparticles, as compared with their larger counterparts, have a greater potential to be inhaled and deposited deep inside the respiratory tract, causing constant lung inflammation (Department of Health and Human Services. Centers for Disease Control and Prevention. National Institute for Occupational Safety and Health 2009; Hoet et al. 2004). It seems that the most evident deteriorative effect of nanoparticles on human health is the long known induction of the aggressive cancer, mesothelioma, by nanorods of blue asbestos (Ramos-Nino et al. 2003). Some nanoparticles, *e.g.* polymeric spheres, cerium oxide and multiwall carbon nanotubes may also induce protein assembly into amyloid fibrils, phenomenon that might lead to many human diseases, such as Alzheimer, Creutzfeld-Jacob and amyloidosis (Linse et al. 2007). Nanosilver may be linked to cytotoxicity, genotoxicity and cell cycle arrest (AshaRani et al. 2009). Polymeric spheres may cause liver oxidative stress (Hoet et al. 2004). Iron oxide particles found in blood may cause an inflammation of endothelial lining and lead to early atherosclerosis (Zhu et al. 2011) In general, some studies predict the potential toxicity of specific nanoparticles to humans by their ability to produce inflammation (Hoet et al. 2004; Seaton et al. 2009).

Environmental impacts of nanoparticles are also extensively studied. A recent review of harmful effects which were assessed for a variety of non-mammalian species, determined that several kinds of nanoparticles, such as nano-titanium, nano-zinc oxide, nano-silver, nano-copper oxide and single- and multi-walled carbon nanotubes, can be defined as harmful to extremely toxic to non-mammalian species which were tested (Kahru and Dubourguier 2010). However, the important finding was that some pesticide nanoparticles show less toxicity towards the non-target organisms compared with bulk or commercial formulations and therefore, higher specificity (Frederiksen et al. 2003). For instance, gamma cyhalothrin encapsulated into solid lipid nanoparticles exhibited reduced toxicity towards fish and daphnia by a factor of 10 and 63, respectively compared with a traditional emulsifiable concentrate formulation. This could result from the encapsulation process.

Hardly any adverse effects are universal enough to be attributed to all kinds of nanoparticles. These effects are foreseen to depend greatly on the chemical structure of the active material, and its physicochemical properties such as size, hydrophobicity, encapsulation material, solubility, surface coating, charge, shape, particle density etc. Therefore, the potential risk of every contact with a particular type of nanoparticle should be specifically assessed. Concentration and exposure routes and duration should be also taken into consideration.

Insecticide nanoparticles are usually applied on a larger scale compared with other nanoparticulate formulations, such as pharmaceutical preparations, and present a much higher environmental load. On the other hand, unlike drugs, most of the currently approved insecticides are unlikely to affect human health and the amount that reaches the final food consumer is usually very small. However, the case-by-case assessment seems to be justified for every new product.

Another implementation of nanotechnology in the field of insect control that should be addressed, but is beyond the scope of this chapter, is the possibility of detection, degradation and removal of insecticide pollutants from ground water. This can be achieved either by adsorption on metallic nanoparticles such as gold and silver, or by catalytic degradation by iron and titania nanoparticles (Nair et al. 2003, 2007; Paknikar et al. 2005; Yu et al. 2007; Senthilnathan and Philip 2010)

6 Summary and Future Perspectives

To sum up, the development of nanoparticle-based formulations for insecticidal applications is very promising in view of the advantages such as decreased amount of active materials due to increased bioavailabilty. There is a significant interest in chemical and agrochemical industry towards developing and producing of effective, environmentally friendly nanometric pesticides. It is anticipated that the new comprehensive regulations on the use of nanoscale pesticides will lead our society to the careful assessment of the risks involved in using those products. Elimination of public environmental and health concerns regarding nanomaterials will be hopefully achieved, and will result in promotion of nanoscale agrochemicals through the acknowledgement of their potential benefits. Although today the large-scale production of nanometric formulations may be relatively expensive, the wide range of preparation methods and the low cost inactive ingredients available in this field are expected to make it more feasible in the near future.

References

Angus D, Duncalf DJ, Elphick JA et al (2007) Improvements relating to nanodispersions. European patent application, EP 2386292 A1

Anjali CH, Khan SS, Margulis-Goshen K et al (2010) Formulation of water-dispersible nanopermethrin for larvicidal applications. Ecotoxicol Environ Safe 73:1932–1936

AshaRani PV, Low Kah Mun G, Hande MP, Valiyaveettil S (2009) Cytotoxicity and genotoxicity of silver nanoparticles in human cells. ACS Nano 3(2):279–290

Boehm AL, Martinon I, Zerrouk R et al (2003) Nanoprecipitation technique for the encapsulation of agrochemical active ingredients. J Microencapsul 20(4):433–441

Cameron NMS, Mitchell ME (2007) Nanoscale: issues and perspectives for the nano century. In: Kimbrell GA (ed) The potential environmental hazards of nanotechnology and the applicability of the existing low, 1st edn. Wiley, Hoboken

Chen J, Wang W, Xu Y, Zhang X (2011) Slow release formulation of a new biological pesticide, pyoluteorin, with mesoporous silica. J Agric Food Chem 59:307–311

Chin CP, Wu HS, Wang SS (2011) New approach to pesticide delivery using nanosuspensions: research and applications. Ind Eng Chem Res 50:7637–7643

Department of Health and Human Services. Centers for Disease Control and Prevention. National Institute for Occupational Safety and Health (2009) Approaches to safe nanotechnology: managing the health and safety concerns associated with engineered nanomaterials. http://www.cdc.gov/niosh/docs/2009-125. Accessed 18 Aug 2011

Desai N, Trieu V, Yao Z et al (2006) Increased antitumor activity, intratumor paclitaxel concentrations, and endothelial cell transport of cremophor-free, albumin-bound paclitaxel, ABI-007, compared with cremophor-based paclitaxel. Clin Cancer Res 12:1317–1324

Directive 98/8/EC of the European Parliament and of the Council of 16 February 1998 concerning the placing of biocidal products on the market (1998). http://ec.europa.eu/environment/biocides/pdf/dir_98_8_biocides.pdf. Accessed 18 Aug 2011

Dunculf DJ, Foster AJ, Long J et al (2008) Improvements relating to biocidal compositions. International Patent Application WO2008006714

Elek N, Hoffman R, Raviv U et al (2010) Novaluron nanoparticles: formation and potential use in controlling agricultural insect pests. Colloid Surf A 372(1–3):66–72

Federal Register/Vol. 76, No. 117/ Friday, June 17, 2011/ Proposed rules. http://www.gpo.gov/fdsys/pkg/FR-2011-06-17/pdf/FR-2011-06-17.pdf. Accessed 18 Aug 2011

Fendler JH, Meldrum FC (1995) The colloid chemical approach to nanostructured materials. Adv Mater 7(7):607–632

Frederiksen HK, Kristensen HG, Pedersen M (2003) Solid lipid microparticle formulations of the pyrethroid gamma-cyhalothrin-incompatibility of the lipid and the pyrethroid and biological properties of the formulations. J Control Release 86(2–3):243–252

Goswami A, Roy I, Sengupta S, Debnath N (2010) Novel applications of solid and liquid formulations of nanoparticles against insect pests and pathogens. Thin Solid Films 519:1252–1257

Hancock BC, Parks M (2000) What is the true solubility advantage for amorphous pharmaceuticals? Pharm Res 17(4):397–404

Hoet PHM, Bruske-Hohlfeld I, Salata OV (2004) Nanoparticles-known and unknown health risks. J Nanobiotechnol 2:12–27

Horn D, Rieger J (2001) Organic nanoparticles in the aqueous phase—theory, experiment, and use. Angew Chem Int Ed 40:4330–4361

Jinno J, Kamada N, Miyake M, Yamada K, Mukai T, Odomi M, Toguchi H, Liversidge GG, Higaki K, Kimura T (2008) In vitro-in vivo correlation for wet-milled tablet of poorly water-soluble cilostazol. J Control Release 130:29–37

Johnson BK, Prud'homme RK (2004) Process and apparatuses for preparing nanoparticle compositions with amphiphilic copolymers and their use. US Patent Application 20040091546

Jung J, Perrut M (2001) Particle design using supercritical fluids: literature and patent survey. J Supercrit Fluids 20:179–219

Kahru A, Dubourguier HC (2010) From ecotoxicology to nanoecotoxicology. Toxicology 269(2–3):105–119

Knapp LF (1921) The solubility of small particles and the stability of colloids. Trans Faraday Soc 17:457–465

Lai F, Wissing SA, Muller RH, Fadda AM (2006) *Artemisia arborescens* L essential oil-loaded solid lipid nanoparticles for potential agricultural application: preparation and characterization. AAPS PharmSciTech 7(1):E1–E9

Levy-Ruso G, Toledano O (2007) Process for the preparation of nanoparticulate pesticidal compositions and compositions obtained there from. US Patent Application 20070197385

Li ZZ, Xu SA, Wen LX et al (2006) Controlled release of avermectin from porous hollow silica nanoparticles: influence of shell thickness on loading efficiency, UV-shielding property and release. J Control Release 111:81–88

Li X, Anton N, Arpagaus C et al (2010a) Nanoparticles by spray drying using innovative new technology: the Büchi nano spray dryer B-90. J Control Release 147(2):304–310

Li F, Pham H, Anderson DG (2010b) Methods to produce polymer nanoparticles and formulations of active ingredients. International Patent Application WO 2010035118

Linse S, Cabaleiro-Lago C, Xue WF et al (2007) Nucleation of protein fibrillation by nanoparticles. PNAS 104(21):8691–8696

Liu F, Wen LX, Li ZZ et al (2006) Porous hollow silica nanoparticles as controlled delivery system for water-soluble pesticide. Mater Res Bull 41(12):2268–2275
Liu Y, Tong Z, Prud'homme RK (2008) Stabilized polymeric nanoparticles for controlled and efficient release of bifenthrin. Pest Manag Sci 64:808–812
Macosko CW, Anaker JL, Hoye TR, Prud'homme RK (2007) Method for producing nanoparticles. US Patent Application 20070122440
Magdassi S, Dayan B, Levy-Ruso G (2008) Pesticide nanoparticles obtained from microemulsions and nanoemulsions. International Patent Application WO2008032328
Margulis-Goshen K, Kesselman E, Danino D, Magdassi S (2010) Formation of celecoxib nanoparticles from volatile microemulsions. Int J Pharm 393:230–237
Martin I, Auweter H, Koltzenburg S et al (2007) Nanoparticulate pesticide formulations. International Patent Application WO 2007093232
Merisko-Liversidge EM, Liversidge GG (2008) Drug nanoparticles: formulating poorly water-soluble compounds. Toxicol Pathol 36(1):43–48
Mihranyan A, Stromme M (2007) Solubility of fractal nanoparticles. Surf Sci 601(2):315–319
Muller RH, Keck CM (2004) Challenges and solutions for the delivery of *biotech* drugs – a review of drug nanocrystal technology and lipid nanoparticles. J Biotechnol 113(1–3):151–170
Nagayasu A, Uchiyama K, Kiwada H (1999) The size of liposomes: a factor which affects their targeting efficiency to tumors and therapeutic activity of liposomal antitumor drugs. Adv Drug Deliv Rev 40:75–87
Nair AS, Tom RT, Pradeep T (2003) Detection and extraction of endosulfan by metal nanoparticles. J Environ Monit 5(2):363–365
Nair AS, Tom RT, Kumar R (2007) Chemical interactions at noble metal nanoparticle surfaces — catalysis, sensors and devices. COSMOS 3(1):103–124
Noyes AA, Whitney WR (1897) The rate of solution of solid substances in their own solutions. J Am Chem Soc 19:930–934
Paknikar KM, Nagpal V, Pethkar AV, Rajwade JM (2005) Degradation of lindane from aqueous solutions using iron sulfide nanoparticles stabilized by biopolymers. Sci Technol Adv Mater 6(3–4):370–374
Ramos-Nino ME, Scapoli L, Martinelli M et al (2003) Microarray analysis and RNA silencing link fra-1 to cd44 and c-met expression in mesothelioma. Cancer Res 63(13):3539–3545
Rogers TL, Andrew CN, Sarkari M et al (2003) Enhanced aqueous dissolution of a poorly water soluble drug by novel particle engineering technology: spray-freezing into liquid with atmospheric freeze-drying. Pharm Res 20(3):485–493
Sap-Iam N, Homklinchan C, Larpudomlert R et al (2010) UV irradiation-induced silver nanoparticles as mosquito larvacides. J Appl Sci 10(23):3132–3136
Sasson Y, Levy-Ruso G, Toledano O, Ishaaya I (2007) Nanosuspensions: emerging Novel agrochemical formulations. In: Ishaaya I, Nauen R, Horowitz AR (eds) Insecticides design using advanced technologies, 1st edn. Springer, Berlin/Heidelberg
Seaton A, Tran L, Aitken R, Donaldson K (2009) Nanoparticles, human health hazard and regulation. J R Soc Interface. doi:10.1098/rsif.2009.0252
Senthilnathan J, Philip L (2010) Removal of mixed pesticides from drinking water system using surfactant-assisted nano -TiO2. Water Air Soil Pollut 210:143–154
Simonelli AP, Mehta SC, Higuchi WI (1970) Inhibition of sulfathiazole crystal growth by polyvinylpyrrolidone. J Pharm Sci 59(5):633–638
Stackelberg PE, Kauffman LJ, Ayers MA, Baehr AL (2001) Frequently co-occurring pesticides and volatile organic compounds in public supply and monitoring wells, southern New Jersey, USA. Environ Toxicol Chem 20(4):853–865
Storm RM, Price DC, Lubetkin SD (2001) Aqueous dispersion of agricultural chemicals. US Patent Application 20010051175
Texter J (2001) Precipitation and condensation of organic particles. J Dispers Sci Technol 22:499–527
Timothy AE, Taylor RAJ, Downer RA, Hall FR (1999) Deposit structure and efficacy of pesticide application. 1: Interactions between deposit size, toxicant concentration and deposit number. Pestic Sci 55:783–792

Whitehouse P, Rannard S (2010) The application of nanodispersions to agriculture. Outlook Pest Manag 21(4):190–192
Wu WJ, Nancollas GH (1998) A new understanding of the relationship between solubility and particle size. J Solut Chem 27(6):521–531
Yang FL, Li XG, Zhu F, Lei CL (2009) Structural characterization of nanoparticles loaded with garlic essential oil and their insecticidal activity against *Tribolium castaneum* (Herbst) (Coleoptera: Tenebrionidae). J Agric Food Chem 57:10156–10162
Yu L (2001) Amorphous pharmaceutical solids: preparation, characterization and stabilization. Adv Drug Deliv Rev 48(1):27–42
Yu B, Zeng J, Gong L et al (2007) Investigation of the photocatalytic degradation of organochlorine pesticides on a nano-TiO2 coated film. Talanta 72:1667–1674
Yuan F, Cao M, Li C et al (2009) Emamectin benzoate solid lipid nanoparticle, and its preparation method and application in insecticide formulation. Chinese Patent Application CN 101692808
Zhao F, Yaqian W (2011) Formula optimization design of pesticide microemulsion. In: Stoytcheva M (ed) Pesticides – formulations, effects, fate. ISBN: 978-953-307-532-7, InTech, Available from: http://www.intechopen.com/articles/show/title/formula-optimization-design-of-pesticide-microemulsion. Accessed 18 Aug 2011
Zhu MT, Wang B, Wang Y et al (2011) Endothelial dysfunction and inflammation induced by iron oxide nanoparticle exposure: risk factors for early atherosclerosis. Toxicol Lett 203:162–171

Index

I. Ishaaya et al. (eds.), *Advanced Technologies for Managing Insect Pests*,
DOI 10.1007/978-94-007-4497-4,

H

I

N

O

P

The manufacturer's authorised representative in the EU is Springer Nature Customer Service Centre GmbH, Europaplatz 3, 69115 Heidelberg, Germany. If you have any concerns regarding our products, please contact ProductSafety@springernature.com

Printed and bound by CPI Group (UK) Ltd, Croydon, CR0 4YY
25/09/2026
02222899-0005